New Trends in Few-Body Systems: a 30th Anniversary Collection

Alejandro Kievsky

Editor

New Trends in Few-Body Systems: a 30th Anniversary Collection

The articles have previously been published with Springer in the journal *Few-Body Systems*, Vols 57–58 (2016–2017)

 Springer

Editor
Alejandro Kievsky
Department of Physics
Istituto Nazionale di Fisica Nucleare
Pisa
Italy

ISBN 978-3-7091-4869-3

Library of Congress Control Number: 2017946033

Printed on acid-free paper

This Springer imprint is published by Springer Nature
The registered company is Springer-Verlag GmbH Austria
The registered company address is: Prinz-Eugen-Str. 8-10, 1040 Wien, Austria

Contents

Few-Body Syst (2017) 58:136
DOI 10.1007/s00601-017-1300-8

A. Kievsky · B. L. G. Bakker · W. Plessas

30th Anniversary of Few-Body Systems

Received: 11 April 2017 / Accepted: 13 April 2017 / Published online: 28 April 2017
© Springer-Verlag Wien 2017

The journal "Few-Body Systems" is thirty years old! It was founded in 1986 after seeking a new role for Acta Physica Austriaca. In the editorial to its first edition the founders, Profs. H. Mitter and W. Plessas, predicted the growing importance and the relevance of systems consisting of few components in many areas of physics. After thirty years of intense work in those areas we can say that they were right. It is interesting to look at the definition of "few-body systems" that is written in the Aims and Scope of the journal and has remained almost unchanged over the years: Conceptually few-body systems are understood as consisting of a small number of well-defined constituent structures. Examples are found in subnuclear, nuclear, atomic, molecular, and condensed-matter physics, etc., and even extend to problems in celestial mechanics. This illustrates the interdisciplinary character of the journal, which is of primary importance. Moreover, mathematical methods developed to obtain rigorous solutions in the description of such systems is an outstanding part of the legacy of the study of few-body systems.

At the very beginning of its appearance Few-Body Systems included, beyond the published articles, also a News Section for few-body physicists providing them with advance information on new preprints and upcoming conferences/meetings, a service that preceded the arXiv. Similarly, in order to foster communication in the few-body-physics community an E-Mail Directory had been published in 1996 and kept further on accessible on the Internet through the Few-Body-Physics Network (FBN).

Few-Body Systems also pioneered electronic publishing as it was the first physics journal to become available electronically on the Internet. Already in 1995, even before the http-protocol had grown into common standard, one could access the issues without relying on the printed versions, using the multi-functional web-tool Hyper-G (later on Hyperwave). For a period of time the regular journal was augmented over the years by the book series "Few-Body-Physics Supplements", which published mainly conference proceedings but also topical volumes of review type. Furthermore, some special anniversary issues of the journal have been devoted to honoring outstanding scientists in the field of few-body physics.

Leafing through the journals issues, the enormous amount of work reported documents all aspects of few-body physics. The characteristic of a few-body physicist emerges as well. She/he is a researcher specializing in a particular area of physics, while his/her expertise allows him/her to engage also in other fields, in which similar phenomena and problems occur. To underline this specificity many journals have similarly developed

This article belongs to the Topical Collection "30th Anniversary of Few-Body Systems".

A. Kievsky (✉)
Istituto Nazionale di Fisica Nucleare, Pisa, Italy
E-mail: alejandro.kievsky@pi.infn.it

B. L. G. Bakker
Department of Physics and Astrophysics, Vrije Universiteit, Amsterdam, The Netherlands

W. Plessas
Institute of Physics, University of Graz, 8010 Graz, Austria

sections dedicated to such kind of research. In these respects Few-Body Systems has led the way and still does so.

Looking at the first issues, the journal published many articles on few-nucleon systems. In the mid eighties, the solution of the three- and four-nucleon problems was a big challenge. The complexity of these problems encompasses many of the characteristics usually present in the description of a few-body system: the simultaneous presence of long and short-range interactions and the large number of coupled channels that need to be included in a suitable method for its rigorous solution. In addition to these general complications the few-nucleon problem deals with a specific characteristic represented by the construction of the nuclear interaction, a long-standing problem in nuclear physics still intensively studied. Common efforts between theoreticians and experimentalists are in progress on this subject and the present volume illustrates these problems with nice examples in the articles by J. Vannase and D.R. Phillips; E. Stephan et al.; H. Wituła et al.; A. Fonseca and A. Deltuva; and S. Deflorian, V.D. Efros and W. Leidemann. Besides the research on few-nucleon systems, many efforts have been devoted also to the few-quark problem. The constituent-quark model was intensively investigated during the eighties and nineties with many interesting results and predictions to be experimentally confirmed (as, e.g., regarding the pentaquark). Hadronic physics studied from a few-body perspective is similarly well represented in this volume with two reviews by J.M. Richard and N. Shevchenko and three articles by C. Roberts; B.L.G. Bakker and C.-R. Ji; and J.H. Alvarenga Nogueira, T. Frederico and O. Lourenço.

One striking property to observe in few-body systems is their universal behaviour when the interaction among the constituents is close to the unitary limit. The Efimov effect, predicted by V. Efimov in 1971 and experimentally established more than two decades after its theoretical prediction, is a particular three-body phenomenon. An intense experimental and theoretical activity was triggered around this problem at the beginning of the nineties after the ability to utilize Feshbach resonances in magnetic traps had been achieved. Many physicists coming from different backgrounds tried and succeeded to give a theoretical context to this phenomenon in terms of effective theories. They opened the door to a new, interdisciplinary world, still waiting for a complete description. In particular, the situation where in a many-body system the few-body dynamics is in the Efimov regime, is a largely uncharted territory. This subject is well represented in the present volume with the articles by U. van Kolck; R. Weiss, E. Pazy and N. Barnea; P. Giannakeas and C.H. Greene; and E.A. Kolganova, A. K. Motovilov and W. Sandhas.

The search for new methods to attack unsolved problems is a common denominator in few-body physics. The basic line is to develop new, analytical or numerical, tools to extend our capability to describe few-body systems. For example the simultaneous treatment of long- and short-range interactions emerged as a challenging difficulty in the description of scattering states. In a different subject the use of a very large expansion basis can be simplified knowing some analytical properties. Three specific examples are given in the present volume by D. Fedorov; E. Garrido, A. Kievsky and M. Viviani; and T. Watanabe, Y. Hiratsuka, S. Oryu, and Y. Togawa.

One feature of the few-body approach to physical problems should not be forgotten, namely the relentless efforts of the experimentalists to perform more and more accurate experiments on as many physical observables as possible. Their achievements urged theoreticians to refine their calculations up to the point where the dynamics used to analyze these data could be put to stringent tests.

We would like to thank the contributors to the present volume for their efforts and for bringing to our attention the modern vision they present in the description of few-body systems. We also hope for a long and a very successful future of the journal.

Enjoy the reading!

Few-Body Syst (2016) 57:1185–1212
DOI 10.1007/s00601-016-1159-0

J.-M. Richard

Exotic Hadrons: Review and Perspectives

Received: 30 June 2016 / Accepted: 16 September 2016 / Published online: 15 October 2016

Abstract The physics of exotic hadrons is revisited and reviewed, with emphasis on flavour configurations which have not yet been investigated. The constituent quark model of multiquark states is discussed in some detail, as it can serve as a guide for more elaborate approaches.

1 Introduction

The physics of hadrons began about one century ago, when it became appealing to build nuclei from protons and neutrons. The experimental search for the neutron was soon resumed along with the search for the particle predicted by Yukawa as being responsible for the tight binding of protons and neutrons. See, e.g., [1–3] for the history of this early period.

When interacting with nucleons or among themselves, the pions produce many resonances. The exchange of meson resonances, and the formation of nucleon resonances in the intermediate states are crucial for our understanding of the nuclear forces beyond the single-pion exchange proposed by Yukawa. For instance, the existence of the vector mesons, nowadays known as ρ and ω, was somewhat anticipated from the need for isospin-dependent spin-orbit components in the nucleon-nucleon interaction. See, e.g., [4] for an introduction into the early literature on this subject. The discovery of numerous meson or baryon resonances stimulated intense efforts to understand the underlying mechanisms that generate the hadrons. For instance, the ρ meson can be described as a $\pi\pi$ resonance in a p-wave, with the s-wave counterpart being the scalar meson responsible for the spin-isospin averaged attraction observed in nuclei. In 1955, Chew and Low described the $\Delta(1232)$ as the result of a strong πN interaction [5], and today, the scattering of pions off nuclei is still pictured as series of Δ-hole formation and annihilation.[1]

An ever more ambitious picture was proposed, called "nuclear democracy" or "bootstrap", in which every hadron consists of all possible sets of hadrons compatible with its quantum numbers [6]. Unfortunately, neither the models based on nucleons and pions as elementary bricks, nor the bootstrap, succeeded in providing an overall picture. For instance, the degeneracy of isospin $I = 0$ and $I = 1$ mesons such as ρ and ω, does not emerge naturally, if one assumes that their content is dominated by the nucleon-antinucleon component [7].

From the very beginning, the attention was focused on hadrons with properties that look unusual, in contrast to the regularity patterns of the rest of the spectrum. The exotic character of a hadron is of course

---

This article belongs to the special issue "30th anniversary of Few-Body Systems".

---

[1] The author thanks G.F. Chew for an enlightening discussion on the subject many years ago.

J.-M. Richard (✉)
Institut de Physique Nucléaire de Lyon, IN2P3-CNRS–UCBL, Université de Lyon, 4, rue Enrico Fermi, 69622 Villeurbanne, France
E-mail: j-m.richard@ipnl.in2p3.fr

a time-dependent concept: once a striking property of a hadron is confirmed by further measurements, and understood within a plausible theoretical framework, the hadron ceases to be exotic.

Among the most famous examples, there is the ϕ meson, a peak of mass 1.02 GeV with a remarkably small width of about 4 MeV [8], decaying preferentially into a pair of kaons, though the decay into pions is energetically much easier. The decay patterns of the ϕ are now understood from its quark content $s\bar{s}$, where s denotes the strange quark, with very little mixing. The suppression of the internal annihilation of $s\bar{s}$ into light mesons has been codified in terms of the so-called "Zweig rule", eventually justified from the asymptotic-freedom limit of Quantum Chromo-Dynamics (QCD).

The process took, however, several years, and was not mature enough when the J/ψ particle was discovered at the end of 1974; it is a meson of mass about 3.1 GeV and width much less than 1 MeV [8]. Why the J/ψ was not immediately identified with $c\bar{c}$? The great surprise was that the Zweig rule works too well! As compared to the $\phi(1020)$, the J/ψ is three times heavier, so its phase-space for decay into light mesons is considerably larger, and yet its width is smaller! For the Υ, the effect is even more pronounced, but when the Υ was discovered in 1977 [8], the lesson was learnt, and the Υ was never considered as exotic.

Today, any attempt to provide with a definition of an "exotic" hadron starts from the quark model. A meson with quantum numbers that cannot be matched by any spin-orbital state of a quark and an antiquark is obviously exotic. States with unusual properties but ordinary quantum numbers are named "crypto-exotic" states. In particular, ordinary baryons can reach any set of values of the spin-parity J^P with half-integer spin J and parity $P = \pm 1$, so a baryon cannot be exotic except for its flavour content. A consensus has thus been reached that a hadron is exotic if it cannot be understood as mainly made of quark-antiquark nor quark-quark-quark, but "mainly" remains vague. While simple quark models will propose a reference with a frozen number of constituents, more advance pictures include an expansion in the Fock space, with additional quark-antiquark pairs in the wave function. The hope is that the expansion converges, and that a clean separation emerges between exotic and non-exotics.

This review is organised as follows. In Sect. 2, we shall review states with unusual properties in the light (Sect. 2.1), open heavy-flavour (Sect. 2.2), double heavy flavour (Sect. 2.3), and hidden heavy-flavour (Sect. 2.4) sectors. The baryon spectrum is discussed in Sect. 3, with, again a separation between the non-exotic light sector (Sect. 3.1 for non strange and Sect. 3.2 for strange baryons), the exotic light pentaquark with positive strangeness in Sect. 3.3, the case of open heavy-flavour, single (Sect. 3.4) or double single (Sect. 3.5), the anticharmed pentaquark in Sect. 3.6, and the case of hidden heavy-flavour (Sect. 3.7) with in, particular, the recent pentaquark states found by the LHCb collaboration. The multibaryon systems will be discussed in Sect. 4. The various theoretical models will be reviewed in Sect. 5, with some emphasis on string dynamics and potential models.

Some limits have to be set. The case of light hadrons will be just outlined, to focus more on the most recent states containing heavy quarks or antiquarks. In the former sector, the literature is enormous, and can be traced back from the conference proceedings that will be cited in the next sections, from some well-known reviews such as [8–11], or from study groups for future facilities [12–15]. For the recent exotics with hidden charm, there exists already excellent and even comprehensive reviews, such as [16–19]. Our aim here is not to compete with these reviews, nor to duplicate them, but rather to discuss critically the recent findings and the models that they inspired. So, this paper will be more subjective. Hopefully, it will trigger some discussion, and, perhaps, attract newcomers into the field, where innovative ideas are welcome.

The bibliography for such a subject cannot be comprehensive. Some past review articles are cited, which give access to the early papers. We apologise in advance in the event where some important contributions are inadvertently omitted.

2 Mesons

2.1 Light Mesons

There are many hot issues on light mesons: why are the pseudoscalars so light? what is the status of the many scalar mesons? why ideal mixing and Zweig rule work so well in some sectors, and not in other sectors? We refer, e.g., to [11] for a critical survey of the claims for exotics, and to [20] for a thorough discussion within QCD and the bibliography, and to the last hadron conferences for an update [21–23].

One of the puzzles arises from the $\gamma\gamma \rightarrow \rho\rho^2$, measured in several experiments, in particular at L3 [24,25]: the comparison of the different charge modes suggest a sizeable isospin $I = 2$ contribution. There

---

[2] I thank Bernard Pire for calling my attention on this channel.

are interesting theoretical studies on the production mechanism [26], but missing is a simple understanding of why there is strong attraction in that channel in terms of quark dynamics.

There is an intense activity in the search of glueballs and light hybrids, either as supernumerary states with ordinary quantum numbers J^{PC}, or as genuine exotics with J^{PC} that cannot be matched by ordinary quantum numbers. In particular, evidence has been found for 1^{-+} states between 1 and $2\,\text{GeV}^2$ by VES at Serpukhov [27], by E852 at Brookhaven [28], and by COMPASS at CERN [29], and the search is resumed actively, in particular by GLUEX at Jlab [30]. For a thorough review of the other contributions, see, e.g., [31].

2.2 Mesons with Single Heavy Flavour

The charmed mesons D, D^* have been discovered shortly after the charmonium, and the mesons with charm and strangeness D_s, D_s^* have followed. Relevant for this review on exotics are the so-called $D_{s,J}$ states. As summarised in [32], the $D_{s,0}^*(2317)$ was discovered at BaBar [33] and the $D_{s,1}^*(2460)$ at CLEO [34], with a mass remarkably lower than predicted in the "bible" [35] and in similar quark-model calculations. Note that the tuning of the spin-dependent forces in such models has been often debated, and no consensus has ever been reached, see, e.g., [36–39]. To explain the low mass and the properties of the $D_{s,J}^*$ states, the coupling to the nearby $D^{(*)}K^{(*)}$ thresholds certainly plays a role. These mesons have even been considered as molecules, for instance, in [40–43].

2.3 Mesons with Double Heavy Flavour

After, the discovery of charm, it has been early recognized that configurations including both light and heavy quarks or antiquarks might be particularly interesting, for instance, in the case of baryon-antibaryon or multi-baryon systems [44,45]. Another pioneering contribution was due to Isgur and Lipkin [46] who considered the flavour exotic $(cs\bar{q}\bar{q})$ configuration.

Shortly after, even more exotic mesons were studied, $(QQ'\bar{q}\bar{q}')$, with two heavy quarks and two light antiquarks. As explained later in this review, their existence is a prediction shared by many theorists using different approaches, which will be discussed in the next sections. Doubly-heavy exotics, which are surprisingly omitted in some "reviews" on multiquarks, have not yet been searched for in any experiment, but will be one of the goals of future experiments, and several theorists estimated the rate of production at hadron [47,48] or electron [49] colliders, or in the decay of hadrons with beauty [50].[3] The production of two charm-anticharm pairs is somewhat disconcerting, with the good surprise of a rather abundant production of double charmonium, leading for instance to the identification of the $\eta_c(2s)$ recoiling against the J/ψ [51], and, on the other hand, the double-charm baryons have not been confirmed in recent experiments, see Sect. 3.5.

2.4 Mesons with Hidden Charm or Beauty

The spectrum is quarkonia, $(c\bar{c})$ and $(b\bar{b})$, is now rather well known and well understood. For a while, the most delicate experimental sector was the one of spin-singlet states, such as $\eta_{c,b}(ns)$, $h_{c,b}(np)$. For instance, a first candidate for the $\eta_c(1s)$ was announced about 300 MeV below the J/ψ, triggering some premature announcements on the quark dynamics, but eventually replaced by a more reasonable $\eta_c(1s)$ about 112 MeV below J/ψ, see, e.g., the reviews [52,53]. A first indication for the h_c was given a by $\bar{p}p$ formation experiment at CERN [54], with another indication in a similar experiment at Fermilab [55]. It was eventually firmly identified at CLEO [56]. The $\eta_c(2s)$ was seen at Belle in B decay [57] and in double-charmonium production [51]. It also took some time until the spin-singlet states of $(b\bar{b})$ were identified [8].

Above the threshold, $D\bar{D}$ for charmonium and $B\bar{B}$ for bottomonium, the spectrum was known to be more intricate, with broader peaks, and an interplay of s- and d-states mixed with the $(Q\bar{q})(\bar{Q}q)$ continuum, and it was a somewhat a daily grind to improve the measurements.

Then came the shock of the discovery of the $X(3872)$ by Belle [58], in the $\pi^+\pi^- J/\psi$ mass spectrum of the reaction $B^\pm \to K^\pm \pi^+\pi^- J/\psi$. Neither the mass nor the decay properties were according to the expectations of the charmonium models, in particular for the 2^3P_2 level. On the other hand, such a state was announced in the

---

[3] I thank Angelo Esposito and Alessandro Pilloni for a reminder about this reference.

molecular models, as discussed in Sect. 5.4.3. Not only the experts on quark- or hadron-dynamics commented this new state, but also some non-experts became attracted to the subject.

As well summarised in a recent review [59], a second shock was the discovery of the $Y(4260)$ by BaBar [60], in e^+e^- with ISR.[4] This 1^{--} sector of charmonium was already thoroughly explored, with radial excitations, with s- and d-waves mixed between them and with the continuum. So this new state was immediately associated with an exotic, or, say crypto-exotic structure. The $Y(4260)$ was seen also at CLEO and Belle [8].

A third shock was the discovery of charged states, with minimal content $c\bar{c}u\bar{d}$ or c.c. The first announcement was the $Z^+(4430)$ by the Belle collaboration, in the $\pi^\pm\psi(2s)$ mass distribution of the decay $B \to K\pi^\pm\psi(2s)$ [61].

It is hardly possible to list all the states of X, Y or Z type without duplicating the very comprehensive reviews such as [18,19]. Let us stress a few points:

- The $X(3940)$ and $X(4160)$ have been seen in double-charmonium production [62,63], which gives a clear signature of their $c\bar{c}$ content,
- Some states have been seen in $\gamma\gamma$, indicating a charge conjugation $C = +1$, and very likely $J^{PC} = 0^{++}$ or 2^{++}. This is the case for $Z(3930)$, $X(3915)$ and $X(4350)$ [64–66].
- Some states, in particular among the above ones, are just candidates for radial excitations of charmonium. If confirmed, they indicate the supernumerary character of other states sharing the same quantum numbers.
- The $Y(4630)$ has been seen as an enhancement in the $\Lambda_c\bar{\Lambda}_c$ spectrum of the ISR reaction $e^+e^- \to \gamma\Lambda_c\bar{\Lambda}_c$ [67]. Many references about this state can be found in a recent paper [68]. As stressed in this literature, the decay into a baryon and an antibaryon might indicate a "baryonium" type of structure, though other scenarios have been proposed [69].
- There is sometimes a clustering of states with similar masses. It might the same state with a peak that depends on the production and decay process.
- Most quark models and, to a lesser extent, molecular models predict an isospin $I = 1$ partner of the $I = 0$ state, analogous to ρ vs. ω. This stimulated searches for charged partner, such as [70].
- While the $X(3872)$ has been seen in many experiments, $Y(4260)$ in a few experiments, some states have been seen in just a single experiment. In between, $Z_c(3900)^\pm$ has been seen by Belle and BESIII [71,72], but these experiments share many collaborators, and used the same reaction $e^+e^- \to Y(4260) \to \pi^\mp Z_c^\pm$, with a slight variant for the confirmation by CLEO [73]. So, the negative result of a search by COMPASS in photoproduction is particularly important for the existence or at least the coupling properties of this state [74].
- The lightest XYZ states have minimal quark content $c\bar{c}q\bar{q}$, where $q = u$, d. In several models, their $c\bar{c}s\bar{s}$ with hidden strangeness are expected. Recently, the LHCB collaboration re-analysed the $J/\psi\,\phi$ mass spectrum from the decay $B \to J/\psi\,\phi K$ [75], and found evidence for $X(4140)$ and $X(4264)$ both with $J^{PC} = 1^{++}$, in disagreement with some models (cited in [75]) and in partial or full agreement with others, in particular the extension by Stancu [76] of the chromomagnetic model proposed for the $X(3872)$ [77].
- The models somewhat differ about the existence and location of the $b\bar{b}$ analogues of the exotic mesons with hidden charm: the heavier b-quark mass, for instance, suppresses the chromomagnetic effect but enhances the chromoelectric ones, while the coupling schemes are sensitive to the gap between bare quarkonia and meson pairs. The list of candidates is rather scarce for the time being, and are due to Belle only. The charged $Z_b^\pm(10610)$ and $Z_b^\pm(10650)$ have been seen in $\Upsilon(5s) \to (b\bar{b})Z_b^\pm\pi^\mp$, where $(b\bar{b})$ stands here for some lower $\Upsilon(ns)$ or $h_b(1p)$ or $h_b(2p)$. Later, Belle also got some candidate for the neutral $Z_b(10610)^0$. On the other hand, a search for the heavy partner of $X(3872)$, decaying into $\omega\Upsilon(1s)$ or $\pi^+\pi^-\Upsilon(1s)$, has been so far unsuccessful.

3 Baryons

The history of the quark model is intimately linked to baryon spectroscopy. One of the motivation for SU(3) flavour symmetry, nowadays denoted SU(3)$_F$, was the observation that the hyperons have properties similar to that of non-strange baryons. The quark model was developed first in the baryon sector, with the pioneering works by Greenberg, Dalitz, etc., and pushed very far by the subsequent works by Dalitz et al., Hey et al., Isgur and Karl, and many others. For a review, see, e.g., [78]. Obviously, the field has been much enriched by the study of baryons with charm or beauty, and baryons carrying two or three heavy flavoured quarks are eagerly awaited.

---

[4] Initial State Radiation, this means that the reaction is $e^+e^- \to \gamma + e^+e^- \to \gamma + Y(4260)$.

3.1 Non Strange Light Baryons

There are many issues there, that have been presented in conferences such as "Baryons", "Nstar", etc. For years, one of the most striking issues was the signature for the "symmetric quark model" with two degrees of freedom, vs. the quark–diquark model. Clearly, the recent progress, as e.g., summarised at the recent conference in Tallahassee [23], confirms the existence of states such as $N(1900)$ which cannot be accommodated in a naive quark-diquark picture with a frozen diquark.

The lightest state with an excitation in each of the Jacobi variables, say, $x = r_2 - r_1$ and $y = (2r_3 - r_1 - r_2)/\sqrt{3}$, has isospin $I = 1/2$, spin $s = 1/2$ and orbital momentum and parity $\ell = 1^+$. In the particular case of the harmonic oscillator, its wave function reads

$$\Psi(x, y) = x \times y \exp(-a(x^2 + y^2)/2) \,, \tag{1}$$

and is antisymmetric under the permutations. Thus, if associated with an antisymmetric colour wave function $(3 \times 3 \times 3 \to 1)$ and an antisymmetric spin-isospin wave function, it fits perfectly the requirement of Fermi statistics. Note that an antisymmetric spin-isospin wave function is possible for $I = 1/2$ (uud), but not for $I = 3/2$ (uuu) nor $I = 0$ (ccc).

Experimentally, the search for doubly-excited baryons is difficult. As often underlined (see, e.g., [79]) photon or meson scattering off a nuclear target favours the production of states with a single quark being excited. And in amplitude analyses, doubly excited baryons, if any, come in a region that is already well populated!

Another hot issue is the Roper resonance, an excitation with the same quantum numbers as the ground state. In potential models, this is mainly a hyper-radial excitation, but in explicit model calculations, it is found higher than the negative-parity excitation, while the experimental Roper state lies degenerate or even slightly below the lowest negative-parity excitation. Chiral dynamics, as implemented in specific models or in lattice simulations, accounts for this property, which sets limits to naive potential models.

3.2 Light Baryons with Negative Strangeness

Here, again, is the problem of missing states, with slow but continuous progress to fill the holes. A very striking property is the low mass of the $\Lambda(1405)$, discussed first by Dalitz [80] and now in an abundant literature (see, e.g., [81] for a recent summary). In the case of the nucleon spectrum, the spin-orbit splittings, such as between $N(1520)$ with $(1/2)^-$ and $N(1535)$ with $(3/2)^-$, are very small, and Isgur and Karl, for instance, have set the spin-orbit potential to zero in their fit [82]. Hence the large $\Lambda(1520) - \Lambda(1405)$ splitting of the strange analogues appears as somewhat a surprise in a pure quark model.

A minimal solution explains the discrepancy by a shift due the coupling to the nearby KN threshold. Similarly, the ψ' is shifted down by the nearby $D\bar{D}$ threshold, and the hyperfine splitting $\psi' - \eta_c(2s)$ is slightly smaller than expected. Similarly, the $2\,^3P_2$ state of charmonium gets some admixture of $D\bar{D}$ + c.c. in conservative models of the $X(3872)$, etc.

Now, a more appealing solution is also suggested, in which there are two poles with $J^P = (1/2)^-$ near $\Lambda(1405)$ [83], a mixing of a $\bar{K}N$ molecule and a quark-model state. As often underlined in the case of mesons [84], the mixing scheme might be more involved than in ordinary quantum mechanics, with the absorptive part, i.e., the coupling the decay channels, playing the major role.

3.3 Light Baryons with Positive Strangeness

For years, there has been debates about baryons carrying positive strangeness. In the 60 and 70s, much effort was devoted to exploit the analyticity properties of the S-matrix, and to reconstruct the amplitudes from the scattering data. This program was rather successful in the case of the πN and $\bar{K}N$ reactions, and led to the identification of a number of baryon resonances with strangeness $S = 0$ and $S = -1$, respectively. The case of KN scattering with $S = +1$ was much more delicate and controversial. Some early analyses, based only on differential cross-sections (and just a few data on the recoil polarisation), suggested the possibility of resonances, see, for instance [85–87]. The resonances were named Z_I or Z_I^*, where $I = 0$ or 1 is the isospin.

Then came data on the analysing power, obtained using a polarised target, as in [88]. Not surprisingly, the uncertainties were narrowed [89], and the Z resonances tended to fade away. For years, the Particle Data

Group (PDG), summarised the state of art, as in the 1980 edition [90], where one finds candidates for several exotic baryons with $S = +1$. There are $Z_0(1780)$ and $Z_0(1865)$ in the isospin $I = 0$ sector, and $Z_1(1900)$, $Z_1(2150)$ and $Z_1(2500)$ for $I = 1$. Then the Z baryon disappeared from the review, and the letter Z was given to another particle. Note that other risky amplitudes analyses were done later, for instance for $\bar{p}p \to \pi^+\pi^-$ or K^+K^- before the advent of data taken on the polarised target, or for the elastic $\bar{p}p$ scattering with just data on the polarisation, while the simpler pp case notoriously requires several spin observables.

Act two of the play is better known. Meanwhile the name "pentaquark" was introduced in the anticharmed sector, as will be discussed shortly, in Sect. 3.6. It reflects that the minimal content of the Z_I baryon is $(\bar{s}qqqq)$. In 2003, Nakano et al. found a narrow peak in the K^+n distribution of the reaction $\gamma n \to K^-K^+n$ [91]. This state, named $\theta^+(1540)$, was predicted in a model by Diakonov et al. [92] based on some developments of the chiral dynamics. It was of course rather remarkable to have a relatively small experiment revealing a major discovery. But the sequels quickly went out of control. On the theory side, a flurry of papers, with hasty calculations and ad-hoc assumptions. On the experimental side, some collaborations discovered that they had data on tape, that were never analysed, and contained in fact the pentaquark! Other positive results came, both for the θ^+, and for other members of the anti-decuplet of $SU(3)_F$ that was predicted. However, the trend became inverted, and experiments with better statistics and particle identification did not confirm the existence of the these light pentaquarks. This is reviewed for instance in [93,94]. The theory part also came under better scrutiny. In Ref. [95], some lattice QCD calculations following the discovery of the θ^+ are criticised. In, e.g., [96,97], some inconsistencies are pointed out in the estimate of the width in the model of Diakonov et al., a crucial property of the putative light pentaquark.[5]

3.4 Baryons with Single Heavy Flavour

In the last 20 years, much experimental progress has been done in the sector of baryons with single charm or beauty. When available, the comparison of b- and c-baryons is compatible with the idea of a flavour-independent confinement. For instance, the first evidence for $\Omega_b(bss)$ came from D0 at Fermilab, with a mass about 6165 MeV, corresponding to a double-strangeness excitation $\Omega_b - \Lambda_b$ substantially larger than its charm analogue. Further measurements by CDF at Fermilab and LHCb at CERN gave a mass around 6050 in closer agreement with the expectations [98–100].

A remarkable feature of charmed baryons is their narrowness, with, e.g., $\Gamma \sim 6\,\text{MeV}$ for $\Lambda_c(2880)$, $\sim 17\,\text{MeV}$ for $\Lambda_c(2940)$, $\sim 4\,\text{MeV}$ for $\Xi_c(3123)$ [8]. Of course, this gives a better view on the spectrum than in the case of overlapping resonances. An optimistic columnist of PDG [101] even stated: "...models of baryon-resonance spectroscopy should now *start* with the narrow charmed baryons, and work back to those broad old resonances." Among the current discussions, there is the suggestion that $\Lambda_c(2980)$ is mainly a D^*p molecule [102].

The same narrowness is observed for b-baryons, as listed in the latest PDG review [8] or revealed in the latest LHCb papers [103,104].

3.5 Baryons with Double Charm

Thirty years ago, double charm baryons were considered as just a straightforward extension of the quark model, to be checked just for the sake of completeness. At most, some experts stressed the co-existence within the same hadron of the slow, adiabatic, charmonium-like motion of the two heavy quarks, and the fast, D-meson-like motion of the light quark around them. In the literature, (ccq) is often described as a quark-diquark [105]; however, the first excitations occur between the two heavy quarks and thus one has to introduce "excited diquarks". An alternative is the Born-Oppenheimer method [106], which, as discussed later, might also be adequate in the XYZ sector and for hybrids.

Some evidence for double-charm baryons was found at Fermilab, in the SELEX experiment [107,108], but unfortunately not confirmed in other experiments at Fermilab or elsewhere [109–112]. This becomes particularly annoying. Double-charm baryons will certainly receive more attention in the near future [113,114].

It is an important question to understand why double-charm baryons have not yet been seen. If the golden modes such as $\Lambda_c + K$ or $\Lambda + D$ are suppressed, the detection scheme should be revised. And, perhaps, new production mechanisms should be exploited.

---

[5] I thank Herbert Weigel for an informative correspondence on this point.

Our ability to detect double-charm baryons and measure their properties is eagerly watched by people waiting for exotic tetraquarks and dibaryons. As pointed out in [115,116], the binding and the correlation of the two charmed quarks in (ccq) gives some idea about $(cc\bar{q}\bar{q})$ and $(ccqqqq)$. And, in most production experiments, they share the same triggers, so that a simultaneous search for double-charm baryons, mesons and dibaryons can be envisaged.

3.6 Anticharmed Baryons

Two simultaneous papers by Gignoux et al. and Lipkin [117,118], in which the word "pentaquark" was first used, suggested the existence of stable or metastable states $(\bar{Q}qqqq)$ with $qqqq$ denoting $uuds, ddsu$ or $ssud$. This pentaquark was searched for in an experiment at Fermilab, which was not conclusive [119,120]. At the time of the light pentaquark, the H1 collaboration at HERA looked at the $\bar{D}^*p$ mass spectrum, i.e., states of minimal content $(\bar{c}uudd)$, without strangeness contrary to the prediction in [117,118], and found a peak near 3.1 GeV with a width of about 12 MeV [121], which was confirmed neither by Zeus [122], also operating at HERA, nor by BaBar [123] at SLAC, nor in the analysis by H1 of their HERA II data.[6] The anticharmed pentaquark was initially one the goals of the COMPASS experiment [124], but it has disappeared from the agenda.

3.7 Baryons with Hidden Charm or Beauty

This is one of the latest findings, inducing an intense phenomenological activity supplementing the discussions about XYZ mesons. The rumour is that, due to the bad reputation of the light pentaquark, the LHCb collaboration tried during months to kill the evidence, with more checks than usual for this type of discovery. The result published in [125] deals with the weak decay

$$\Lambda_b \to J/\psi + p + K , \tag{2}$$

of the lowest baryon with beauty. The Dalitz plot, as well as the $(p, J/\psi)$ mass spectrum, indicates two peaks, one with a mass of $4380 \pm 8 \pm 29$ MeV and a width of $205 \pm 18 \pm 86$ MeV, an another with a mass of $4449.8 \pm 1.7 \pm 2.5$ MeV and a width of $39 \pm 5 \pm 19$ MeV. As the Λ_b mass is always reconstructed, some ambiguities are removed. The background is fitted as $J/\psi + \Lambda^{(*)}$, with many Λ^* excitations, up to high mass and high spin. When two resonances are added, the preferred quantum numbers are spin $J = 3/2$ and $5/2$, and opposite parities, namely $(3/2)^\pm$ and $(5/2)^\mp$. See, also, [126].

The process is schematically described in Fig. 1, as well as the companion reaction (Cabibbo suppressed) $\Lambda_b \to J/\psi + p + \pi$.

This process $\Lambda_b \to J/\psi + p + \pi$ has been measured by LHCb and published before the discovery of hidden-charm pentaquarks [127], but the statistics was not sufficient to identify peaks. This reaction has been discussed in a few papers, as well as other weak decays involving the hidden-charm pentaquark [128,129], shown in Fig. 2. A more recent analysis of the reaction $\Lambda_b \to J/\psi + p + \pi$ provides some support for the contribution of hidden-charm pentaquarks [130].

It has been proposed [131–135] to check the existence of the LHCb pentaquarks and provide more information on their properties by studying the photoproduction reaction $\gamma + p \to J/\psi + p$, and the experiment will be done at Jlab.

4 Multibaryons

4.1 Dibaryons

The situation in the non-strange sector has been recently reviewed at the Conference Baryon 2016 [136]. There are regularly some claims for resonant behaviour of some partial waves [137–140]. Negative results include: [141] in the isospin $I = 2$ sector, [142] for multineutrons bound by a few π^-. The most serious candidates are the state of mass 2380 MeV and width 70 MeV by the WASA collaboration, in the $^3D_3 {}^3G_3$ coupled partial waves, corresponding to 3^+ [143,144], which was expected by various models [145], and the more recent state at 2.15 GeV which is under investigation at Jlab [136].

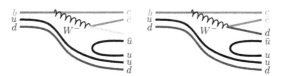

Fig. 1 $\Lambda_b \to J/\psi + p + K$ decay (*left*) and $\Lambda_b \to J/\psi + p + \pi$ decay (*right*)

Fig. 2 Other weak decays of b-hadrons producing a hidden-charm pentaquark

In the sector of single strangeness $S = -1$, there are recurrent discussions about the possibility of dibaryons. See, e.g., [146,147] and refs. there. Note that a study of the atomic (K^-, p, p) system could provide some indirect information. In the two-body case, say (a^+, b^-), the atomic spectrum is much modified if the strong-interaction (a, b) potential generates a bound state near threshold. This was realised by Zel'dovich in the 50's and has been much studied [148–150]. The case of three-body exotic atoms with a combination of Coulomb interaction and short-range hadronic interaction remains to be clarified, but probably deserves a detailed study.[7]

The most advertised case deals with strangeness $S = -2$, following the work by Jaffe[151] on the coherent chromomagnetic attraction in $H = (uuddss)$ with isospin and spin $I = J = 0$. More than 20 experiments have searched without success for the H, ranging from the absence of H in the decay of double hypernuclei [152] to the mass spectra emerging from heavy-ion collisions [153]. New experiments are currently under construction. The status of theoretical studies will be reviewed in Sect. 5.9.1.

The speculations have been extended to the $S = -3$ sector, see, e.g., [154,155].

4.2 Light Hypernuclei, Neutron-Rich Hypernuclei

If the H, considered as $\Lambda\Lambda$ system, is close to binding, as well as some other baryon-baryon systems, it is natural to guess that some multi-baryon configurations will be bound, as an illustration of the quantum phenomenon of Borromean binding [156–158]. There are for instance debates about the existence of (Λ, Λ, N) or (Λ, Λ, n, n) [159–162], as well as $(NN\Xi)$ or $(NN\Xi\Xi)$ [163]. The field of hypernuclear physics is rather active, with new facilities contributing, and is reviewed regularly at dedicated conferences.

Almost immediately after the discovery of charm, the possibility of new species of hypernuclei was envisaged, with hyperons replaced by charmed baryons [44].

5 Theory

In atomic physics, the study of multi-electron systems looks at first rather straightforward: the same non relativistic dynamics is at work, and one has "just" to solve the N-body equations with increasing N. Even there, the progress is slow, and the calculations, which are very delicate, have to be cross-checked. For instance, the last bound state of H_2^+, the excitations of the positronium molecule or the existence of Borromean molecules were discovered only recently [164–167]. In nuclear physics, the situation is even more delicate, as the basic two-body interaction is presumably modified by the presence of other nucleons, and supplemented by 3-body forces.

In QCD, the situation is even worse. If one uses simple constituent models to describe mesons and baryons, their extension to multiquark states is hazardous. If one uses more sophisticated approaches to ordinary hadrons, their application to higher configurations becomes technically very difficult. Before discussing in some details a few specific schemes, we review the early history of exotic hadrons.

---

[6] I thank Katja Krüger and Stefan Schmitt for a clarifying correspondence.

[7] I thank Claude Fayard for several discussions on this topics.

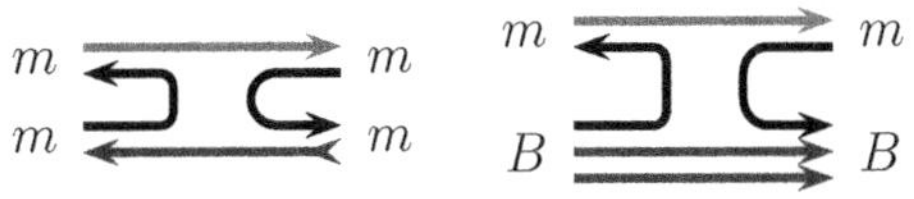

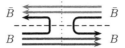

Fig. 3 Duality diagram for meson-meson scattering (*left*) and meson-baryon scattering (*right*)

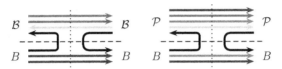

Fig. 4 Duality diagram for baryon-antibaryon scattering. The intermediate state in the *s*-channel (intersecting the *dotted line*) contains two quarks and two antiquarks. The intermediate state in the *t*-channel (crossing the *dashed line*) is made of a quark and antiquark

Fig. 5 Pandora box syndrome for duality. *Left* pentaquarks (four quarks and an antiquark in the *s*-channel) dual of mesons (quark-antiquark in the *t*-channel) in baryonium-baryon scattering. *Right* dibaryon (six quarks in the *s*-channel) dual of mesons (quark-antiquark in the *t*-channel) in pentaquark-baryon scattering

5.1 Duality

In the 60s, much attention was paid to *crossing symmetry*, which relates, e.g., $a + b \rightarrow c + d$ (s) to $a + \bar{c} \rightarrow \bar{b} + d$ (t). The reaction (t) describes what is exchanged to monitor the direct reaction (s). It is reasonable to assume an equivalence between the content of the two reactions, as well as that of the third reaction, $a + \bar{d} \rightarrow \bar{b} + c$ (u). This principle of *duality* was often used as a constraint in building models. For instance, if one describes πN elastic scattering with several $N^{(*)}$ and $\Delta^{(*)}$ resonances, duality dictates to refrain from exchanging σ, ρ, ..., in the *t*-channel.

A consequence of duality is that is the reaction (s) is weak, i.e., without resonances, it does not benefit from coherent *t*-channel exchanges. Though duality was first formulated at the hadron level, it is nowadays explained in terms of quark diagrams. We refer to [168] for a review and references. For instance, in the case of $\pi\pi$ scattering with isospin $I = 0$, the diagram of Fig. 3 (left) shows that the possibility of annihilating a quar-antiquark pair opens the way for $(q\bar{q})$ exchange in the *t*-channel and $(q\bar{q})$ resonances in the *s*-channel. This is why there is a strong $\pi\pi$ interaction with $I = 0$. On the other hand, for isospin $I = 2$, there is s-channel resonance, and the *t*-channel dynamics suffers from a more intricate topology.

Similarly, $\bar{K}N$ scattering, shown in Fig. 3 (right), benefits from both (qqq) resonances in the *s*-channel and $(q\bar{q})$ exchanges in the *t*-channel. This is not the case for KN scattering which is notoriously less strong.

This led Rosner to address the problem of baryon-antibaryon scattering [169], which is depicted in Fig. 4. The interaction is very strong, and benefits, indeed, from the exchanges of mesons in the *t*-channel. Hence there should be resonances in the *s*-channel. This is how baryonium was predicted.

As noted by Roy [168], once one accepts baryonium, in a *Gedankenexperiment* colliding a baryonium against a baryon, meson-exchange is dual of a pentaquark. In turn, the analysis of pentaquark-baryon scattering implies the existence of dibaryons! This is illustrated in Fig. 5.

5.2 String Dynamics

In line with duality, a string model was developed for hadrons, explaining why the squared masses increase linearly with the spin J. This model was echoed by the elongated version of the bag model, the flux-tube model, or some potential models with multi-body forces which will be mentioned later in this section.

The string model was first rather empirical, but later received a support from some topological and non-perturbative properties of QCD, as reviewed recently [170]. The mesons are pictured as a string linking a quark to antiquark, see Fig. 6 (left). The picture of baryons was suggested by Artru, and rediscovered in several approaches [171–176]. It consists of three strings linking each quark to a junction J, which plays

Fig. 6 String model of ordinary mesons (*left*) and baryons (*right*)

Fig. 7 String model of baryonium

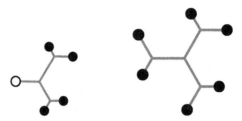

Fig. 8 Possible string structures for pentaquarks (*left*) and dibaryons (*right*)

somewhat the role of a fourth constituent of each baryon, and, in particular, is submitted to an extension of the Zweig rule, which prohibits any internal junction-junction annihilation. In this scheme, the baryonium of Rosner contains two junctions, see Fig. 7 and thus widely differs from a set of two disconnected mesons. The decay of baryonium occurs by string breaking via pair creation: in the case of an external leg, it corresponds to the pionic cascade $\mathcal{B}^* \rightarrow \mathcal{B} + \pi$, while the breaking of the central string corresponds to the decay into baryon and antibaryon. Several experimental states were good candidates for baryonium [9,177], but they were not confirmed in experiments using better antiproton beams. Perhaps the nucleon-antinucleon interaction suffers from too strong an annihilation, while the heavy quark sector gives better opportunities.

The model can be extended to more complicated structures, as to include pentaquarks and dibaryons. Some examples are given in Fig. 8.

5.3 Colour Chemistry

The above string pictures of baryonium was challenged by other schemes: the nuclear model which is reviewed in the next section, and some quark models. One of them describes baryonium states as a diquark and an antidiquark separated by an orbital barrier which suppresses the decay into mesons [178]. In colour chemistry, there is also the possibility of colour-sextet diquark, and thus of $(qq) - (\bar{q}\bar{q})$ states, named 'mock baryonia" which are predicted to be rather narrow [179]. This was, indeed, an interesting challenge to identify spectroscopic signatures of the colour degree of freedom.

5.4 Molecular Binding

The name has evolved. It used to be *quasi-nuclear*, and became *molecular* model. It starts with the observation that the meson-exchange interaction of Yukawa, which successfully binds the deuteron and other nuclei, holds for other hadrons, in particular these containing light quarks.

5.4.1 Baryon-Antibaryon

In the 60s, attempts have been made to describe mesons as baryon-antibaryon states, a simplified version of the bootstrap. Fermi and Yang [180] had noticed that the $N\bar{N}$ potential is often more attractive than the NN

one, as the cancellations in the latter become coherences in the former. But this approach does not account for some important properties of the meson spectrum [7], such as exchange degeneracy, e.g., the ρ and ω having the same mass.

In the late 70s and early 80s, another point of view was adopted, namely that baryon-antibaryon bound states and resonances correspond to new type of mesons, lying higher in mass than ordinary mesons, and preferentially coupled to the baryon-antibaryon channel. In practice, the $N\bar{N}$ potential is deduced from the NN one by the G-parity rule: if the exchange of a meson m (or set of mesons m) contributes as $V_I(m)$ to the NN potential in isospin I, then its contributes as $G_m V_I(m)$ to the $N\bar{N}$ in isospin I [180]. When annihilation is neglected, a rich spectrum of bound states and resonances is found [181,182], with the hope than some of them will not become too broad when annihilation is switched on. Before the discovery of the antiproton, annihilation was believed to be be a short-range correction. However, the annihilation part of the $N\bar{N}$ integrated cross-section was measured to be about twice the elastic one. To fit the cross-sections, one needs an annihilation acting up to about 1 fm, and this makes most quasi-nuclear baryonium states very broad. The molecular model might have better chances to predict exotics in channels with heavy hadrons whose interaction is less absorptive.

5.4.2 Binding of Heavy Hadrons, First Attempts

The Yukawa model made a come back in the heavy quark sector. The $D^{(*)}D^{(*)}$ or $D^{(*)}\bar{D}^{(*)}$ potential, where $D^{(*)} = (c\bar{q})$, is perhaps weaker than the NN potential, but being experienced by heavier particles, it can produce binding as well. The first papers were by Okun and Voloshin and Glashow et al. [183,184]. In the latter case, the idea was to explain some intriguing properties of some high-lying charmonium states, but it was eventually realised that these properties were due to the nodes of the wave function when the states are described as radial excitations [185–187].

5.4.3 Binding of Heavy Hadrons, XYZ

The best known application of the molecular model deals with XYZ states. Almost simultaneously, a few papers came, stressing the attraction due to one-pion-exchange in some $D^{(*)}D^{(*)}$ or $D^{(*)}\bar{D}^{(*)}$ channels [188–191]. Of course, the $X(3872)$ discovery at Belle was greeted as a success of this approach. Today, the $X(3872)$ is generally considered as having both a quarkonium and a multiquark component [192]. For a clarification of the molecular approach, see [16,193–195].

5.4.4 Molecular Pentaquark

The anticharmed pentaquark has been studied with molecular dynamics, as a light baryon linked to an anticharmed meson. See, for instance, [196]. Note that the *uuds* or *ddsu* and *ssud* content of the light sector in the chromomagnetic model is not necessarily the same when mesons are exchanged between a $\bar{D}^{(*)}$ or a $\bar{D}_s^{(*)}$ and a baryon.

After the publication of the LHCb results on the hidden-charm pentaquark [125], it was reminded or discovered that there were some predictions of bound states of a charmed baryon and an anticharmed meson, in particular by Hofman and Lutz [197], Oset and collaborators [198], and Yang et al. [199]. The game was even initiated earlier for the case of hidden-strangeness [200]. Of course, the model has been refined recently and compared to the LHCb results. The literature is comprehensively reviewed in [19].

One might wonder why the hidden-charm pentaquark was not searched for before, given the above predictions.

1. There are many predictions vs. just a few experiments, which are not primarily devoted to spectroscopy.
2. In the molecular model, there are many states of this type, and it is not easy to focus on the configuration giving the best signature for the molecular dynamics. For instance, the binding of a charmed baryon and an anticharmed antibaryon was already envisaged in 1977 [45], shortly after the meson-meson case [183,184], but he meson-baryon case was unfortunately overlooked in the 70s.
3. There is rather generally more effort and more interest in the meson than in the baryon sector.

The negative parity states, either $(3/2)^-$ or $(5/2)^-$, can be accommodated in the molecular model, with a main s-wave between a meson and a baryon, and a small admixture of d-wave, as in the more familiar case of the deuteron. The orbital excitation with positive parity is probably much higher in mass.

5.4.5 Binding Other Heavy Hadrons

In $D^{(*)}\bar{D}^{(*)}$, the annihilation and creation of light quarks opens a coupling to $c\bar{c}$, leading to description of some isoscalar XYZ states as a mixture of molecule and quarkonium states. Isovector states do not mix with quarkonium, except for some minor effects of isospin violation. The sector of double-charm molecules has also received much attention, from the very beginning. See for instance, [189,201,202].

It is immediately realised that if the exchange of mesons bind $D^{(*)}D^{(*)}$, $D^{(*)}\bar{D}^{(*)}$, and $\Sigma_c\bar{D}^{(*)}$ their beauty or mixed charm-beauty analogues and other configurations could as well be formed by this mechanism, such as some meson-baryon and baryon-baryon states [203].

We already mentioned the prediction of hidden-charm pentaquarks as molecules such as $\bar{D}^*\Sigma_c$, before their discovery by LHCb. One could also include $\bar{D}N$ or $\bar{D}$-nuclei states, see, e.g., [204], where the possibility of $\bar{D}NN$ states is also envisaged. The non-strange heavy pentaquark $\bar{D}N$ is related to the non-confirmed state claimed by H1 [121].

In particular, there are many studies of bound states of charmed or doubly-charmed baryons with nucleons or with other singly or doubly charmed baryons [205–210]. One can even speculate about a new periodic table, where the proton is replaced by Ξ_{cc}^{++} and the neutron by Ξ_{cc}^{+} [211]. Of course some rearrangement into triple charm and single charm should be taken into account.

In several model calculations, many interesting hadron-hadron systems are estimated at the edge between binding and non-binding. This means that 3-body or larger systems are probably bound as long as they are not suppressed by the Fermi statistics. So, accounting for the XYZ states implies many other states.

5.5 Gluonium

In QCD, the gluon field carries a colour charge and coupled to itself. This lead to speculations about the existence of new neutral states, made essentially of gluons, except for $g \leftrightarrow q\bar{q}$ loop corrections. This was sometimes hotly debated. I remember for instance that in his talk at the Thessaloniki conference on antiprotons, Van Hove expressed doubts about the necessity of glueballs within QCD, though the written version was somewhat softened [212].

The estimates for the mass and even the quantum numbers of the lightest glueball or gluonium has somewhat changed during the years. The situation is now stabilised, and is reviewed, e.g., in [213]. Note that the lowest glueballs with unusual quantum numbers, such as 0^{+-}, are sometimes found with rather high masses [214].

5.6 Hybrids

Another sector that used to be rather fashionable is the one of hybrids. Very schematically, hybrids combine quark(s), antiquark(s) and gluons. There are even explicit constituent models in which the constituent gluon is given a mass, and interacts via an effective potential, see, e.g., [215].

The first approaches describes hybrids as an excitation of the string or, say, of the gluon field linking the quarks. Some pioneering attempts can be found in [216,217]. In [218], a Born-Oppenheimer treatment was made, with a variant of the bag model used to estimate the energy of the gluon field. For a given interquark separation, the shape of the bag is adjusted as to give the lowest gluon energy. The ground state of the gluon field provide with the usual quarkonium potential. The first excited state of the gluon field is linked to the effective $Q\bar{Q}$ potential within a hybrid. The first $(c\bar{c}g)$ hybrid was predicted near 4.0 GeV, and this estimate has been much refined with the bag model replaced by more elaborate lattice calculations [219,220], or effective field theory [221].

Note that in the late 70s and early 80s, the motivation was that the hybrids would show up clearly atop a clean and well-understood charmonium spectrum! With the proliferation of XYZ states, we realise today that identifying hybrids is a little more challenging.

5.7 QCD Sum Rules

The method of QCD sum rules (QCDSR) is a very ambitious and efficient approach, initiated by Shifman, Vainshtein and Zakharov [222–224], allowing an interpolation between the rigorous perturbative regime and the long-range regime. For a review, see, e.g., [20]. It was first applied to ground-state mesons. There is no limitation, in principle, but the numerical stability become more and more delicate for excitations with the same quantum numbers as the ground state (say, radial excitations). For baryons, the choice of the most appropriate operator is not unique, and triggered some debates. This is of course even more delicate for multiquarks. Hence, the choice of the strategy is often dictated by models, such a diquark-antidiquark.

A few groups have applied QCDSR to exotic states, in particular at Montpellier, São Paulo and Peking, see, e.g., [225, 226] for examples of early attempts, and [17, 19, 227] for some reviews of the state of art over the years. Needless to say that QCDSR calculations of multiquark exotics are extremely delicate and require weeks of cross-checks. So any overnight analysis published just after the release of a new experimental state should have been anticipated long in advance.

5.8 Lattice QCD

Lattice QCD (LQCD) is well known and well advertised, and hardly needs to be presented. For an introduction to the recent literature, see, e.g., [228]. Dramatic progress have been accomplished on various aspects of strong interaction. There are interesting recent developments, for instance to study how the results behave numerically and have to depend theoretically on the pion mass. Another extension deals with *ab-initio* nuclear physics, at least for light nuclei, and multiquark physics [229–236], with in particular, a renewed interest in the study of the H dibaryon.

In the sector of heavy quark spectroscopy, LQCD has confirmed many important results obtained by simpler, but empirical, tools. For instance, the $c\bar{c}$ interaction of charmonium, including its spin dependence, has been estimated, in agreement with the current potential models. Latter, the charmonium spectrum was calculated directly.

As already mentioned, some decisive contributions were made for the study of hybrid mesons with hidden heavy flavour [219, 220]. A Born-Oppenheimer potential can be computed for other configurations, e.g., double-charm baryons, and LQCD is very suited for the treatment of the light degrees of freedom.

For multiquarks, pure *ab-initio* studies are preceded by preliminary investigations inspired by models. For instance, in the sector of doubly-heavy mesons, an effective QQ potential or an effective $D^{(*)}D^{(*)}$ or $B^{(*)}B^{(*)}$ interaction has been studied see, e.g., [237–240]. Nowadays, multiquark states are often studied directly on the lattice, without the intermediate step of an effective potential. Needless to say, LQCD faces difficulties as QCDSR, for the study of multiquark states. In the early days, the quantum numbers were not always clearly identified, and the genuine states were cleanly separated from the artefacts due to lattice boundaries. Nowadays, the field is mature, and an impressive program of multiquark and multihadron physics has been developed by several groups [229–236].

5.9 Constituent Models

In this section, we briefly survey some phenomenological models which have been used to predict some exotic hadrons or describe them after their discovery. It should be stressed, indeed, that if these models are more empirical than QCDSR and LQCD, they can be adapted more flexibly to describe the world of ordinary hadrons and used to extrapolate towards novel structures.

5.9.1 Chromomagnetic Binding

In the 70s, the hyperfine splitting among hadron multiplet was interpreted as the QCD analogue of the Breit-Fermi interaction in QED, namely the spin dependent part of the one-gluon exchange [241]. The corresponding hyperfine interaction reads

$$V_{\text{hyp}} = -\sum_{i<j} \frac{C}{m_i\, m_j} \tilde{\lambda}_i.\tilde{\lambda}_j\, \boldsymbol{\sigma}_i.\boldsymbol{\sigma}_j\, \delta^{(3)}(\boldsymbol{r}_{ij})\,, \tag{3}$$

to be treated at first order, otherwise it requires some regularisation in the channels where it is attractive. A tensor term also contributes beyond s-waves.

In 1977, Jaffe applied this Hamiltonian (3) to the configuration $H = (uudss)$ with $I = J = 0$ assuming *i)* SU(3)$_{\text{F}}$ flavour symmetry, and *ii)* that the short-range correlation $\langle\delta^{(3)}(\boldsymbol{r}_{ij})\rangle$ has the same value as for ordinary baryons [151]. He found a remarkable coherence in the spin-colour operator, with $\langle\sum\tilde{\lambda}_i.\tilde{\lambda}_j\,\boldsymbol{\sigma}_i.\boldsymbol{\sigma}_j\rangle = 24$ for H, to be compared to 8 for N or Λ, i.e., 16 for the $\Lambda\Lambda$ threshold. This means that the H is bound below the $\Lambda\Lambda$ threshold by half the Δ-N splitting, about 150 MeV. The H was searched for in many experiments, without success so far. For instance, the double hypernucleus $_{\Lambda\Lambda}^{6}$He [242] was not seen decaying into $H + \alpha$.

The model has been reexamined thoroughly [243–246], with the result that any correction tends to reduce the binding, and eventually, the H is pushed in the continuum. These corrections are: SU(3)$_{\rm F}$ breaking, more realistic estimate of the short-short correlation (a dibaryon is more dilute than a baryon), and proper inclusion of the full Hamiltonian (kinetic energy and spin-independent confinement besides $V_{\rm hyp}$). Of course, some additional attraction such a scalar-isoscalar exchange between the two Λ can restore the binding.

Meanwhile, some other configurations were found, for which the chromomagnetic interaction, if alone, induces binding. In 1987, the Grenoble group and H.J. Lipkin [117,118] proposed independently and simultaneously a heavy pentaquark $P = (\bar{Q}qqqq)$, where $qqqq = uuds$, $ddsu$ or $ssud$. With the same hypotheses as Jaffe's for the H, and assuming an infinite mass for the heavy antiquark $\bar{Q}$, they found an optimal $\langle \sum \tilde{\lambda}_i.\tilde{\lambda}_j \, \sigma_i.\sigma_j \rangle = 16$, to be compared to 8 for the threshold $(\bar{Q}q) + (qqq)$, i.e., the same binding energy 150 MeV. In a series of papers [247–249], Leandri and Silvestre-Brac applied the chromomagnetic operator

$$- \sum_{i<j} \frac{C}{m_i \, m_j} \tilde{\lambda}_i.\tilde{\lambda}_j \, \sigma_i.\sigma_j \,, \tag{4}$$

with suitable changes for antiquarks, to a variety of configurations, and found several cases, for which the chromomagnetic interaction, if alone, induces binding. As noted by Lichtenberg [250], the $(m_i \, m_j)^{-1}$ dependence is somewhat extreme, as heavy quarks give more compact wave-functions with larger short-range correlation factor. This led some authors to adopt a more empirical prescription,

$$- \sum_{i<j} c_{ij} \, \tilde{\lambda}_i.\tilde{\lambda}_j \, \sigma_i.\sigma_j \,, \tag{5}$$

where the coefficients are taken from a fit to ordinary mesons and baryons [76,77,251]. For instance, in the case of $(c\bar{c}q\bar{q})$ with $J^{PC} = 1^{++}$, the diagonalisation of (5), supplemented by effective constituent masses, gives a state which corresponds nicely to the mass and properties of the $X(3872)$. Its narrowness comes from being almost a pure octet-octet state of colour in the channels $(c\bar{c}) - (q\bar{q})$. As most approaches based on quark dynamics, the chromomagnetic model predicts an isospin $I = 1$ partner of $X(3872)$, with $J = 1$ and $C = +1$ for the neutral, which is not yet seen experimentally.

5.9.2 Chromoelectric Binding

The role of flavour independence In constituent models, one can switch off the spin-dependent part of the interaction, and study the limit of a pure central potential. Its main property is *flavour independence*, with some interesting consequences, which are independent of the detailed shape of the interaction. In particular, if a constituent mass increases, the kinetic energy decreases, and one gets a lower energy, or, say, more binding. This explains why the Υ family has more bound states than the Ψ family below the open-flavour threshold.

Another well-studied consequence deals with doubly-heavy tetraquarks $(QQ\bar{q}\bar{q})$. The 4-body system benefits from the heavy-heavy interaction, which is absent in the $(Q\bar{q}) + (Q\bar{q})$ threshold. Hence, for a large enough Q-to-q mass ratio, the system becomes stable. Explicit calculations with a colour-additive model

$$V = -\frac{16}{3} \sum_{i<j} \tilde{\lambda}_i.\tilde{\lambda}_j \, v(r_{ij}) \,, \tag{6}$$

normalised to the quarkonium potential $v(r)$, indicates that in such spin-independent interaction, $(cc\bar{q}\bar{q})$ is probably unbound, but $(bb\bar{q}\bar{q})$ stable [252–255].

One hardly finds similar configurations. If, for instance, one considers $(QQqqqq)$ within (6), there is no obvious gain with respect to the best threshold made of two baryons, but flavour independence dictates that before any spin corrections, the threshold $(QQq) + (qqq)$ is lower than $2(Qqq)$.

The stability of $(QQ\bar{q}\bar{q})$ for large mass ratios is intimately related to the observation that the hydrogen molecule is more deeply bound (in units of the threshold binding energy) than the equal-mass molecules such as Ps$_2$. The Hamiltonian for (M^+, M^+, m^-, m^-) reads

$$H = H_0 + H_1 = \left[\left(\frac{1}{4M} + \frac{1}{4m} \right) \sum p_i^2 + V \right]$$
$$+ \left(\frac{1}{4M} - \frac{1}{4m} \right) \left(p_1^2 + p_2^2 - p_3^2 - p_4^2 \right) \,. \tag{7}$$

Its C-breaking term H_1 lowers the ground-state energy with respect to the C-even part H_0. Since H_0, a rescaled version of the Ps_2 molecule, and H, have the same threshold, and since Ps_2 is stable, H_2 is even more stable. Clearly, the Coulomb character of V hardly matters in this reasoning. The key property is that the potential does not change when the masses are modified. In the case of a simple spin-independent potential with the colour-additive rule (6), one does not get binding for $M = m$, but binding shows up when M/m increases.

Note that, if one breaks the symmetry of Ps_2 differently and considers instead (M^+, m^+, M^-, m^-) configurations, stability is lost for $M/m \gtrsim 2.2$, or, of course, $M/m \lesssim 1/2.2$ [256, 257]. A protonium atom cannot polarise strongly enough a positronium atom and attach it. Similarly a chromoelectric potential cannot produce a bound $(Qq\bar{Q}\bar{q})$ tetraquark: one needs chromomagnetism or long-range Yukawa interaction.

More detailed comparison of Ps_2 and tetraquark Another problem is to understand why the positronium molecule Ps_2 (neglecting annihilation) lies slightly below its dissociation threshold, while a chromo-electric model, associated with the colour-additive rule (6), does not bind (at least according to most computations, an exception being [258]). This is due to a larger disorder in the colour coefficients than in the electrostatic strength factors $q_i q_j$ entering the Coulomb potential in Ps_2. Consider, indeed, a 4-body Hamiltonian scaled to

$$H[G] = p_1^2 + p_2^2 + p_3^2 + p_4^2 + \sum_{i<j} g_{ij}\, v(r_{ij}), \quad G = \{g_{ij}\}, \tag{8}$$

with $\sum_{i<j} g_{ij} = 2$, having at least one bound state in the symmetric case $G = G_0$ for which $g_{ij} = 1/3 \; \forall i, j$. The variational principle, with the solution of this symmetric case as a trial function, implies that

$$\min(H[G]) \le \min(H[G_0]), \tag{9}$$

so that, schematically, the more asymmetric the distribution G, the lower the ground-state energy. For the cases of interest, one could be more precise, as the distribution is of the type $G(\lambda) = (1 - \lambda)\, G_0 + \lambda\, G_1$. The ground-state energy $E(\lambda)$ is convex as a function of λ, and, since maximal at $\lambda = 0$, it is convexly decreasing as a function of $|\lambda|$ for both $\lambda \ge 0$ and $\lambda \le 0$. If one compares two values of λ with opposite sign, then it is a good guess, but not a rigorous statement, to say that the larger $|\lambda|$, the lower the energy, as $E(\lambda)$ is nearly parabolic near $\lambda = 0$.

Now consider the actual values in Table 1.

The asymmetry parameter $|\lambda|$ is much larger for Ps_2 than for $Ps + Ps$. This explains why the positronium molecule is bound.[8]

A tetraquark with a pure $\bar{3}3$ colour structure has its asymmetry parameter smaller than that of the threshold and has the same sign, so this toy model is rigorously unbound. For $6\bar{6}$, the signs are different, and the magnitude are comparable, so the analysis is more delicate. We note that $|\lambda| = 2$ for Ps_2 results in a fragile binding, so it is not surprising that $|\lambda| = 7/8$ leaves $6\bar{6}$ unbound in numerical studies.

To summarise, the tetraquark is penalised by the non-abelian character of the colour algebra in the simple model (6), and its stability cannot rely on the asymmetries of the potential energy. It should use other asymmetries, in particular through the masses entering the kinetic energy, or spin effects, or mixing of $\bar{3}3$ and $6\bar{6}$, or the coupling to decay channels, etc.

String dynamics The prescription (6) is of course rather crude, as it treats colour as a global variable similar to isospin, and assumes arbitrarily the choice of a pairwise interaction. It can be justified in the limiting cases were $(N - 1)$ quarks are clustered and the last quark isolated. So it might provide a reasonable interpolation for other spatial configurations.

As mentioned earlier, an alternative is the Y-shape interaction of Fig. 6, now used to estimate a potential which reads

$$V_Y^{\text{conf}} = \lambda \, \min_J (r_{1J} + r_{2J} + r_{3J}), \tag{10}$$

if $\lambda\, r_{12}$ is the linear confinement of mesons. Note that (10) features the minimal path linking three points, an old problem first raised by Fermat and Torricelli.[9] For baryons alone, solving the 3-quark problem with (10)

---

[8] Of course, this is not exactly what is written in the textbooks on quantum chemistry, but a large $|\lambda|$ means that Ps_2 benefits from correlation and anticorrelation effects in the various pairs.

[9] This minimal distance is also intimately linked to a theorem by Napoleon: by adding external equilateral triangle on each side of the quark triangle, one easily construct the location of the junction J and the value of the minimal sum of distances [259].

Table 1 Strength coefficients for the positronium molecule and the tetraquark states with colour wave-function $\bar{3}3$ or $6\bar{6}$, as well as for their threshold

$(abcd)$	$v(r)$	g'	g''	λ
Ps + Ps	$-1/r$	1	0	1
Ps$_2$	$-1/r$	-1	1	-2
$(q\bar{q}) + (q\bar{q})$	$-1/r, r$	1	0	1
$[(qq)_{\bar{3}}(\bar{q}\bar{q})_3]$	$-1/r, r$	1/2	1/4	1/4
$[(qq)_6(\bar{q}\bar{q})_{\bar{6}}]$	$-1/r, r$	$-1/4$	5/8	$-7/8$

The list of coefficients is of the type $\{g', g', g'', g'', g'', g''\}$ after suitable renumbering, and summarised as $G = (1-\lambda)\,G_0 + \lambda\,G_1$, where G_0 is the symmetric case $g' = g'' = 1/3$

Fig. 9 Potential for the tetraquark in a simple string model. The minimum is taken of the connected diagram (*right*) and the two possible meson-meson diagrams (*left*)

Fig. 10 Some contributions to the pentaquark potential: for each set of quark positions, the minimum is taken of the connected diagrams such as the one on the *right* and disconnected diagrams such as the one on the *left*

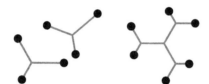

Fig. 11 Some contributions to the hexaquark potential: for each set of quark positions, the minimum is taken of the connected diagrams such as the one on the *right* and disconnected diagrams such as the one on the *left*

instead of the perimeter-based confinement

$$V_P^{\text{conf}} = \frac{\lambda}{2}\,(r_{12} + r_{23} + r_{31})\,, \tag{11}$$

is technically more involved, but does not change the patterns of the excitation spectrum.

For tetraquarks, a precursor of non-pairwise potential was the flip-flop model [260], with quadratic confinement. In the more recent linear version, one takes the minimum of the two possible meson-meson pairing configurations and the connected diagram shown in Fig. 9.

The good surprise is that this potential is more attractive than the one deduced from the colour-additive rule (6) and thus is more favourable for binding [259,261]. The same property is observed for the pentaquark [262], Fig. 10, the dibaryon, Fig. 11 and the system made of 3 quarks and 3 antiquarks [263], Fig. 12. However, this string potential corresponds to a kind of Born-Oppenheimer approximation for the colour fluxes at given quark positions. It mixes colour $\bar{3}$ and 6 for any pair of quark, to optimise the overall energy. Thus it holds for quarks bearing different flavours. In [264], an attempt is proposed to combine quark statistics and string dynamics in the case of tetraquarks with identical quarks or antiquarks: not surprisingly, the possibility of binding is reduced.

Note that in the above potential models, there is no explicit junction à la Rossi and Veneziano: the minimal length is the only guidance, and nothing refrain a transition from a disconnected flip-flop to a connected string. It is observed, however, that the ground-state uses mainly the flip-flop term, so there are probably two classes of states: the lowest have the same number of junctions as the thrshold, and thus are rather broad if the fall-apart decay is energetically allowed. The second class might be higher in mass, but benefits from a kind of metastability due to its junction content.

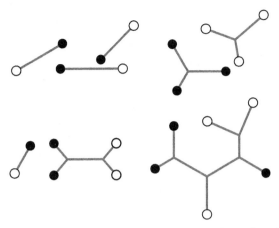

Fig. 12 Some contributions to the potential for 3 quarks and 3 antiquarks: for each set of quark positions, the minimum is taken of connected diagrams (*bottom, right*) and disconnected diagrams (*top* and *bottom left*)

These string-inspired potentials, however simplistic, have motivated some studies. For instance, the literature on Steiner trees helps in optimising the location of the junctions and the estimate of the minimal strength [259,265–267].

5.9.3 Combining Chromoelectric and Chromomagnetic Effects

As discussed at length in Sect. 5.9.2, the tetraquark $(QQ\bar{q}\bar{q})$ might become stable by the sole effect of chromomagnetic forces, but the stability requires a very large M/m mass ratio. It was also noticed [268,269], that some configurations such as $J^{PC} = 1^{++}$, there is a favourable spin-spin interaction in the light sector that benefits the tetraquark and not the threshold. A calculation by Rosina et al. [269], using a potential tuned to fit the ordinary hadrons, and the colour-additive rule, demonstrated stability for $(cc\bar{u}\bar{d})$, and this was confirmed by Vijande et al. [270].

There are not too many similar configurations, in which both chromoelectric and chromomagnetic effects are favourable. For instance, for the doubly-flavoured dibaryons $(QQqqqq)$, the two thresholds $(QQq)+(qqq)$ and $(Qqq) + (Qqq)$ are nearly degenerate, at least in case the charm, $Q = c$. The former threshold benefits from the chromoelectric attraction between the two heavy quarks, while the second can receive a favourable chromomagnetic interaction. It is, however, difficult for the 6-quark configuration to combine astutely the two effects, and double-charm baryons seem hardly stable in constituent models [271].

5.9.4 Solving the Schrödinger Equation

Once a model is adopted, typically

$$V = -\frac{16}{3} \sum_{i<j} \tilde{\lambda}_i.\tilde{\lambda}_j \left[v(r_{ij}) + v_{ss}(r_{ij})\, \boldsymbol{\sigma}_i.\boldsymbol{\sigma}_j \right] , \tag{12}$$

or a more complicated model involving multi-body forces, one is left with solving the wave equation to obtain the energies and the wavefunctions.

In the case of quarkonium, there are dozens of reliable numerical methods, starting with the one proposed by Hartree [272], consisting of matching a regular outgoing solution and a regular incoming solution of the radial equation. Most methods can be extended to the case of coupled equations in the unnatural-parity channels, similar to the $s - d$ coupling of the deuteron in nuclear physics. Many rigorous results have been revisited and developed for studies related to the quarkonium: scaling law, virial-like theorems, level order, value of wave-function at the origin for the successive radial excitations, etc. This is reviewed in [273–275].

The 3-body problem one has to solve for baryons is often a psychological barrier that refrain to envisage a detailed and rigorous treatment beyond the harmonic oscillator, and there is sometimes a confusion between the "constituent model of baryons" and the "harmonic oscillator model of baryons". For instance, the negative-parity excitation being half-way of the positive-parity excitation is a property of the latter, but not necessarily of the former.

For baryons, there are just a few rigorous results, for instance dealing the level order of states when the degeneracy obtained in the harmonic oscillator or hyperscalar limit are lifted at first order [276,277]. As for the numerical solution, various methods have been used, such as hyperspherical expansion, Faddeev equations, or variational methods, with a comparison in [278]. Note that among the latter, the expansion on a harmonic-oscillator basis is less and less adopted, as it does not give good convergence for the short-range correlation parameters which enter some decays and some production mechanisms.

For multiquark systems, the hyperspherical expansion works rather well [270], especially for deeply bound states, if any, for which the various r.m.s. radii are of the same order of magnitude. The tendency is towards the method of Gaussian expansion, already used in quantum chemistry. If $\tilde{\boldsymbol{x}}^{\dagger} = \{\boldsymbol{x}_1, \ldots, \boldsymbol{x}_{N-1}\}$ are Jacobi variable describing the relative motion of the N-body system, then the wave-function is sought as

$$\Psi = \sum_{a,i} \gamma_{a,i} \exp[-\tilde{\boldsymbol{x}}^{\dagger}.A_{a,i}.\tilde{\boldsymbol{x}}/2] |a\rangle , \tag{13}$$

where $|a\rangle$ denotes the basis of internal colour-spin-flavour states needed to construct the most general colour singlet of given overall spin and flavour, and $A_{a,i}$ is a $(N-1) \times (N-1)$ definite positive symmetric matrix. For a given set of matrices $A_{a,i}$, the weight factors $\gamma_{a,i}$ are given by a generalised eigenvalue equation. Then the range parameters entering the $A_{a,i}$ are minimised numerically. This latter task is not obvious, and sophisticated techniques have been developed, such as the stochastic search of Mitroy et al. [279], and the method of Hiyama, Kino and Kamimura [280], in which a specific choice of Jacobi variables is chosen in each channel. The variational energies below the threshold are interpreted as bound states. The scaling properties of the states above the threshold are scrutinised to decide whether they are genuine resonances or artefacts of the finite variational expansion. This is somewhat similar to the analyses of states above the threshold in QCDSR or LQCD.

5.9.5 Limits and Extensions

Diquarks Hadron spectroscopy with diquarks has ups and downs, or, say, pros and cons. The art of physics is to identity effective degrees of freedom at a certain scale, and any simplification is welcome. The concept of diquark is very useful to describe many phenomenons such a multiple production [281], and might be legitimated by boson-fermion symmetry [282]. In spectroscopy, the diquark model provides a simple physical picture that might be obscured in complicated three-body calculations, and it provides an efficient tool for investigating exotics [283]. Some reservations are however expressed about the abundant literature on diquarks, which is sometimes confusing, as some recent papers do refer to the old ones:

- Explicit quark model calculations find diquark correlations, but the effect depends on a somewhat arbitrary regularisation of the chromomagnetic interaction [284]. In other studies, no diquark clustering is found for the ground state [285]. This means that the diquark picture of low-lying baryons implies a change of dynamics. On the other hand, a dynamical diquark formation is obtained for highly excited baryons: if the total angular momentum L increases, it is first shared equally among the 3 quarks, but at larger L the third quark takes $\ell_y \sim L$ while the two first are mainly in a relative s-wave (of course modulo permutations) [286].
- The diquark model becomes a good approximation for the ground state of double-charm baryons (QQq) provided it is handled and interpreted properly, since a part of the QQ interaction comes from the third quark. In the case of the harmonic oscillator, with the usual definition of the Jacobi variables $\boldsymbol{x} = \boldsymbol{r}_2 - \boldsymbol{r}_1$ and $\boldsymbol{y} = (2\boldsymbol{r}_3 - \boldsymbol{r}_1 - \boldsymbol{r}_2)/\sqrt{3}$, one notices that $\sum r_{ij}^2 = (3/2)(x^2 + y^2)$, so the direct QQ interaction x^2 gets a 50 % increase from the light quark. Moreover, the first excitations of (QQq) (besides spin) occur with (QQ), so a new diquark has to be estimated for each level. In fact a Born-Oppenheimer treatment à la H_2^+ is more appropriate.

Born–Oppenheimer Along this review, we stressed the importance of the Born-Oppenheimer point of view: the quarkonium potential is just an effective interaction integrating out the gluon and light-quark degrees of freedom surrounding a heavy quark and a heavy antiquark; doubly-heavy baryons (QQq) have typically two scales of energy for its spectrum, the excitation energy of the QQ pair, and the excitation energy of the light quark, and this comes very naturally in the Born-Oppenheimer treatment, as it does in atomic physics for H_2^+.

We also mentioned that hybrid mesons, schematically noted ($Q\bar{Q}g$) have been studied by estimating an "excited quarkonium potential": for a given $Q\bar{Q}$ separation, the energy of the excited gluon field is computed either from a simple bag model, or from LQCD. A Born-Oppenheimer treatment of $(QQ\bar{q}\bar{q})$ would also be feasible, to cross-check the 4-body calculations discussed in Sect. 5.9.2.

Recently, a very comprehensive and unifying Born-Oppenheimer treatment of the XYZ states was presented [287], with a series of effective $Q\bar{Q}$ potentials adapted to the quantum states of the light constituents (gluon and light quarks). This work could of course be generalised as to include qqq instead of $q\bar{q}$ in the light sector, and applied to the LHCb pentaquarks.

There are interesting alternatives for the coupling of $Q\bar{Q}$ to light quarks and antiquarks. For instance, a relatively weak interaction between J/ψ or η_c with nuclei has been proposed, but, once more, this potential is probed by a particle which is heavy enough to get bound. In *hadrocharmonium*, the compact $Q\bar{Q}$ is trapped inside an extended light hadron. See [195, 288, 289].

6 Super-Exotics

In the future, the field of exotic hadrons might be extended very far, if some new constituents are discovered, for instance at LHC. As stressed by many authors (see, for instance, [290]), some anomalies in the Higgs branching ratios, and the other recent observations at CERN might indicate the existence of heavy scalars. Among the many possible scenarios, there is the possibility of scalar quarks with colour charge 3, having the same colour-mediated interaction as the ordinary quarks, except of course for spin and statistics. Perhaps the lowest of theses scalars will live long enough to hadronise, and give rise to new hadrons.

The lifetime of the hypothetical scalar quarks would of course be crucial. Remember that for many years the potential model of charmonium and bottomonium was extrapolated towards topomonium. Then it was realised that, when the mass of t quark is too large, its weak decay becomes too fast for hadronisation to take place [291]. If this is the case for any new super-heavy quark or squark, then the realm of exotic hadrons will remain the privilege of low-energy hadron physics.

Let us consider a scenario with heavy scalar quarks of colour 3, as predicted by supersymmetry or any other BSM (Beyond Standard Model) theory. (There are speculations about colour 6 or other gauge groups.) We note them "squarks" without necessarily endorsing supersymmetry. Then, one gets squarkonia $(\varsigma\bar{\varsigma})$, and perhaps the scalar ground state could be responsible for the excess tentatively seen at 750 GeV. See, e.g., [292]. I apologise for the lack of space needed to cite the other $\gtrsim 1000$ relevant references, and the lack of expertise to separate the really innovative papers from the "ambulance chasing" opportunities [293].

The lightest charged hadron is likely $\underline{d}u$ for $\varsigma = \underline{d}$ with charge $-1/3$ or $\underline{u}ud$ of Λ_Q type, but spin 0, if $\varsigma = \underline{u}$ with charge $2/3$.

The baryon sector becomes entertaining if several squarks are put together. For instance $(\varsigma\varsigma q)$ has quantum numbers $(1/2)^-$ as the $\varsigma\varsigma$ disquark contains one unit of orbital excitation, to fulfill the Bose statistics. The triply-superexotic $(\varsigma\varsigma\varsigma)$ requires a fully antisymmetric wave function $\propto \boldsymbol{\rho} \times \boldsymbol{\lambda}$, as per Eq. (1) that was proposed 50 years ago for the last member of the $N = 2$ multiplet for light baryons, the so-called $[20, 1^+]$ multiplet in the notation of, e.g., [294]. Fifty years after the first baryon conference, the physics has evolved, but one uses the same wave function (besides some scaling).

7 Outlook

Exotic hadrons are coming up and down, or back and forth if one prefers. For outsiders, there is somewhat a discredit on a field in which many states regularly show up rapidly and then disappear within a few years, with many papers hastily written to explain the new hadrons, and never any deep understanding on why the models fail.

This is of course, a superficial view. The truth is that the experiments are very difficult, due to a low statistics for a signal above an intricate background. The situation is even worse for broad resonances. As often stressed [295, 296], the location of the peak might depend on the production process and on the final state, so that there is often the question on whether two or more candidates correspond to the same "state". So far, several mass ranges and flavour sectors have been investigated with already a few good candidates for exotics such as $X(3872)$. The main lesson of the last decades is clearly the richness of the configurations mixing heavy and light constituents. The hidden-charm sector is already much explored, and even if the recent discoveries have to be checked, refined and completed, there is clearly an urgent need for investigating the double charm and other flavor exotic sectors. Hopefully, there will be an emulation between electron-positron machines, proton colliders, heavy-ion collisions and intense hadron facilities to develop this program.

On the theory side, the fundamental approaches such as LQCD or QCDSR involve lengthy calculations and usually come second, to clarify some predictions. First appear the speculations based on more empirical models, inspired by QCD.

In phenomenological QCD, *Arx tarpeia Capitoli proxima*:[10] Physicists inventing molecules or quark clusters that account for the latest exotics are praised for their inspiration, invited to many workshops and conferences, hired in committees, etc., but, when the states disappear, the same model-builders are pointed out as lacking rigour and corrupting the field. To be honest, it often happens that the pioneers have written their predictions rather carefully, with many warnings about the width of states in the continuum, but that the followers have developed the models beyond any reasonable control, with tables and tables of states which simply fall part.

For one of the models of multiquarks, a warning was set [297] that only a few states will show up as peaks, the others, being very broad, just paving the continuum, but the warning was often forgotten. This *Pandora box syndrome* affects practically all approaches. For instance, the molecular picture was first focused on one-pion exchange, with strict selection rules, that makes the interaction repulsive in any second channel. With the exchange of vector mesons, and the old-fashioned potentials replaced by effective theories, the molecular picture has gained some respectability, but has dramatically widened the spectrum of predictions. In the diquark model, Frederikson and Jandel [298] discussed the existence of possible dibaryons, and the problem was rediscovered recently [299], that the picture of XYZ as diquark-antidiquark states probably implies higher configurations. This is why genuinely exotic states, lying below their dissociation threshold, will always play a privileged role, and the searches in the sector of double heavy flavour should be given the highest priority.

In any case, the studies on the mechanisms of confinement and the resulting nuclear forces should be pursued. The strong-coupling limit of QCD and lattice simulations explain how a linear confinement is developed between a quark and an antiquark. It is already more delicate to understand its extension to baryons, with possible interplay between Δ-shape and Y-shape configurations for the gluon density, and even more speculative to write down a confining Hamiltonian for multiquarks. What remain to be further investigated, based on the already rich literature is how clustering sometimes occur in baryons and higher multiquarks, and how the linear confinement breaks to lead to the actual decay of resonances and the virtual hadron-hadron components, i.e., to operate the matching of the quark dynamics and the long-range hadron-hadron interaction.

Acknowledgments This review benefited from stimulating discussions with my past and present collaborators and with many colleagues. It was completed during a visit at RIKEN, in the framework of a JSPS-CNRS agreement, and finalised during a visit at ECT*. The hospitality provided by Emiko Hiyama, by Jochen Wambach, and by their colleagues is gratefully acknowledged. Thanks are also due to M. Asghar for very interesting comments on the manuscript.

References

1. Segrè, E.: From X-rays to quarks: modern physicists and their discoveries (Freeman, San Francisco, CA, 1980) trans. of: Personaggi e scoperte nella fisica contemporanea. Milano: Mondadori (1976)
2. Pais, A.: Inward Bound: of Matter and Forces in the Physical World. Clarendon Press, Oxford (1986)
3. Ezhela, V.V., Filimonov, B.B., Lugovsky, S.B., Polishchuk, B.V., Striganov, S.I., Stroganov, Y.G., Armstrong, B., Barnett, R.M., Groom, D.E., Gee, P.S., Trippe, T.G., Wohl, C.G., Jackson, J.D.: Particle Physics, One Hundred Years of Discoveries: An Annotated Chronological Bibliography. AIP, New York (1996)
4. Breit, G.: Aspects of nucleon-nucleon scattering theory. Rev. Mod. Phys. **34**, 766–812 (1962)
5. Chew, G.F., Low, F.E.: Effective range approach to the low-energy p wave pion-nucleon interaction. Phys. Rev. **101**, 1570–1579 (1956)
6. Chew, G.F.: The Analytic S-Matrix: a Basis for Nuclear Democracy. Benjamin, New York (1966)
7. Ball, J.S., Scotti, A., Wong, D.Y.: One-boson-exchange model of NN and $N\bar{N}$ interaction. Phys. Rev. **142**, 1000–1012 (1966)
8. Olive, K.A., et al.: (Particle Data Group), Review of particle physics. Chin. Phys. C **38**, 090001 (2014)
9. Montanet, L., Rossi, G.C., Veneziano, G.: Baryonium physics. Phys. Rep. **63**, 149–222 (1980)
10. Amsler, C., Tornqvist, N.A.: Mesons beyond the naive quark model. Phys. Rep. **389**, 61–117 (2004)
11. Klempt, E., Zaitsev, A.: Glueballs, hybrids, multiquarks. Experimental facts versus QCD inspired concepts. Phys. Rep. **454**, 1–202 (2007). arXiv:0708.4016 [hep-ph]

---

[10] "The Tarpeian Rock is close to the Capitol". In ancient Rom, some prominent wining generals had a triumph procession, the Fifth-Avenue parade of that time, ending to the Capitol Temple, and the worse criminals were plunged to death from the nearby Tarpeian Rock. This Latin sentence is probably posterior to the Roman period. I thank the historian François Richard for a correspondence on this point.

12. Dalpiaz, P., Klapisch, R., Lefevre, P., Macri, M., Montanet, L., Mohl, D., Martin, A., Richard, J.-M., Pirner, H.J.L.: Tecchio, Physics at SuperLEAR. In: Les Houches. Workshop **1987**(0414), 0414 (1987)

13. Amsler, C., et al.: Perspectives in hadron and quark dynamics, prospects of hadron and quark physics with electromagnetic probes. In: Proceedings, 2nd ELFE Workshop on Hadronic Physics, Saint Malo, France, Sept 23–27, 1996. Nucl. Phys. **A622**, 315C–354C (1997)

14. von Harrach, D.: (COMPASS), The COMPASS experiment at CERN, Quark lepton nuclear physics. In: Proceedings, International Conference, QULEN'97, Osaka, Japan, May 20–23, 1997. Nucl. Phys. **A629**, 245C–254C (1998)

15. Destefanis, M.: (PANDA) The PANDA experiment at FAIR. In: Proceedings, 7th Joint International Hadron Structure'13 Conference (HS 13. Nucl. Phys. Proc. Suppl. **245**, 199–206 (2013)

16. Swanson, E.S.: The new heavy mesons: a status report. Phys. Rep. **429**, 243–405 (2006). arXiv:hep-ph/0601110

17. Nielsen, M., Navarra, F.S., Lee, S.H.: New charmonium states in QCD sum rules: a concise review. Phys. Rep. **497**, 41–83 (2010). arXiv:0911.1958 [hep-ph]

18. Hambrock, C.: Exotic heavy quark spectroscopy: theory interpretation vs. data. In: Proceedings, 14th International Conference on B-Physics at Hadron Machines (Beauty 2013), PoS Beauty **2013**, 044 (2013), arXiv:1306.0695 [hep-ph]

19. Chen, H.-X., Chen, W., Liu, X., Zhu, S.-L.: The hidden-charm pentaquark and tetraquark states (2016). doi:10.1016/j.physrep.2016.05.004, arXiv:1601.02092 [hep-ph]

20. Narison, S.: QCD as a theory of Hadrons: from partons to confinement. In: Nuclear Physics and Cosmology. Cambridge University Press, Cambridge Monographs on Particle Physics (2004)

21. Pennington, M.R. (ed.): Proceedings, 16th International Conference on Hadron Spectroscopy (Hadron 2015), vol. 1735 (2016)

22. Bicudo, P., Dubnicka, S., Giacosa, F., Kaminski, R., Marinkovic, M.K. (eds.) Proceedings, International Meeting of Excited QCD 2015, vol. 8 (2015)

23. https://baryons2016.physics.fsu.edu/indico/event/0/contributions

24. Achard, P., et al.: (L3), Measurement of exclusive $\rho^+\rho^-$ production in mid-virtuality two-photon interactions and study of the $\gamma\gamma^* \to \rho\rho$ process at LEP. Phys. Lett. B **615**, 19–30 (2005). arXiv:hep-ex/0504016 [hep-ex]

25. Achard, P., et al.: (L3), Measurement of exclusive $\rho^0\rho^0$ production in two photon collisions at high Q^2 at LEP. Phys. Lett. B **568**, 11–22 (2003). arXiv:hep-ex/0305082 [hep-ex]

26. Anikin, I.V., Pire, B., Teryaev, O.V.: Search for isotensor exotic meson and twist-4 contribution to $\gamma^*\gamma \to \rho\rho$. Phys. Lett. B **626**, 86–94 (2005). arXiv:hep-ph/0506277 [hep-ph]

27. Dorofeev, V. et al.: (VES), The $J^{PC} = 1^{-+}$ hunting season at VES. In: Proceedings, 9th International Conference on Hadron Spectroscopy (Hadron 2001), AIP Conference Proceedings **619**, 143–154 (2002), [143(2001)], arXiv:hep-ex/0110075 [hep-ex]

28. Thompson, D.R., et al.: (E852), Evidence for exotic meson production in the reaction pi-p-> eta pi-p at 18-GeV/c. Phys. Rev. Lett. **79**, 1630–1633 (1997). arXiv:hep-ex/9705011 [hep-ex]

29. Alekseev, M., et al.: (COMPASS), Observation of a $J^{PC} = 1^{-+}$ exotic resonance in diffractive dissociation of 190 GeV/c π^- into $\pi^-\pi^-\pi^+$. Phys. Rev. Lett. **104**, 241803 (2010). arXiv:0910.5842 [hep-ex]

30. Al Ghoul H., et al.: (GlueX), First results from the GlueX experiment. In: 16th International Conference on Hadron Spectroscopy (Hadron 2015) Newport News, Virginia, USA, Sept 13–18, 2015 (2015) arXiv:1512.03699 [nucl-ex]

31. Meyer, C.A., Swanson, E.S.: Hybrid mesons. Prog. Part. Nucl. Phys. **82**, 21–58 (2015). arXiv:1502.07276 [hep-ph]

32. Amsler, C., Hanhart, C.: Non-$q\bar{q}$ mesons, in [6]

33. Aubert, B., et al.: (BaBar), Observation of a narrow meson decaying to $D_s^+\pi^0$ at a mass of 2.32 GeV/c^2. Phys. Rev. Lett. **90**, 242001 (2003). arXiv:hep-ex/0304021 [hep-ex]

34. Besson, D., et al.: (CLEO), Observation of a narrow resonance of mass 2.46 GeV/c^2 decaying to $D*+_s\pi^0$ and confirmation of the $D_{sJ}^*(2317)$ state. Phys. Rev. D **68**, 032002 (2003). [Erratum: Phys. Rev. **D75**, 119908 (2007)]

35. Godfrey, S., Isgur, N.: Mesons in a relativized quark model with chromodynamics. Phys. Rev. D **32**, 189–231 (1985)

36. Pignon, D., Piketty, C.A.: Charmed meson spectra. Phys. Lett. B **81**, 334–338 (1979)

37. Schnitzer, H.J.: Spin dependence and glueball mixing with $\theta(1640)$ in ordinary meson spectroscopy. Nucl. Phys. B **207**, 131–156 (1982)

38. Cahn, R.N., Jackson, J.D.: Spin-orbit and tensor forces in heavy-quark light-quark mesons: implications of the new D_s state at 2.32 GeV. Phys. Rev. D **68**, 037502 (2003). arXiv:hep-ph/0305012

39. Matsuki, T., Morii, T., Sudoh, K.: New heavy-light mesons $Q\bar{q}$. Prog. Theor. Phys. **117**, 1077–1098 (2007). arXiv:hep-ph/0605019 [hep-ph]

40. Nielsen, M., D'Elia, R.M., Navarra, F.S., Bracco, M.E.: Testing the nature of the $D_{sJ}^+(2317)$ and $X(3782)$ states using QCD sum rules. In: Proceedings, 18th International IUPAP Conference on Few-Body Problems in Physics (FB18). Nucl. Phys. **A790**, 526–529 (2007)

41. Vijande, J., Fernandez, F., Valcarce, A.: The puzzle of the D and D_s mesons, quarks and nuclear physics. In: Proceedings, 4th International Conference, QNP 2006, Madrid, Spain, June 5–10, 2006, Eur. Phys. J. **A31**, 722–724 (2007a)

42. Segovia, J., Entem, D.R., Fernandez, F.: Charmed-strange Meson spectrum: old and new problems. Phys. Rev. D **91**, 094020 (2015). arXiv:1502.03827 [hep-ph]

43. Godfrey, S., Moats, K.: Properties of excited charm and charm-strange Mesons. Phys. Rev. D **93**, 034035 (2016). arXiv:1510.08305 [hep-ph]

44. Dover, C.B., Kahana, S.H.: Possibility of charmed hypernuclei. Phys. Rev. Lett. **39**, 1506–1509 (1977)

45. Dover, C.B., Kahana, S.H., Trueman, T.L.: Bound states of Charmed Baryons and anti-Baryons. Phys. Rev. D **16**, 799–815 (1977)

46. Isgur, N., Lipkin, H.J.: On the possible existence of stable four quark scalar mesons with charm and strangeness. Phys. Lett. B **99**, 151 (1981)

47. Cho, S., et al.: (ExHIC), Multi-quark hadrons from Heavy Ion Collisions. Phys. Rev. Lett. **106**, 212001 (2011). arXiv:1011.0852 [nucl-th]

48. Yu-qi, C., Su-zhi, W.: Production of four-quark states with double heavy quarks at LHC. Phys. Lett. B **705**, 93–97 (2011). arXiv:1101.4568 [hep-ph]

49. Hyodo, T., Liu, Y.-R., Oka, M., Sudoh, K., Yasui, S.: Production of doubly charmed tetraquarks with exotic color configurations in electron-positron collisions (2012). arXiv:1209.6207 [hep-ph]

50. Esposito, A., Papinutto, M., Pilloni, A., Polosa, A.D., Tantalo, N.: Doubly charmed tetraquarks in B_c and Ξ_{bc} decays. Phys. Rev. D **88**, 054029 (2013). arXiv:1307.2873 [hep-ph]

51. Abe, K., et al.: (Belle), study of double charmonium production in e+ e− annihilation at $\sqrt{s} \sim 10.6$ GeV. Phys. Rev. D **70**, 071102 (2004). arXiv:hep-ex/0407009 [hep-ex]

52. Martin, A., Richard, J.M.: The eventful story of charmonium singlet states. CERN Cour. **43**, 17–18 (2003)

53. Barnes, T., Olsen, S.L.: Charmonium spectroscopy. Int. J. Mod. Phys. A **24**(S1), 305–325 (2009)

54. Baglin, C., et al.: (R704, Annecy(LAPP)-CERN-Genoa-Lyon-Oslo-Rome-Strasbourg-Turin), search for the p wave singlet charmonium state in $\bar{p}p$ annihilations at the CERN intersecting storage rings. Phys. Lett. B **171**, 135–141 (1986)

55. Armstrong, T.A., et al.: Observation of the p wave singlet state of charmonium. Phys. Rev. Lett. **69**, 2337–2340 (1992)

56. Rubin, P., et al.: (CLEO), Observation of the $^1P_1$ state of charmonium. Phys. Rev. D **72**, 092004 (2005). arXiv:hep-ex/0508037 [hep-ex]

57. Choi, S.K., et al.: (Belle), Observation of the $\eta_c(2s)$ in exclusive $B \rightarrow K K_s K^- \pi^+$ decays. Phys. Rev. Lett. **89**, 102001 (2002), [Erratum: Phys. Rev. Lett. **89**, 129901 (2002)], arXiv:hep-ex/0206002 [hep-ex]

58. Choi, S.K., et al.: (Belle), Observation of a new narrow charmonium state in exclusive $B^\pm \rightarrow K^\pm \pi^+ \pi^- J/\psi$ decays. Phys. Rev. Lett. **91**, 262001 (2003). arXiv: hep-ex/0309032

59. Lebed, R.F.: Exotic discoveries in familiar places: theory of the onia and exotics. In: 16th International Conference on B-Physics at Frontier Machines (Beauty 2016), 2–6 May 2016. Marseille, France, to appear in PoS, arXiv:1605.07975 [hep-ph]

60. Aubert, B., et al.: (BABAR), Observation of a broad structure in the $\pi^+\pi^- J/\psi$ mass spectrum around 4.26 GeV/c^2. Phys. Rev. Lett. **95**, 142001 (2005a). arXiv:hep-ex/0506081

61. Choi, S.K., et al.: (BELLE Collaboration), Observation of a resonance-like structure in the $\pi^\pm \psi'$ mass distribution in exclusive $B \rightarrow K\pi^\pm \psi'$ decays. Phys. Rev. Lett. **100**, 142001 (2008). arXiv:0708.1790 [hep-ex]

62. Abe, K., et al.: (Belle), Observation of a new charmonium state in double charmonium production in e^+e^- annihilation at $\sqrt{s} \sim 10.6$ GeV. Phys. Rev. Lett. **98**, 082001 (2007). arXiv:hep-ex/0507019 [hep-ex]

63. Pakhlov, P., et al.: (Belle), Production of new charmoniumlike states in $e^+e^- \rightarrow J/\psi D^{(*)} \bar{D}^{(*)}$ at $\sqrt{s} \sim 10$ GeV. Phys. Rev. Lett. **100**, 202001 (2008). arXiv:0708.3812 [hep-ex]

64. Uehara, S., et al.: (Belle), Observation of a χ'_2 candidate in $\gamma\gamma \rightarrow D\bar{D}$ production at BELLE. Phys. Rev. Lett. **96**, 082003 (2006). arXiv:hep-ex/0512035 [hep-ex]

65. Uehara, S., et al.: (Belle), Observation of a charmonium-like enhancement in the $\gamma\gamma \rightarrow \omega J/\psi$ process. Phys. Rev. Lett. **104**, 092001 (2010). arXiv:0912.4451 [hep-ex]

66. Shen, C.P., et al.: (Belle), Evidence for a new resonance and search for the $Y(4140)$ in the $\gamma\gamma \rightarrow \phi J/\psi$ process. Phys. Rev. Lett. **104**, 112004 (2010). arXiv:0912.2383 [hep-ex]

67. Pakhlova, G., et al.: (Belle), Observation of a near-threshold enhancement in the $e^+e^- \rightarrow \Lambda_c^+ \bar{\Lambda}_c^-$ cross section using initial-state radiation. Phys. Rev. Lett. **101**, 172001 (2008). arXiv:0807.4458 [hep-ex]

68. Sonnenschein, J., Weissman, D: A tetraquark or not a tetraquark: a holography inspired stringy hadron (HISH) perspective (2016). arXiv:1606.02732 [hep-ph]

69. Guo, F.-K., Haidenbauer, J., Hanhart, C., Meissner, U.-G.: Reconciling the X(4630) with the Y(4660). Phys. Rev. D **82**, 094008 (2010). arXiv:1005.2055 [hep-ph]

70. Aubert, B., et al.: (BaBar), Search for a charged partner of the X(3872) in the B meson decay $B \rightarrow X^- K, X^- \rightarrow J/\psi \pi^- \pi^0$. Phys. Rev. D **71**, 031501 (2005b). arXiv:hep-ex/0412051 [hep-ex]

71. Ablikim, M., et al.: (BESIII Collaboration), Observation of a charged charmoniumlike structure in $e^+e^- \rightarrow \pi^+\pi^- J/\psi$ at $\sqrt{s} = 4.26$ GeV. Phys. Rev. Lett. **110**, 252001 (2013). arXiv:1303.5949 [hep-ex]

72. Liu, Z.Q., et al.: (Belle Collaboration), Study of $e^+e^- \rightarrow \pi^+\pi^- J/\psi$ and observation of a charged charmonium-like state at Belle. Phys. Rev. Lett. **110**, 252002 (2013). arXiv:1304.0121 [hep-ex]

73. Xiao, T., Dobbs, S., Tomaradze, A., Seth, K.K.: Observation of the charged Hadron $Z_c^\pm(3900)$ and evidence for the neutral $Z_c^0(3900)$ in $e^+e^- \rightarrow \pi\pi J/\psi$ at $\sqrt{s} = 4170$ MeV. Phys. Lett. B **727**, 366–370 (2013a). arXiv:1304.3036 [hep-ex]

74. Nerling, F.: (COMPASS), Highlights from the COMPASS experiment at CERN—Hadron spectroscopy and excitations. In: 4th International Conference on New Frontiers in Physics (ICNFP 2015) Kolymbari, Greece, Aug 23–30, 2015 (2016), arXiv:1601.05025 [hep-ex]

75. Roel, A. et al.: (LHCb), Observation of $J/\psi\phi$ structures consistent with exotic states from amplitude analysis of $B^+ \rightarrow J/\psi\phi K^+$ decays (2016a) arXiv:1606.07895 [hep-ex]

76. Stancu, F.: Can $Y(4140)$ be a $c\bar{c}s\bar{s}$ tetraquark? J. Phys. G**37**, 075017 (2010). arXiv:0906.2485 [hep-ph]

77. Høgåsen, H., Richard, J.-M., Sorba, P.: A chromomagnetic mechanism for the X(3872) resonance. Phys. Rev. D **73**, 054013 (2006). arXiv:hep-ph/0511039

78. Crede, V., Roberts, W.: Progress towards understanding baryon resonances. Rep. Prog. Phys. **76**, 076301 (2013). arXiv:1302.7299 [nucl-ex]

79. Isgur, N.: An introduction to the quark model for baryons. Int. J. Mod. Phys. E **1**, 465–490 (1992)

80. Dalitz, R.H., Tuan, S.F.: The phenomenological description of K-nucleon reaction processes. Ann. Phys. **10**, 307–351 (1960)

81. Oset, E.: in [21]

82. Isgur, N., Karl, G.: P-wave Baryons in the quark model. Phys. Rev. D **18**, 4187 (1978)

83. Kamiya, Y., Miyahara, K., Ohnishi, S., Ikeda, Y., Hyodo, T., Oset, E., Weise, W.: Antikaon-nucleon interaction and $\Lambda(1405)$ in chiral SU(3) dynamics. Nucl. Phys. A **954**, 41–57 (2016). arXiv:1602.08852 [hep-ph]

84. Close, F.E.: Light hadron spectroscopy: theory and experiment, Lepton and photon interactions at high energies. In: Proceedings, 20th International Symposium, LP 2001, Rome, Italy, July 23–28, 2001, Int. J. Mod. Phys. A17, 3239–3258 (2002), [327(2001)], arXiv:hep-ph/0110081 [hep-ph]

85. Carter, A.A.: Evidence for an $S = +1$, $I = 0$ resonance in the K^+-nucleon system. Phys. Rev. Lett. **18**, 801–803 (1967)

86. Lea, A.T., Martin, B.R., Oades, G.C.: $K^+ p$ phase-shift analysis below 1500 MeV/c. Phys. Rev. **165**, 1770–1786 (1968)

87. Wilson, B.C., et al.: Phase-shift analysis of the K^+-nucleon interaction in the $I = 0$ state up to 1.5 GeV/c. Nucl. Phys. **B42**, 445–453 (1972)

88. Albrow, M.G., Andersson-Almehed, S., Bosnjakovic, B., Daum, C., Erne, F.C., Kimura, Y., Lagnaux, J.P., Sens, J.C., Udo, F., Wagner, F.: Elastic scattering of positive kaons on polarized protons between 0.87 and 2.74 Gev/c. Results and phase-shift analysis. Nucl. Phys. **B30**, 273–305 (1971)

89. Martin, B.R.: Kaon-nucleon partial wave amplitudes below 1.5 GeV/c for $I = 0$ and 1. Nucl. Phys. **B94**, 413–430 (1975)

90. Kelly, R.L., et al.: (Particle Data Group), Review of particle properties. Particle data group. Rev. Mod. Phys. **52**, S1–S286 (1980)

91. Nakano, T., et al.: (LEPS), Evidence for narrow $S = +1$ Baryon resonance in photo-production from neutron. Phys. Rev. Lett. **91**, 012002 (2003). arXiv:hep-ex/0301020

92. Diakonov, D., Petrov, V., Polyakov, M.V.: Exotic anti-decuplet of baryons: prediction from chiral solitons. Z. Phys. A **359**, 305–314 (1997). arXiv:hep-ph/9703373

93. Wohl, C.G., Pentaquarks (2008)

94. Hicks, K.H.: On the conundrum of the pentaquark. Eur. Phys. J. **H37**, 1–31 (2012)

95. Tariq, A.S.B.: Revisiting the pentaquark episode for lattice QCD. In: Proceedings, 25th International Symposium on Lattice field theory (Lattice 2007), PoS LAT2007, 136 (2007), arXiv:0711.0566 [hep-lat]

96. Jaffe, R.L.: The width of the theta+ exotic baryon in the chiral soliton model. Eur. Phys. J. C **35**, 221–222 (2004). arXiv:hep-ph/0401187 [hep-ph]

97. Weigel, H.: Axial current matrix elements and pentaquark decay widths in chiral soliton models. Phys. Rev. D **75**, 114018 (2007). arXiv:hep-ph/0703072 [hep-ph]

98. Abazov, V.M., et al.: (D0), Observation of the doubly strange b baryon Ω_b^-. Phys. Rev. Lett. **101**, 232002 (2008). arXiv:0808.4142 [hep-ex]

99. Aaltonen, T., et al.: (CDF), Observation of the Ω_b^- and measurement of the properties of the Ξ_b^- and Ω_b^-. Phys. Rev. D **80**, 072003 (2009). arXiv:0905.3123 [hep-ex]

100. Aaij, R., et al.: (LHCb), Measurement of the Λ_b^0, Ξ_b^- and Ω_b^- baryon masses. Phys. Rev. Lett. **110**, 182001 (2013a). arXiv:1302.1072 [hep-ex]

101. Amsler, C., et al.: (Particle Data Group), Review of particle physics. Phys. Lett. B **667**, 1–1340 (2008a)

102. Dong, Y., Faessler, A., Gutsche, T., Kumano, S., Lyubovitskij, V.E.: Strong three-body decays of $\Lambda_c(2940)^+$. Phys. Rev. D **83**, 094005 (2011). arXiv:1103.4762 [hep-ph]

103. Aaij, R., et al.: (LHCb), Observation of two new Ξ_b^- baryon resonances. Phys. Rev. Lett. **114**, 062004 (2015a). arXiv:1411.4849 [hep-ex]

104. Aaij, R. et al. (LHCb), Measurement of the properties of the Ξ_b^{*0} baryon, (2016b). arXiv:1604.03896 [hep-ex]

105. Ebert, D., Faustov, R.N., Galkin, V.O., Martynenko, A.P.: Mass spectra of doubly heavy baryons in the relativistic quark model. Phys. Rev. D **66**, 014008 (2002). arXiv:hep-ph/0201217 [hep-ph]

106. Fleck, S., Richard, J.M.: Baryons with double charm. Prog. Theor. Phys. **82**, 760–774 (1989)

107. Mattson, M., et al.: (SELEX), First observation of the doubly charmed baryon Ξ_{cc}^+. Phys. Rev. Lett. **89**, 112001 (2002). arXiv:hep-ex/0208014

108. Ocherashvili, A., et al.: (SELEX), Confirmation of the double charm baryon $\Xi_{cc}^+(3520)$ via its decay to pD^+K^-. Phys. Lett. B **628**, 18–24 (2005). arXiv: hep-ex/0406033

109. Ratti, S.P.: New results on c-baryons and a search for cc-baryons in FOCUS, Proceedings, 5th International Conference on Hyperons, charm and beauty hadrons (BEACH 2002), Nucl. Phys. Proc. Suppl. 115, 33–36 (2003), [33(2003)]

110. Aubert, B., et al.: (BaBar), Search for doubly charmed baryons Ξ_{cc}^+ and Ξ_{cc}^{++} in BaBar. Phys. Rev. D **74**, 011103 (2006a). arXiv:hep-ex/0605075 [hep-ex]

111. Kato, Y., et al.: (Belle), Search for doubly charmed baryons and study of charmed strange baryons at Belle. Phys. Rev. D **89**, 052003 (2014). arXiv:1312.1026 [hep-ex]

112. Aaij, R., et al.: (LHCb), Search for the doubly charmed baryon Ξ_{cc}^+. JHEP **12**, 090 (2013b). arXiv:1310.2538 [hep-ex]

113. Brodsky, S., de Teramond, G., Karliner, M.: Puzzles in hadronic physics and novel quantum chromodynamics phenomenology. Ann. Rev. Nucl. Part. Sci. **62**, 1–35 (2012), [Ann. Rev. Nucl. Part. Sci. 62, 2082(2011)], arXiv:1302.5684 [hep-ph]

114. Karliner, M., Rosner, J.L.: Baryons with two heavy quarks: masses, production, decays, and detection. Phys. Rev. D **90**, 094007 (2014). arXiv:1408.5877 [hep-ph]

115. Gelman, B.A.: Nussinov, S.: Does a narrow tetraquark $(cc\bar{u}\bar{d})$ state exist? Phys. Lett. B **551**, 296–304 (2003). arXiv:hep-ph/0209095 [hep-ph]

116. Cohen, T.D.: Doubly heavy hadrons and the domain of validity of doubly heavy diquark-anti-quark symmetry. Phys. Rev. D **74**, 094003 (2006). arXiv:hep-ph/0606084 [hep-ph]

117. Gignoux, C., Silvestre-Brac, B., Richard, J.M.: Possibility of stable multi-quark baryons. Phys. Lett. B **193**, 323 (1987)

118. Lipkin, H.J.: New new possibilities for exotic hadrons: anticharmed strange Baryons. Phys. Lett. B **195**, 484 (1987)

119. Aitala, E.M., et al.: (E791), Search for the pentaquark via the $P^0(\bar{c}s)$ decay. Phys. Rev. Lett. **81**, 44–48 (1998). arXiv:hep-ex/9709013 [hep-ex]

120. Aitala, E.M., et al.: (E791), Search for the pentaquark via the $P^0(\bar{c}s) \rightarrow K^{0,*}K^- p$ decay. Phys. Lett. B **448**, 303–310 (1999)

121. Aktas, A., et al.: (H1), Evidence for a narrow anti-charmed baryon state. Phys. Lett. B **588**, 17 (2004). arXiv:hep-ex/0403017 [hep-ex]

122. Chekanov, S., et al.: (ZEUS), Search for a narrow charmed baryonic state decaying to $D^{*\pm} p^{\mp}$ in ep collisions at HERA. Eur. Phys. J. C **38**, 29–41 (2004). arXiv:hep-ex/0409033 [hep-ex]

123. Aubert, B., et al.: (BaBar), Search for the charmed pentaquark candidate $\Theta_c(3100)$ in $e^+ e^-$ annihilations at $\sqrt{s} = 10.58$ GeV. Phys. Rev. D **73**, 091101 (2006b). arXiv:hep-ex/0604006 [hep-ex]

124. Baum, G., Kyynäräinen, J., Tripet, A.: (NA58 Collaboration), COMPASS: a proposal for a common muon and proton apparatus for structure and spectroscopy, Tech. Rep. CERN-SPSLC-96-14. SPSLC-P-297 (CERN, Geneva, 1996)

125. Aaij, R., et al.: (LHCb), Observation of $J/\psi p$ resonances consistent with pentaquark states in $\Lambda_b^0 \to J/\psi K^- p$ decays. Phys. Rev. Lett. **115**, 072001 (2015b). arXiv:1507.03414 [hep-ex]

126. Aaij, R. et al. (LHCb), Model-independent evidence for $J/\psi p$ contributions to $\Lambda_b^0 \to J/\psi p K^-$ decays, (2016c). arXiv:1604.05708 [hep-ex]

127. Aaij, R., et al.: (LHCb), Observation of the $\Lambda_b^0 \to J/\psi p \pi^-$ decay. JHEP **07**, 103 (2014). arXiv:1406.0755 [hep-ex]

128. Burns, T.J.: Phenomenology of $P_c(4380)^+$, $P_c(4450)^+$ and related states. Eur. Phys. J. A **51**, 152 (2015). arXiv:1509.02460 [hep-ph]

129. Wang, E., Chen, H.-X., Geng, L.-S., Li, D.-M., Oset, E.: A hidden-charm pentaquark state in $\Lambda_b^0 \to J/\psi p \pi^-$ decay. Phys. Rev. D **93**, 094001 (2016). arXiv:1512.01959 [hep-ph]

130. Aaij, R. et al. (LHCb), Evidence for exotic hadron contributions to $\Lambda_b^0 \to J/\psi p \pi^-$ decays, Phys. Rev. Lett. 117, 082003 (2016d), [Addendum: Phys. Rev. Lett. **117**(10), 109902 (2016)], arXiv:1606.06999 [hep-ex]

131. Kubarovsky, V., Voloshin, M.B.: Formation of hidden-charm pentaquarks in photon-nucleon collisions. Phys. Rev. D **92**, 031502 (2015). arXiv:1508.00888 [hep-ph]

132. Kubarovsky, V., Voloshin, M.B.: Search for hidden-charm pentaquark with CLAS12, (2016). arXiv:1609.00050 [hep-ph]

133. Blin, H., Fernández-Ramírez, C., Jackura, A., Mathieu, V., Mokeev, V.I., Pilloni, A., Szczepaniak, A.P.: Studying the $P_c(4450)$ resonance in J/ψ photoproduction off protons. Phys. Rev. D **94**, 034002 (2016). arXiv:1606.08912 [hep-ph]

134. Karliner, M., Rosner, J.L.: Photoproduction of exotic Baryon resonances. Phys. Lett. B **752**, 329–332 (2016). arXiv:1508.01496 [hep-ph]

135. Meziani, Z.E., et al.: A search for the LHCb charmed 'Pentaquark' using photo-production of J/ψ at threshold in hall C at Jefferson Lab, (2016). arXiv:1609.00676 [hep-ex]

136. Mattione, P.: in [21]

137. Arndt, R.A., Hyslop III, J.S., Roper, L.D.: Nucleon-nucleon partial wave analysis to 1100 MeV. Phys. Rev. D **35**, 128 (1987)

138. Arndt, R.A., Strakovsky, I.I., Workman, R.L., Bugg, D.V.: Analysis of the reaction $\pi^+ d \to pp$ to 500 MeV. Phys. Rev. C **48**, 1926–1938 (1993)

139. Arndt, R.A., Strakovsky, I.I., Workman, R.L.: Analysis of πd elastic scattering data to 500 MeV. Phys. Rev. C **50**, 1796–1806 (1994). arXiv:nucl-th/9407032 [nucl-th]

140. Tatischeff, B., et al.: Evidence for narrow dibaryons at 2050 MeV, 2122 MeV, and 2150 MeV observed in inelastic pp scattering. Phys. Rev. C **59**, 1878–1889 (1999)

141. Parker, B., Seth, K.K., Ginsburg, C.M., O'Reilly, B., Sarmiento, M.: Search for a $T = 2$ Dibaryon. Phys. Rev. Lett. **63**, 1570 (1989)

142. De Boer, F.W.N., et al.: Search for bound states of neutrons and negative pions. Phys. Rev. Lett. **53**, 423–426 (1984)

143. Adlarson, P., et al.: (WASA-at-COSY) Evidence for a new resonance from polarized neutron-proton scattering. Phys. Rev. Lett. **112**, 202301 (2014a). arXiv:1402.6844 [nucl-ex]

144. Adlarson, P., et al.: (WASA-at-COSY), Neutron-proton scattering in the context of the d* (2380) resonance. Phys. Rev. C **90**, 035204 (2014b). arXiv:1408.4928 [nucl-ex]

145. Goldman, T., Maltman, K., Stephenson Jr., G.J., Schmidt, K.E., Wang, F.: An 'inevitable' nonstrange dibaryon. Phys. Rev. C **39**, 1889–1895 (1989)

146. Ohnishi, S., Ikeda, Y., Kamano, H., Sato, T.: Signature of strange dibaryon in kaon-induced reaction. In: Proceedings, 5th asia-pacific conference on few-body problems in physics 2011 (APFB2011). Few Body Syst. **54**, 347–351 (2013), arXiv:1109.4724 [nucl-th]

147. Hashimoto, T., et al.: (J-PARC E15), Search for the deeply bound $K^- pp$ state from the semi-inclusive forward-neutron spectrum in the in-flight K^- reaction on ^{3}He. PTEP 2015, 061D01 (2014), arXiv:1408.5637 [nucl-ex]

148. Zel'dovich, Y.B.: Energy levels in a distorted Coulomb field. Sov. Phys. Solid State **1**, 1497–1501 (1960)

149. Gal, A., Friedman, E., Batty, C.J.: On the interplay between Coulomb and nuclear states in exotic atoms. Nucl. Phys. A **606**, 283–291 (1996)

150. Combescure, M., Khare, A., Raina, A., Richard, J.-M., Weydert, C.: Level rearrangement in exotic atoms and quantum dots. Int. J. Mod. Phys. B **21**, 3765–3781 (2007). arXiv:cond-mat/0701006 [cond-mat]

151. Jaffe, R.L.: Perhaps a stable dihyperon, Phys. Rev. Lett. **38**, 195–198 (1977), [Erratum: Phys. Rev. Lett. **38**, 617(1977)]

152. Dalitz, R.H., Davis, D.H., Fowler, P.H., Montwill, A., Pniewski, J., Zakrzewski, J.A.: The Identified $\Lambda\Lambda$ Hypernuclei and the Predicted H Particle. Proc. R. Soc. Lond. **A426**, 1–17 (1989)

153. Kuhn, C., Hippolyte, B., Coffin, J.P., Baudot, J., Belikov, I., Dietrich, D., Germain, M., Suire, C.: Search for strange dibaryons in STAR and ALICE, Strange quarks in matter. Proceedings, 6th International Conference, SQM 2001, Frankfurt, Germany, September 24–29, 2001. J. Phys. **G28**, 1707–1714 (2002)

154. Goldman, T., Maltman, K., Stephenson Jr., G.J., Schmidt, K.E., Wang, F.: Strangeness -3 dibaryons. Phys. Rev. Lett. **59**, 627 (1987)

155. Huang, H., Jialun, P., Fan, W.: Further study of the $N\Omega$ dibaryon within constituent quark models. Phys. Rev. C **92**, 065202 (2015). arXiv:1507.07124 [hep-ph]

156. Thomas, L.H.: The interaction between a neutron and a proton and the structure of ^{3}H. Phys. Rev. **47**, 903–909 (1935)

157. Zhukov, M.V., Danilin, B.V., Fedorov, D.V., Bang, J.M., Thompson, I.J., Vaagen, J.S.: Bound state properties of Borromean Halo nuclei: ^{6}He and ^{11}Li. Phys. Rep. **231**, 151–199 (1993)

158. Richard, J.-M.: Borromean binding (2003). http://theor.jinr.ru/Few-body/Belyaev-70/, arXiv:nucl-th/0305076 [nucl-th]

159. Hiyama, E.: Few-body aspects of hypernuclear physics. Few Body Syst. **53**, 189–236 (2012)

160. Filikhin, I.N., Gal, A.: Light $\Lambda\Lambda$ hypernuclei and the onset of stability for $\Lambda\Xi$ hypernuclei. Phys. Rev. C **65**, 041001 (2002). arXiv:nucl-th/0110008 [nucl-th]

161. Garcilazo, H., Valcarce, A., Carames, T.F.: The $N\Lambda\Lambda - \Xi NN$ bound state problem with $N\Lambda\Sigma$ and $N\Sigma\Sigma$ channels. J. Phys. **G42**, 025103 (2015)

162. Richard, J.-M., Wang, Q., Zhao, Q.: Lightest neutral hypernuclei with strangeness -1 and -2. Phys. Rev. C **91**, 014003 (2015). arXiv:1404.3473 [nucl-th]

163. Garcilazo, H., Valcarce, A., Vijande, J.: Maximal isospin few-body systems of nucleons and Ξ hyperons. Phys. Rev. C **94**, 024002 (2016). arXiv:1608.05192 [nucl-th]

164. Carbonell, J., Lazauskas, R., Delande, D., Hilico, L., Kiliç, S.: A new vibrational level of the H_2^+ molecular ion. EPL (Europhysics Letters) **64**, 316–322 (2003). arXiv:physics/0207007

165. Varga, K., Usukura, J., Suzuki, Y.: Second bound state of the positronium molecule and biexcitons. Phys. Rev. Lett. **80**, 1876–1879 (1998). arXiv:cond-mat/9802261

166. Czarnecki, A.: Positronium and polyelectrons. Nucl. Phys. A **827**, 541–543c (2009)

167. Richard, J.-M.: Critically bound four-body molecules. Phys. Rev. A **67**, 034702 (2003). arXiv:physics/0302004

168. Roy, D.P.: History of exotic meson (4-quark) and baryon (5-quark) states. J. Phys. **G30**, R113 (2004). arXiv:hep-ph/0311207

169. Rosner, J.L.: Possibility of baryon-anti-baryon enhancements with unusual quantum numbers. Phys. Rev. Lett. **21**, 950–952 (1968)

170. Rossi, G., Veneziano, G.: The string-junction picture of multiquark states: an update. JHEP **06**, 041 (2016). arXiv:1603.05830 [hep-th]

171. Artru, X.: String model with baryons: topology, classical motion. Nucl. Phys. B **85**, 442 (1975)

172. Muller, V.F.: On composite hadrons in nonabelian lattice gauge theories. Nucl. Phys. B **116**, 470 (1976)

173. Hasenfratz, P., Horgan, R.R., Kuti, J., Richard, J.M.: Heavy Baryon Spectroscopy in the QCD bag model. Phys. Lett. B **94**, 401–404 (1980a)

174. Carlson, J., Kogut, J.B., Pandharipande, V.R.: A quark model for baryons based on quantum chromodynamics. Phys. Rev. D **27**, 233 (1983)

175. Bagan, E., Latorre, J.I., Merkurev, S.P., Tarrach, R.: The baryon loop tension from continuum QCD. Phys. Lett. B **158**, 145 (1985)

176. de la Ripelle, M.F., Lassaut, M.: Transformation of a three-body interaction into a sum of pairwise potentials: application to the quark-string-junction and the Urbana potentials. Few Body Syst. **23**, 75–86 (1997)

177. Evangelista, C., et al.: Evidence for a narrow width Boson of Mass 2.95 GeV. Phys. Lett. **B72**, 139 (1977)

178. Jaffe, R.L.: $Q^2\bar{Q}^2$ resonances in the Baryon-antibaryon system. Phys. Rev. D **17**, 1444 (1978)

179. Chan, H.-M., Fukugita, M., Hansson, T.H., Hoffman, H.J., Konishi, K., Høgåsen, H.: Tsun Tsou, S.: Color chemistry: a study of metastable multi-quark molecules. Phys. Lett. **B76**, 634–640 (1978)

180. Fermi, E., Yang, C.-N.: Are mesons elementary particles? Phys. Rev. **76**, 1739–1743 (1949)

181. Shapiro, I.S.: The physics of nucleon-anti-nucleon systems. Phys. Rep. **35**, 129–185 (1978)

182. Buck, W.W., Dover, C.B., Richard, J.-M.: The interaction of nucleons with anti-nucleons. 1. General features of the $\bar{N}N$ spectrum in potential models. Ann. Phys. **121**, 47 (1979)

183. Voloshin, M.B., Okun, L.B.: Hadron molecules and charmonium atom. JETP Lett. **23**, 333–336 (1976)

184. De Rujula, A., Georgi, H., Glashow, S.L.: Molecular charmonium: a new spectroscopy? Phys. Rev. Lett. **38**, 317 (1977)

185. Le Yaouanc, A., Oliver, L., Pene, O., Raynal, J.C.: Strong decays of ψ (4.028) as a radial excitation of charmonium. Phys. Lett. **B71**, 397 (1977a)

186. Le Yaouanc, A., Oliver, L., Pene, O., Raynal, J.C.: Why is ψ''' (4.414) so narrow? Phys. Lett. **B72**, 57 (1977b)

187. Eichten, E., Gottfried, K., Kinoshita, T., Lane, K.D., Yan, T.-M.: Charmonium: comparison with experiment. Phys. Rev. D **21**, 203 (1980)

188. Törnqvist, N.A.: Possible large deuteron-like meson meson states bound by pions. Phys. Rev. Lett. **67**, 556–559 (1991)

189. Manohar, A.V., Wise, M.B.: Exotic $QQ\bar{q}\bar{q}$ states in QCD. Nucl. Phys. B **399**, 17–33 (1993). arXiv:hep-ph/9212236

190. Ericson, T.E.O.: Strength of pion exchange in hadronic molecules. Phys. Lett. B **309**, 426–430 (1993)

191. Törnqvist, N.A.: From the deuteron to deusons, an analysis of deuteron - like meson meson bound states. Z. Phys. C **61**, 525–537 (1994). arXiv:hep-ph/9310247 [hep-ph]

192. Ferretti, J., Galatà, G., Santopinto, E.: Interpretation of the X(3872) as a charmonium state plus an extra component due to the coupling to the meson-meson continuum. Phys. Rev. C **88**, 015207 (2013). arXiv:1302.6857 [hep-ph]

193. Thomas, C.E., Close, F.E.: Is X(3872) a molecule? Phys. Rev. D **78**, 034007 (2008). arXiv:0805.3653 [hep-ph]

194. Ohkoda, S., Yamaguchi, Y., Yasui, S., Sudoh, K., Hosaka, A.: Exotic mesons with hidden bottom near thresholds. Phys. Rev. D **86**, 014004 (2012a). arXiv:1111.2921 [hep-ph]

195. Cleven, M., Guo, F.-K., Hanhart, C., Wang, Q., Zhao, Q.: Employing spin symmetry to disentangle different models for the XYZ states. Phys. Rev. D **92**, 014005 (2015). arXiv:1505.01771 [hep-ph]

196. Hofmann, J., Lutz, M.F.M.: Coupled-channel study of baryon resonances with charm. In: Proceedings, 11th International Conference on Hadron spectroscopy (Hadron 2005) (2005) arXiv:nucl-th/0510091 [nucl-th]

197. Hofmann, J., Lutz, M.F.M.: Coupled-channel study of crypto-exotic baryons with charm. Nucl. Phys. A **763**, 90–139 (2005b). arXiv:hep-ph/0507071 [hep-ph]

198. Xiao, C.W., Nieves, J., Oset, E.: Combining heavy quark spin and local hidden gauge symmetries in the dynamical generation of hidden charm baryons. Phys. Rev. D **88**, 056012 (2013b). arXiv:1304.5368 [hep-ph]

199. Yang, Z.-C., Sun, Z.-F., He, J., Liu, X., Zhu, S.-L.: The possible hidden-charm molecular baryons composed of anti-charmed meson and charmed baryon. Chin. Phys. C **36**, 6–13 (2012). arXiv:1105.2901 [hep-ph]

200. Landsberg, L.G.: Do the narrow cryptoexotic baryon resonances exist? Phys. Atom. Nucl. **57**, 2127–2131 (1994), [Yad. Fiz.57N12,2210(1994)]

201. Wong, C.-Y.: Molecular states of heavy quark mesons. Phys. Rev. C **69**, 055202 (2004). arXiv:hep-ph/0311088 [hep-ph]

202. Ohkoda, S., Yamaguchi, Y., Yasui, S., Sudoh, K., Hosaka, A.: Exotic mesons with double charm and bottom flavor. Phys. Rev. D **86**, 034019 (2012b). arXiv:1202.0760 [hep-ph]

203. Karliner, M., Rosner, J.L.: New exotic Meson and Baryon resonances from doubly-heavy hadronic molecules. Phys. Rev. Lett. **115**, 122001 (2015). arXiv:1506.06386 [hep-ph]

204. Yasui, S., Sudoh, K.: Exotic nuclei with open heavy flavor mesons. Phys. Rev. D **80**, 034008 (2009). arXiv:0906.1452 [hep-ph]

205. Froemel, F., Julia-Diaz, B., Riska, D.O.: Bound states of double flavor hyperons. Nucl. Phys. A **750**, 337–356 (2005). arXiv:nucl-th/0410034 [nucl-th]

206. Liu, Y.-R., Oka, M.: $\Lambda_c N$ bound states revisited. Phys. Rev. D **85**, 014015 (2012). arXiv:1103.4624 [hep-ph]

207. Meguro, W., Liu, Y.-R., Oka, M.: Possible $\Lambda_c \Lambda_c$ molecular bound state. Phys. Lett. B **704**, 547–550 (2011). arXiv:1105.3693 [hep-ph]

208. Huang, H., Ping, J., Wang, F.: Possible H-like dibaryon states with heavy quarks. Phys. Rev. C **89**, 035201 (2014). arXiv:1311.4732 [hep-ph]

209. Lee, N., Luo, Z.-G., Chen, X.-L., Zhu, S.-L.: Possible deuteron-like molecular states composed of heavy baryons. Phys. Rev. D **84**, 014031 (2011). arXiv:1104.4257 [hep-ph]

210. Li, N., Zhu, S.-L.: Hadronic molecular states composed of heavy flavor baryons. Phys. Rev. D **86**, 014020 (2012). arXiv:1204.3364 [hep-ph]

211. Julia-Diaz, B., Riska, D.O.: Nuclei of double charm hyperons, the structure of Baryons. In: Proceedings, 10th international conference, Baryons'04, Palaiseau, France, Oct 25–29, 2004. Nucl. Phys. **A755**, 431–434 (2005). arXiv:nucl-th/0405061 [nucl-th]

212. Van Hove, L.: Future prospects of particle physics. In: 7th European symposium on nucleon antinucleon interactions: antiproton 86, Thessaloniki, Greece, Sept 1–5, 1986 (1986)

213. Ochs, W.: The status of glueballs. J. Phys. **G40**, 043001 (2013). arXiv:1301.5183 [hep-ph]

214. Tang, L., Qiao, C.-F.: Mass spectra of 0^{+-}, 1^{-+}, and 2^{+-} exotic glueballs. Nucl. Phys. B **904**, 282–296 (2016). arXiv:1509.00305 [hep-ph]

215. Buisseret, F., Semay, C., Mathieu, V., Silvestre-Brac, B.: Excited string and constituent gluon descriptions of hybrid mesons. In: Proceedings, 20th European Conference on Few-Body Problems in Physics (EFB20), Few Body Syst. 44, 87–89 (2008)

216. Giles, R., Tye, S.H.H.: The application of the quark-confining string to the psi spectroscopy. Phys. Rev. D **16**, 1079 (1977)

217. Horn, D., Mandula, J.: A model of mesons with constituent gluons. Phys. Rev. D **17**, 898 (1978)

218. Hasenfratz, P., Horgan, R.R., Kuti, J., Richard, J.M.: The effects of colored glue in the QCD motivated bag of heavy quark-anti-quark systems. Phys. Lett. B **95**, 299 (1980b)

219. Mandula, J.E.: Quark-anti-quark-gluon exotic fields for lattice QCD. Phys. Lett. B **135**, 155–158 (1984)

220. Juge, K.J., Kuti, J., Morningstar, C.: The heavy-quark hybrid meson spectrum in lattice qcd. AIP Conf. Proc. **688**, 193–207 (2004). arXiv:nucl-th/0307116

221. Berwein, M., Vairo, A., Brambilla, N., Castella, J.T.: Quarkonium hybrids with non-relativistic filed theory, Phys. Rev. D 62, 114019 (2015). arXiv:1511.04299 [hep-ph]

222. Shifman, M.A., Vainshtein, A.I., Zakharov, V.I.: Theoretical foundations. Nucl. Phys. **B147**, 385–447 (1979a)

223. Shifman, M.A., Vainshtein, A.I., Zakharov, V.I.: QCD and resonance physics. Applications. Nucl. Phys. **B147**, 448–518 (1979b)

224. Shifman, M.A., Vainshtein, A.I., Zakharov, V.I.: QCD and resonance physics. The rho-omega mixing. Nucl. Phys. **B147**, 519–534 (1979c)

225. Matheus, R.D., Narison, S., Nielsen, M., Richard, J.M.: Can the $X(3872)$ be a 1^{++} four-quark state? Phys. Rev. D **75**, 014005 (2007). arXiv:hep-ph/0608297

226. Chen, H.-X., Liu, X., Hosaka, A., Zhu, S.-L.: The Y(2175) state in the QCD sum rule. Phys. Rev. D **78**, 034012 (2008). arXiv:0801.4603 [hep-ph]

227. Narison, S., Navarra, F.S., Nielsen, M.: Investigating different structures for the $X(3872)$. In: Proceedings, 15th High-Energy Physics International Conference on Quantum Chromodynamics (QCD 10), Nucl. Phys. Proc. Suppl. **207–208**, 249–252 (2010), arXiv:1007.4575 [hep-ph]

228. DeTar, C.: LQCD: Flavor Physics and Spectroscopy, (2015), Lepton-Photon 2015, Ljubljana, Slovenia, to appear in PoS, arXiv:1511.06884 [hep-lat]

229. Fiebig, H.R., Rabitsch, K., Markum, H., Mihaly, A.: Exploring the $\pi^+\pi^+$ interaction in lattice qcd. Few Body Syst. **29**, 95–120 (2000). arXiv:hep-lat/9906002

230. Inoue, T., et al.: (HAL QCD), Bound H-dibaryon in Flavor SU(3) Limit of Lattice QCD. Phys. Rev. Lett. **106**, 162002 (2011). arXiv:1012.5928 [hep-lat]

231. Beane, S.R., et al.: (NPLQCD), Evidence for a bound H-dibaryon from Lattice QCD. Phys. Rev. Lett. **106**, 162001 (2011). arXiv:1012.3812 [hep-lat]

232. Suganuma, H., Iritani, T., Okiharu, F., Takahashi, T.T., Yamamoto, A.: Lattice QCD study for confinement in Hadrons. AIP Conf. Proc. **1388**, 195–201 (2011). arXiv:1103.4015 [hep-lat]

233. Aoki, S., Doi, T., Hatsuda, T., Ikeda, Y., Inoue, T., Ishii, N., Murano, K., Nemura, H., Sasaki, K.: (HAL QCD), Lattice QCD approach to nuclear physics. PTEP **2012**, 01A105 (2012). arXiv:1206.5088 [hep-lat]

234. Kirscher, J., Barnea, N., Gazit, D., Pederiva, F., van Kolck, U.: Spectra and scattering of light latticenuclei from effective field theory. Phys. Rev. C **92**, 054002 (2015). arXiv:1506.09048 [nucl-th]

235. Dudek, J.J.: Hadron scattering and resonances in QCD. In: Proceedings, 16th International Conference on Hadron Spectroscopy (Hadron 2015), AIP Conference Proceedings **1735**, 020014 (2016)

236. Francis, A., Hudspith, R.J., Lewis, R., Maltman, K.: Doubly bottom strong-interaction stable tetraquarks from lattice QCD (2016). arXiv:1607.05214 [hep-lat]

237. Green, A.M., Pennanen, P.: An interquark potential model for multiquark systems. Phys. Rev. C **57**, 3384–3391 (1998). arXiv:hep-lat/9804003 [hep-lat]

238. Michael, C., Pennanen, P.: (UKQCD), Two heavy-light mesons on a lattice. Phys. Rev. D **60**, 054012 (1999). arXiv:hep-lat/9901007 [hep-lat]

239. Bali, G., Hetzenegger, M.: Static-light meson-meson potentials. PoS LATTICE **2010**, 142 (2010). arXiv:1011.0571 [hep-lat]

240. Bicudo, P., Cichy, K., Peters, A., Wagner, M.: BB interactions with static bottom quarks from lattice QCD. Phys. Rev. D **93**, 034501 (2016). arXiv:1510.03441 [hep-lat]
241. De Rújula, A., Georgi, H., Glashow, S.L.: Hadron masses in a gauge theory. Phys. Rev. D **12**, 147–162 (1975)
242. Takahashi, H., et al.: Observation of a $^6_{\Lambda\Lambda}$He double hypernucleus. Phys. Rev. Lett. **87**, 212502 (2001)
243. Oka, M., Shimizu, K., Yazaki, K.: The dihyperon state in the quark cluster model. Phys. Lett. B **130**, 365 (1983)
244. Rosner, J.L.: SU(3) breaking and the H dibaryon. Phys. Rev. D **33**, 2043 (1986)
245. Karl, G., Zenczykowski, P.: H dibaryon spectrocopy. Phys. Rev. D **36**, 2079 (1987)
246. Faessler, A., Straub, U.: Baryon baryon interaction in the quark model and the H dibaryon. In: Erice 1989, Proceedings, the Nature of Hadrons and Nuclei by Electron Scattering, 323–332. Prog. Part. Nucl. Phys. **24**, 323–332 (1990)
247. Leandri, J., Silvestre-Brac, B.: Systematics of $\bar{Q}Q^4$ systems with a pure chromomagnetic interaction. Phys. Rev. D **40**, 2340–2352 (1989)
248. Silvestre-Brac, B., Leandri, J.: Systematics of q^6 systems in a simple chromomagnetic model. Phys. Rev. D **45**, 4221–4239 (1992)
249. Leandri, J., Silvestre-Brac, B.: Dibaryon states containing two different types of heavy quarks. Phys. Rev. D **51**, 3628–3637 (1995)
250. Lichtenberg, D.B., Roncaglia, R.: Colormagnetic interaction, diquarks, and exotic hadrons. In: International Workshop on Diquarks, Turin, Italy, Nov 2–4, 1992 (1992)
251. Buccella, F., Høgåsen, H., Richard, J.-M., Sorba, P.: Chromomagnetism, flavour symmetry breaking and S-wave tetraquarks. Eur. Phys. J. C **49**, 743–754 (2007). arXiv:hep-ph/0608001
252. Ader, J.P., Richard, J.M., Taxil, P.: Do narrow heavy multiquark states exist? Phys. Rev. D **25**, 2370 (1982)
253. Heller, L., Tjon, J.A.: On the existence of stable dimesons. Phys. Rev. D **35**, 969 (1987)
254. Zouzou, S., Silvestre-Brac, B., Gignoux, C., Richard, J.M.: Four quark bound states. Z. Phys. C **30**, 457 (1986)
255. Carlson, J., Heller, L., Tjon, J.A.: Stability of dimesons. Phys. Rev. D **37**, 744 (1988)
256. Bressanini, D., Mella, M., Morosi, G.: Stability of four-body systems in three and two dimensions: A theoretical and quantum monte carlo study of biexciton molecules. Phys. Rev. A **57**, 4956–4959 (1998)
257. Varga, K., Usukura, J., Suzuki, Y.: Recent applications of the stochastic variational method. In: Desplanques, B., Protasov, K., Silvestre-Brac, B., Carbonell, J. (eds.) Few-Body Problems in Physics '98 (1999) p. 11
258. Lloyd, R.J., Vary, J.P.: All charm tetraquarks. Phys. Rev. D **70**, 014009 (2004). arXiv:hep-ph/0311179 [hep-ph]
259. Cafer, A.Y., Richard, J.-M., Rubinstein, J.H.: Stability of asymmetric tetraquarks in the minimal-path linear potential. Phys. Lett. B **674**, 227–231 (2009). arXiv:0901.3022 [math-ph]
260. Lenz, F., Londergan, J.T., Moniz, E.J., Rosenfelder, R., Stingl, M., Yazaki, K.: Quark confinement and hadronic interactions. Ann. Phys. **170**, 65 (1986)
261. Vijande, J., Valcarce, A., Richard, J.M.: Stability of multiquarks in a simple string model. Phys. Rev. D **76**, 114013 (2007b). arXiv:0707.3996 [hep-ph]
262. Richard, J.-M.: Stability of the pentaquark in a naive string model. Phys. Rev. C **81**, 015205 (2010). arXiv:0908.2944 [hep-ph]
263. Vijande, J., Valcarce, A., Richard, J.-M.: Stability of hexaquarks in the string limit of confinement. Phys. Rev. D **85**, 014019 (2012). arXiv:1111.5921 [hep-ph]
264. Vijande, J., Valcarce, A., Richard, J.-M.: Adiabaticity and color mixing in tetraquark spectroscopy. Phys. Rev. D **87**, 034040 (2013). arXiv:1301.6212 [hep-ph]
265. Silvestre-Brac, B., Semay, C., Narodetskii, I.M., Veselov, A.I.: The Baryonic Y-shape confining potential energy and its approximants. Eur. Phys. J. C **32**, 385–397 (2003). arXiv:hep-ph/0309247 [hep-ph]
266. Bicudo, P., Cardoso, M.: Iterative method to compute the Fermat points and Fermat distances of multiquarks. Phys. Lett. B **674**, 98–102 (2009). arXiv:0812.0777 [physics.comp-ph]
267. Dmitrašinović, V., Sato, T., Šuvakov, M.: Low-lying spectrum of the Y-string three-quark potential using hyper-spherical coordinates. Eur. Phys. J. C **62**, 383–397 (2009). arXiv:0906.2327 [hep-ph]
268. Semay, C., Silvestre-Brac, B.: Diquonia and potential models. Z. Phys. C **61**, 271–275 (1994)
269. Janc, D., Rosina, M.: The $T_{cc} = DD^*$ molecular state. Few Body Syst. **35**, 175–196 (2004). arXiv:hep-ph/0405208
270. Vijande, J., Weissman, E., Valcarce, A., Barnea, N.: Are there compact heavy four-quark bound states? Phys. Rev. D **76**, 094027 (2007c). arXiv:0710.2516 [hep-ph]
271. Vijande, J., Valcarce, A., Richard, J.M., Sorba, P.: Search for doubly-heavy dibaryons in a quark model. Phys. Rev. D **94**, 034038 (2016). arXiv:1608.03982 [hep-ph]
272. Hartree, D.R.: The Calculation of Atomic Structures. Wiley, New York (1957)
273. Quigg, C., Rosner, J.L.: Quantum mechanics with applications to quarkonium. Phys. Rep. **56**, 167–235 (1979)
274. Grosse, H., Martin, A.: Exact results on potential models for quarkonium systems. Phys. Rep. **60**, 341 (1980)
275. Grosse, H., Martin, A.: Particle physics and the Schrödinger Equation; new ed., Cambridge monographs on particle physics, nuclear physics, and cosmology (Cambridge Univ., Cambridge, 2005)
276. Richard, J.-M., Taxil, P.: The ordering of low lying bound states of three identical particles. Nucl. Phys. B **329**, 310–326 (1990)
277. Stancu, F., Stassart, P.: Negative parity nonstrange baryons. Phys. Lett. B **269**, 243–246 (1991)
278. Richard, J.M.: The nonrelativistic three-body problem for baryons. Phys. Rep. **212**, 1–76 (1992)
279. Mitroy, J., et al.: Theory and application of explicitly correlated Gaussians. Rev. Mod. Phys. 85, 693 (2013)
280. Hiyama, E., Kino, Y., Kamimura, M.: Gaussian expansion method for few-body systems. Prog. Part. Nucl. Phys. **51**, 223–307 (2003)
281. Anselmino, M., Predazzi, E., Ekelin, S., Sverker, F., Lichtenberg, D.B.: Diquarks. Rev. Mod. Phys. **65**, 1199–1234 (1993)
282. Brodsky, S.J., de Téramond, G.F., Dosch, H., Lorcé, C.: Meson/Baryon/Tetraquark Supersymmetry from Superconformal Algebra and Light-Front Holography, Conference on New Physics at the Large Hadron Collider Singapore, Singapore, February 29-March 4, 2016. Int. J. Mod. Phys. **A31**, 1630029 (2016), arXiv:1606.04638 [hep-ph]

283. Lichtenberg, D.B., Roncaglia, R., Predazzi, E.: Predicting exotic hadron masses from supersymmetry and a quark-diquark model. J. Phys. **G23**, 865–874 (1997)
284. Grach, I.L., Narodetsky, I.M.: Diquark correlations in the proton. Few Body Syst. **16**, 151–163 (1994)
285. Fleck, S., Silvestre-Brac, B., Richard, J.M.: Search for diquark clustering in Baryons. Phys. Rev. D **38**, 1519–1529 (1988)
286. Martin, A.: Regge trajectories in the quark model. Z. Phys. C **32**, 359 (1986)
287. Braaten, E., Langmack, C., Hudson, S.D.: Born-Oppenheimer approximation for the XYZ mesons. Phys. Rev. D **90**, 014044 (2014). arXiv:1402.0438 [hep-ph]
288. Dubynskiy, S., Voloshin, M.B.: Hadro-Charmonium. Phys. Lett. B **666**, 344–346 (2008). arXiv:0803.2224 [hep-ph]
289. Voloshin, M.B.: $Z_c(3900)$; What is inside? Phys. Rev. D **87**, 091501 (2013). arXiv:1304.0380 [hep-ph]
290. Wu, T.T., Wu, S.L.: Augmented standard model and the simplest scenario. Int. J. Mod. Phys. A **30**, 1550201 (2015)
291. Ikaros, I.Y., Bigi, D., Yuri, L., Khoze, V.A., Kuhn, J.H., Zerwas, P.M.: Production and decay properties of ultraheavy quarks. Phys. Lett. B **181**, 157–163 (1986)
292. Luo, M., Wang, K., Tao, X., Zhang, L., Zhu, G.: Squarkonium, diquarkonium, and octetonium at the LHC and their diphoton decays. Phys. Rev. D **93**, 055042 (2016). arXiv:1512.06670 [hep-ph]
293. Backović, M.: A Theory of Ambulance Chasing (2016). arXiv:1603.01204 [physics.soc-ph]
294. Hey, A.J.G., Kelly, R.L.: Baryon spectroscopy. Phys. Rep. **96**, 71 (1983)
295. Dytman, S.A.: Impact of recent data on N* structure. In: ICTP 4th International Conference on Perspectives in Hadronic Physics Trieste, Italy, May 12–16, 2003, Eur. Phys. J. A **19**(Suppl 1), 61–65 (2004), [61 (2004)]
296. Briceno, R.A., et al.: Issues and opportunities in exotic hadrons. Chin. Phys. C **40**, 042001 (2016). arXiv:1511.06779 [hep-ph]
297. Jaffe, R.L., Low, F.E.: The connection between quark model eigenstates and low-energy scattering. Phys. Rev. D **19**, 2105–2118 (1979)
298. Fredriksson, S., Jändel, M.: The diquark deuteron. Phys. Rev. Lett. **48**, 14 (1982)
299. Maiani, L., Polosa, A.D., Riquer, V.: From pentaquarks to dibaryons in $\Lambda_b(5620)$ decays. Phys. Lett. B **750**, 37–38 (2015). arXiv:1508.04459 [hep-ph]

Few-Body Syst (2017) 58:1–25
DOI 10.1007/s00601-016-1170-5

N. V. Shevchenko

Three-Body Antikaon–Nucleon Systems

Received: 31 July 2016 / Accepted: 3 November 2016 / Published online: 9 December 2016
© Springer-Verlag Wien 2016

Abstract The paper contains a review of the exact or accurate results achieved in the field of the three-body antikaon–nucleon physics. Different states and processes in $\bar{K}NN$ and $\bar{K}\bar{K}N$ systems are considered. In particular, quasi-bound states in K^-pp and K^-K^-p systems were investigated together with antikaonic deuterium atom. Near-threshold scattering of antikaons on deuteron, including the K^-d scattering length, and applications of the scattering amplitudes are also discussed. All exact three-body results were calculated using some form of Faddeev equations. Different versions of $\bar{K}N$, ΣN, $\bar{K}\bar{K}$, and NN potentials, specially constructed for the calculations, allowed investigation of the dependence of the three-body results on the two-body input. Special attention is paid to the antikaon–nucleon interaction, being the most important for the three-body systems. Approximate calculations, performed additionally to the exact ones, demonstrate accuracy of the commonly used approaches.

1 Introduction

An interest to the exotic systems, which consist of antikaons and nucleons, rose recently after the statement [1,2], that deep and narrow quasi-bound states can exist in $\bar{K}NN$ and $\bar{K}NNN$ systems. Due to this, several calculations of the quasi-bound state in the lightest $\bar{K}NN$ system with $J^P = 0^-$ quantum numbers, that is K^-pp, were performed. Among all calculations those using Faddeev-type equations are the most accurate ones. The first results of the two accurate calculations [3–5] confirmed the existence of the quasi-bound state in the K^-pp, but the evaluated binding energies and widths are far different from those predicted in [1,2]. The results of the two groups [3–5] also differ one from the other, and the main reason for this is a choice of the antikaon–nucleon interaction, being an input for the three-body calculations.

The question of the quasi-bound state in the K^-pp system is far from being settled from experimental point of view as well. The first experimental evidence of the K^-pp quasi-bound state existence occurred in the FINUDA experiment [6] at the DAΦNE e^+e collider. Recently performed new analyses of old experiments, such as OBELIX [7] at CERN and DISTO [8] at SATURNE also claimed the observation of the state. However, there are some doubts, whether the observed structure corresponds to the quasi-bound state. The experimental results also differ from each other, moreover, their binding energies and widths are far from all theoretical predictions. Since the question of the possible existence of the quasi-bound state in the K^-pp system is still highly uncertain, new experiments are being planned and performed by HADES [9] and LEPS [10] Collaborations, and in J-PARC E15 [11] and E27 [12] experiments.

The work was supported by the Czech CACR Grant 15-04301S.

This article belongs to the special issue "30th anniversary of Few-Body Systems".

N. V. Shevchenko (✉)
Nuclear Physics Institute, 25068 Řež, Czech Republic
E-mail: shevchenko@ujf.cas.cz

It was demonstrated [4] that the $\bar{K}N$ interaction plays a crucial role in the $\bar{K}NN$ calculations. It is much more important than the nucleon–nucleon one, but is far less known. Therefore, a model of the antikaon–nucleon interaction, which is more accurate in reproducing experimental data than that from [3,4], was necessary to construct. The experimental data on $\bar{K}N$ interaction, which can be used for fitting parameters of the potential, are: near-threshold cross-sections of K^-p scattering, their threshold branching ratios, and shift and width of $1s$ level of kaonic hydrogen (which should be more accurately called "antikaonic hydrogen"). The last observable has quite interesting experimental history and finally was measured quite accurately. As for the theoretical description of kaonic hydrogen, many authors used approximate Deser-type formulas, which connect $1s$ level shift of an hadronic atom with the scattering length, given by strong interaction in the pair. The question was, how accurate are the approximate formulas, derived for the pion–nucleon interaction, for the antikaon–nucleon system. It was demonstrated in [13,14] and later in [15] that Deser-type formulas has low accuracy for the antikaon–nucleon system.

Another question of antikaon–nucleon interaction is a structure of the $\Lambda(1405)$ resonance, which couples the $\bar{K}N$ system to the lower $\pi\Sigma$ channel. $\Lambda(1405)$ is usually assumed as a quasi-bound state in the higher $\bar{K}N$ channel and a resonance state in the lower $\pi\Sigma$ channel. But it was found in [16] and later in other papers that a chirally motivated model of the interaction lead to a two-pole structure of the resonance. Keeping these two points of view in mind, two phenomenological $\bar{K}N$ potentials with one- and two-pole structure of the $\Lambda(1405)$ resonance were constructed in [14]. Their parameters were fitted to the experimental data, and $1s$ level shift and width of kaonic hydrogen were calculated directly, without any approximate formulas.

It turned out [14] that it is possible to construct phenomenological potentials with one- and two-pole $\Lambda(1405)$ resonance which describe existing low-energy experimental data with the same level of accuracy. Due to this, the two $\bar{K}N$ potentials were used as an input in calculations of the low-energy elastic K^-d scattering in [17]. But after the publication of the K^-d results, SIDDHARTA collaboration reported results of their measurement of kaonic hydrogen characteristics [18]. The results turned out to be quite different from the previously measured results of DEAR experiment [19] and compatible with older KEK data [20]. Due to this, the $\bar{K}N$ potentials were refitted in such a way, that they reproduce the most recent experimental data on the $1s$ level shift and width of kaonic hydrogen. The calculations of the low-energy elastic kaon-deuteron scattering were repeated in [21] with the new potentials. In addition, an approximate calculation of the $1s$ level shift and width of kaonic deuterium was performed. It was done approximately using a complex $K^- - d$ potential, reproducing the elastic three-body K^-d amplitudes.

It was found that the three-body K^-d system also does not allow to make preference to one of the two phenomenological $\bar{K}N$ potentials and by this to solve the question of the number of $\Lambda(1405)$ poles. In order to support the statement, one more, a chirally motivated $\bar{K}N$ potential was constructed. As other chiral models, it has two poles forming the $\Lambda(1405)$ resonance. Parameters of this potential were also fitted to the low-energy experimental data on K^-p scattering and kaonic hydrogen, the chirally motivated $\bar{K}N$ potential reproduces all antikaon–nucleon data with the same accuracy as the two phenomenological models.

Another way of investigation of the $\Lambda(1405)$ resonance was suggested and realised in [22], were low-energy breakup of the K^-d system was considered. The idea was that the resonance should be seen as a bump in so called deviation spectrum of neutrons in the final state of the reaction. However, $\Lambda(1405)$ is so broad that it was seen as a bump in some cases only.

Finally, the calculations of the three-body $\bar{K}NN$ system with different quantum numbers were repeated in [23,24] using all three models of the $\bar{K}N$ interaction. In particular, the binding energy and width of the K^-pp quasi-bound state were evaluated, the low-energy K^-d amplitudes were calculated and the $1s$ level shift and width of kaonic deuterium were predicted. A search of the quasi-bound state in the K^-d system was also performed, but the results are negative.

After the approximate calculations of the characteristics of deuterium, the exact calculations were performed in [25]. Namely, Faddeev-type equations with strong plus Coulomb interactions, suggested in [26], were solved. It was the first time, when the equations [26], initially written and used for a system with Coulomb interaction being a correction to a strong potential, were used for investigations of an hadronic atom, where Coulomb potential plays the main role. Since the equations are much more complicated than "usual" AGS ones (containing short-range potentials only), the calculations were performed with simple complex $\bar{K}N$ potentials, reproducing only some of the experimental K^-p data. Comparison of the dynamically exact three-body results with the previous approximate ones shown that the approximation of the kaonic deuterium as a two-body system is quite accurate for this task.

Another three-body exotic system, consisting of two antikaons and one nucleon, was studied in [27]. It was expected that a quasi-bound state can exist in the $\bar{K}\bar{K}N$ system too. The three $\bar{K}N$ potentials were used, and a quasi-bound state was found with smaller binding energy than in the $K^- pp$ and larger width. It is interesting, that the parameters of the state allow to associate it with a Ξ state mentioned in the Particle Data Group [28].

The paper summarises results of the series of exact or accurate calculations [3,4,14,17,21,23–25,27]. The next section contains information about the two-body interactions, necessary for the three-body calculations. Faddeev-type Alt–Grassberger–Sandhas equations with coupled channels, which were used for three-body calculations with strong interactions, are described in Sect. 3. Section 4 is devoted to the quasi-bound states in the $K^- pp$, $K^- d$, and $K^- K^- p$ systems. The near-threshold $K^- d$ scattering is considered in Sect. 5, the kaonic deuterium - in Sect. 6. The last section summarises the results.

2 Two-Body Interactions

In order to investigate some three-body system it is necessary to know the interactions of all the pairs of the particles. The interactions, necessary for investigations of the $\bar{K}NN - \pi\Sigma N$ and $\bar{K}\bar{K}N - \bar{K}\pi\Sigma$ systems, are $\bar{K}N$ and ΣN with other channels coupled to them, and the one-channel NN and $\bar{K}\bar{K}$ interactions (the rest of them were omitted in the three-body calculations). All potentials, except one of the NN potentials, were specially constructed for the calculations. They have a separable form and $\mathbb{N}$-term structure

$$V_{i,II'}^{\alpha\beta} = \sum_{m=1}^{\mathbb{N}_i^{\alpha}} \lambda_{i(m),II'}^{\alpha\beta} |g_{i(m),I}^{\alpha}\rangle \langle g_{i(m),I'}^{\beta}|, \tag{1}$$

which leads to a separable T-matrix

$$T_{i,II'}^{\alpha\beta} = \sum_{m,n=1}^{\mathbb{N}_i^{\alpha}} |g_{i(m),I}^{\alpha}\rangle \tau_{i(mn),II'}^{\alpha\beta} \langle g_{i(n),I'}^{\beta}|. \tag{2}$$

$\mathbb{N}_i^{\alpha}$ in Eqs. (1) and (2) is a number of terms of the separable potential, λ is a strength constant, while g is a form-factor. The two-body isospin I in general is not conserved. In particular, the two-body isospin is conserved in the phenomenological $\bar{K}N - \pi\Sigma$ potentials, but not in the corresponding T-matrices due to Coulomb interaction and the physical masses, taken into account. The chirally motivated $\bar{K}N - \pi\Sigma - \pi\Lambda$ potential does not conserve the isospin due to its energy and mass dependence.

The separable potentials are simpler than other models of interactions. However, the potentials, entering the equations for the antikaon–nucleon systems, were constructed in such a way that they reproduce the low-energy experimental data for every subsystem very accurately. From this point of view they are not worse than other models of the antikaon–nucleon or the Σ-nucleon interaction (in fact, they are even better than some chiral models). The one-term NN potential does not have a repulsive part, but the two-term model is repulsive at short distances. Finally, all three-body observables described in the present paper turned out to be dependent on the NN and ΣN interactions very weakly, therefore the most important is the accuracy of the $\bar{K}N$ potential.

The antikaon–nucleon interaction is the most important one for the three-body systems under consideration. There are several models of the $\bar{K}N$ interaction, some of them are "stand-alone" ones having the only aim to reproduce experimental data, others were used in few- of many-body calculations. The problem is that the first ones are too complicated to be used in few-body calculations, while the models from the second group are too simple to reproduce all the experimental data properly. Due to this, the $\bar{K}N$ potentials, which are simple enough for using in Faddeev-type three-body equations and at the same time reproduce all low-energy antikaon–nucleon experimental data, were constructed.

2.1 Antikaon–Nucleon Interaction, Experimental Data

$\Lambda(1405)$ *Resonance*

The $\Lambda(1405)$ resonance is a manifestation of the attractive nature of the antikaon–nucleon interaction in isospin zero state, it couples $\bar{K}N$ to the lower $\pi\Sigma$ channel. Not only position and width, but the nature of the resonance itself are opened questions. A usual assumption is that $\Lambda(1405)$ is a resonance in the $\pi\Sigma$ channel and a quasi-bound state in the $\bar{K}N$ channel. According to the most recent Particle Data Group issue [28], the resonance has mass $1405.1^{+1.3}_{-1.0}$ MeV and width 50.5 ± 2.0 MeV. There is also an assumption suggested in [16] and supported by other chiral models, that the bump, which is usually understood as the $\Lambda(1405)$ resonance, is an effect of two poles. Due to this, the two different phenomenological models of the antikaon–nucleon interaction with one- or two-pole structure of the $\Lambda(1405)$ were constructed. The third model is a chirally motivated potential, which has two poles by construction.

Extraction of the resonance parameters from experimental data is complicated for two reasons. First, it cannot be studied in a two-body reaction and can be seen in a final state of some few- or many-body process. Second, its width is large, so the corresponding peak could be blurred.

A theoretical paper [22] was devoted to the possibility of tracing the $\Lambda(1405)$ resonance in the neutron spectrum of a K^-d breakup reaction. The neutron spectra of the $K^-d \to \pi \Sigma n$ reaction were calculated in center of mass energy range 0–50 MeV. The three-body system with coupled $\bar{K}NN$ and $\pi \Sigma N$ channels was studied using the Faddeev-type AGS equations, described in Sect. 3, with four phenomenological $\bar{K}N$ potentials with one- or two-pole structure of $\Lambda(1405)$. It was found that kinematic effects completely mask the peak corresponding to the $\Lambda(1405)$ resonance. Therefore, comparison of eventual experimental data on the low-energy $K^-d \to \pi \Sigma n$ reaction with theoretical results hardly can give an answer to the question of the number of $\Lambda(1405)$ poles.

Later, similar calculations of the same process were performed for initial kaon momentum 1 GeV in [29,30]. Coupled-channel AGS equations were solved as well with energy-dependent and -independent $\bar{K}N$ potentials. The authors predict a pronounced maximum in the double-differential cross section with a forward emitted neutron at $\pi \Sigma$ invariant mass 1.45 GeV. However, applicability of the $\bar{K}N$ potentials, fitted to the near-threshold data, and of the nonrelativistic Faddeev equations for such high energies is quite doubtful.

Several arguments, suggested in support to the idea of the two-pole structure of the $\Lambda(1405)$ resonance, were checked in [17] using the one- and two-pole phenomenological models of the antikaon–nucleon interaction. One of the arguments is the difference between the $\pi \Sigma$ cross-sections with different charge combinations, which is seen in experiments, e.g. in CLAS [31]. The elastic $\pi^+ \Sigma^-$, $\pi^- \Sigma^+$, and $\pi^0 \Sigma^0$ cross-sections were plotted to check the assumption, that the difference is caused by the two-pole structure. However, it turned out that the cross sections are different and their maxima are shifted one from each other for both one- and two-pole versions of the $\bar{K}N$ potential (see Fig. 5 of [17]). Therefore, the effect is not a proof of the two-pole structure, but a manifestation of the isospin non-conservation and differences in the background.

Another argument for the two-pole structure comes from the fact, that the poles in a two-pole model are coupled to different channels. Indeed, a gradual switching off of the coupling between the $\bar{K}N$ and $\pi \Sigma$ channels turns the upper pole into a real bound state in $\bar{K}N$, while the lower one becomes a resonance in the uncoupled $\pi \Sigma$ channel (see e.g. Fig. 2 of [23]). Consequently, it was suggested, that the poles of a two-body model manifest themselves in different reactions. In particular, the $\bar{K}N - \bar{K}N$, $\bar{K}N - \pi \Sigma$, and $\pi \Sigma - \pi \Sigma$ amplitudes should "feel" only one of the two poles. The hypothesis was also checked in [17], and indeed, the real parts of the $\bar{K}N - \bar{K}N$, $\bar{K}N - \pi \Sigma$, and $\pi \Sigma - \pi \Sigma$ amplitudes in $I = 0$ state cross the real axis at different energies. But it is true for the both: the one- and the two-pole versions of the potential (see Fig. 6 of [17]). This effect must be caused by different background contributions in the reactions. Therefore, a proof of the two-pole structure of the $\bar{K}N$ interaction does not exist.

Cross-Sections and Threshold Branching Ratios

Three threshold branching ratios of the K^-p scattering

$$\gamma = \frac{\Gamma\left(K^-p \to \pi^+ \Sigma^-\right)}{\Gamma\left(K^-p \to \pi^- \Sigma^+\right)} = 2.36 \pm 0.04, \tag{3}$$

$$R_c = \frac{\Gamma\left(K^-p \to \pi^+ \Sigma^-, \pi^- \Sigma^+\right)}{\Gamma\left(K^-p \to \text{all inelastic channels}\right)} = 0.664 \pm 0.011, \tag{4}$$

$$R_n = \frac{\Gamma\left(K^-p \to \pi^0 \Lambda\right)}{\Gamma\left(K^-p \to \text{neutral states}\right)} = 0.189 \pm 0.015 \tag{5}$$

were measured rather accurately in [32,33]. Since the phenomenological $\bar{K}N$ potentials, used in the calculations, take the lowest $\pi\Lambda$ channel into account indirectly, a new ratio

$$R_{\pi\Sigma} = \frac{\Gamma\left(K^-p \to \pi^+\Sigma^-\right) + \Gamma\left(K^-p \to \pi^-\Sigma^+\right)}{\Gamma\left(K^-p \to \pi^+\Sigma^-\right) + \Gamma\left(K^-p \to \pi^-\Sigma^+\right) + \Gamma\left(K^-p \to \pi^0\Sigma^0\right)}, \tag{6}$$

which contains the measured R_c and R_n and has an "experimental" value

$$R_{\pi\Sigma} = \frac{R_c}{1 - R_n\left(1 - R_c\right)} = 0.709 \pm 0.011 \tag{7}$$

was constructed and used.

In contrast to the branching ratios, the elastic and inelastic total cross sections with K^-p in the initial state [35–40] were measured not so accurately, see Fig. 1.

Kaonic Hydrogen

The most promising source of knowledge about the $\bar{K}N$ interaction is kaonic hydrogen atom (which correctly should be called "antikaonic hydrogen"). The atom has rich experimental history, several experiments measured its $1s$ level shift

$$\Delta E_{1s} = E_{1s}^{Coul} - \mathrm{Re}\left(E_{1s}^{Coul+Strong}\right) \tag{8}$$

and width Γ_{1s}, caused by the strong $\bar{K}N$ interaction in comparison to pure Coulomb case, with quite different results. The most recent measurement was performed by SIDDHARTA collaboration [18], their results are:

$$\Delta E_{1s}^{SIDD} = -283 \pm 36 \pm 6 \text{ eV}, \quad \Gamma_{1s}^{SIDD} = 541 \pm 89 \pm 22 \text{ eV}. \tag{9}$$

Paradoxically, the directly measurable observables are not reproduced in the same way in the most of the theoretical works devoted to the antikaon–nucleon interaction. Some approximate formula are usually used for reproducing the $1s$ level shift. The most popular is a "corrected Deser" formula [34], which connects the shift with the scattering length a_{K^-p} of the K^-p system:

$$\Delta E^{cD} - i\frac{\Gamma^{cD}}{2} = -2\alpha^3\mu_{K^-d}^2 a_{K^-p}$$
$$\times \left[1 - 2\alpha\mu_{K^-p}a_{K^-p}\left(\ln\alpha - 1\right)\right]. \tag{10}$$

The formula is one of quite a few versions of the original formula, derived by Deser for the pion–nucleon system. It differs from the original one by the second term in the brackets. However, it was shown (e.g. in [14] and other papers) that for the antikaon–nucleon system the formula is not accurate, it gives $\sim 10\%$ error.

2.2 Phenomenological and Chirally Motivated $\bar{K}N$ Potentials

The constructed phenomenological models of antikaon–nucleon interaction with one- or two-pole structure of the $\Lambda(1405)$ resonance together with the chirally motivated model reproduce the $1s$ level shift and width of kaonic hydrogen, measured by SIDDHARTA collaboration, directly, without using approximate formulas. The potentials also reproduce the experimental data on the K^-p scattering and the threshold branching ratios, described in the previous subsection. All three potentials are suitable for using in accurate few-body equations.

The problem of two particles interacting by the strong and Coulomb potentials, considered on the equal basis, was solved. The method of solution of Lippmann–Schwinger equation for a system with Coulomb plus a separable strong potential is based on the fact that the full T-matrix of the problem can be written as a sum $T = T^c + T^{sc}$. Here T^c is the pure Coulomb transition matrix and T^{sc} is the Coulomb-modified strong T-matrix. It was necessary to extend the formalism to describe the system of the coupled $\bar{K}N$, $\pi\Sigma$ (and $\pi\Lambda$ for the chirally motivated potential) channels. The physical masses of the particles were used in the equation, therefore, the two-body isospin of the system is not conserved. More details on the formalism can be found in [14].

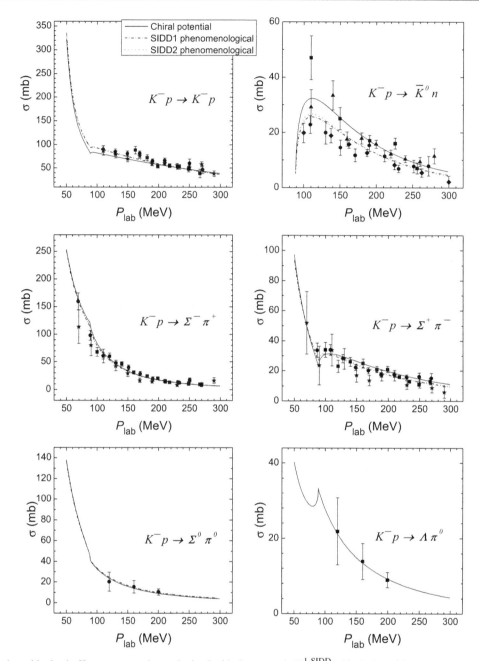

Fig. 1 Elastic and inelastic $K^- p$ cross-sections, obtained with the one-pole $V^{1,\text{SIDD}}_{\bar{K}N-\pi\Sigma}$ (*dash-dotted line*), the two-pole $V^{2,\text{SIDD}}_{\bar{K}N-\pi\Sigma}$ (*dotted line*) phenomenological potentials, and the chirally motivated $V^{Chiral}_{\bar{K}N-\pi\Sigma-\pi\Lambda}$ (*solid line*) potential. The experimental data are taken from [35–40]

The phenomenological potentials describing the $\bar{K}N$ system with coupled $\pi\Sigma$ channel are the one-term separable ones defined by Eq. (1). In momentum representation they have a form

$$V_I^{\bar{\alpha}\bar{\beta}}\left(k^{\bar{\alpha}}, k'^{\bar{\beta}}\right) = \lambda_I^{\bar{\alpha}\bar{\beta}} \, g^{\bar{\alpha}}\left(k^{\bar{\alpha}}\right) g^{\bar{\beta}}\left(k'^{\bar{\beta}}\right), \tag{11}$$

where indices $\bar{\alpha}, \bar{\beta} = 1, 2$ denote the $\bar{K}N$ or $\pi\Sigma$ channel respectively, and I is a two-body isospin. Different form-factors were used for the one- and two-pole versions of the phenomenological potential. While for the one-pole version Yamaguchi form-factors

$$g^{\bar{\alpha}}\left(k^{\bar{\alpha}}\right) = \frac{1}{\left(k^{\bar{\alpha}}\right)^2 + \left(\beta^{\bar{\alpha}}\right)^2} \tag{12}$$

were used, the two-pole version has slightly more complicated form-factors in the $\pi\Sigma$ channel

$$g^{\bar{\alpha}}\left(k^{\bar{\alpha}}\right) = \frac{1}{\left(k^{\bar{\alpha}}\right)^2 + \left(\beta^{\bar{\alpha}}\right)^2} + \frac{s\left(\beta^{\bar{\alpha}}\right)^2}{\left[\left(k^{\bar{\alpha}}\right)^2 + \left(\beta^{\bar{\alpha}}\right)^2\right]^2}. \tag{13}$$

In the $\bar{K}N$ channel the form-factor of the two-pole version is also of Yamaguchi form Eq. (12).

Range parameters $\beta^{\bar{\alpha}}$, strength parameters $\lambda_I^{\bar{\alpha}\bar{\beta}}$ and an additional parameter s of the two-pole version were obtained by fitting to the experimental data described in the previous subsection. They are: the elastic and inelastic K^-p cross-sections, the threshold branching ratios and the $1s$ level shift and width of kaonic hydrogen. The first versions of the potentials, presented in [14] and used in [17], were fitted to the KEK data [20] on kaonic hydrogen. The actual versions of the phenomenological potentials were fitted to the most recent experimental data of SIDDHARTA collaboration [18]. The parameters of the one- and two-pole versions of the phenomenological potentials fitted to SIDDHARTA data can be found in [21].

All fits were performed directly to the experimental values except the threshold branching ratios R_c and R_n. The reason is that the ratios contain data on scattering of K^-p into all inelastic channels including $\pi\Lambda$, which is taken by the phenomenological potentials into account only indirectly through imaginary part of one of the λ parameters. Due to this the phenomenological potentials were fitted to the new ratio $R_{\pi\Sigma}$, defined in Eq. (7).

The third model of the antikaon–nucleon interaction is the chirally motivated potential. It connects all three open channels: $\bar{K}N$, $\pi\Sigma$ and $\pi\Lambda$, and has a form

$$V_{II'}^{\alpha\beta}\left(k^\alpha, k'^\beta; \sqrt{s}\right) = g_I^\alpha\left(k^\alpha\right)\bar{V}_{II'}^{\alpha\beta}\left(\sqrt{s}\right)g_{I'}^\beta\left(k'^\beta\right), \tag{14}$$

where $\bar{V}_{II'}^{\alpha\beta}(\sqrt{s})$ is the energy dependent part of the potential in isospin basis. In particle basis the energy dependent part has a form

$$\bar{V}^{ab}(\sqrt{s}) = \sqrt{\frac{M_a}{2\omega_a E_a}}\frac{C^{ab}(\sqrt{s})}{(2\pi)^3 f_a f_b}\sqrt{\frac{M_b}{2\omega_b E_b}}. \tag{15}$$

Indices a, b here denote the particle channels $a, b = K^-p$, $\bar{K}^0n$, $\pi^+\Sigma^-$, $\pi^0\Sigma^0$, $\pi^-\Sigma^+$ and $\pi^0\Lambda$. The square roots with baryon mass M_a, baryon energy E_a and meson energy w_a of the channel a ensure proper normalization of the corresponding amplitude. SU(3) Clebsh–Gordan coefficients C_I^{WT} enter the non-relativistic form of the leading order Weinberg–Tomozawa interaction

$$C^{ab}\left(\sqrt{s}\right) = -C^{WT}\left(2\sqrt{s} - M_a - M_b\right). \tag{16}$$

Since, as in the case of the phenomenological potentials, the physical masses of the particles were used, the two-body isospin I in Eq. (14) is not conserved. It is different from the phenomenological potentials situation since in that case the potentials conserve the two-body isospin (but the corresponding T-matrices do not). Another feature, which distinguish the chirally motivated potential from the phenomenological ones, is the isospin dependence of its form-factors:

$$g_I^\alpha\left(k^\alpha\right) = \frac{\left(\beta_I^\alpha\right)^2}{(k^\alpha)^2 + \left(\beta_I^\alpha\right)^2}. \tag{17}$$

Besides, they are dimensionless due to the additional factor $(\beta_I^\alpha)^2$ in the numerator.

The pseudo-scalar meson decay constants f_π, f_K and the isospin dependent range parameters β_I^α are free parameters of the chirally motivated potential. They also were found by fitting the potential to the experimental data in the same way as in the phenomenological potentials case. The chirally-motivated $\bar{K}N - \pi\Sigma - \pi\Lambda$ potential reproduces the elastic and inelastic K^-p cross-sections, SIDDHARTA $1s$ level shift and width of kaonic hydrogen. In contrast to the phenomenological potentials, the chirally-motivated one directly reproduces all three K^-p branching ratios: γ, R_c and R_n. The parameters of the potential can be found in [23].

Table 1 Physical characteristics of the three antikaon nucleon potentials: phenomenological $V_{\bar{K}N-\pi\Sigma}^{1,SIDD}$ and $V_{\bar{K}N-\pi\Sigma}^{2,SIDD}$ with one- and two-pole structure of the $\Lambda(1405)$ resonance respectively, and the chirally motivated $V_{\bar{K}N-\pi\Sigma-\pi\Lambda}^{Chiral}$ potential: $1s$ level shift $\Delta E_{1s}^{K^-p}$ (eV) and width $\Gamma_{1s}^{K^-p}$ (eV) of kaonic hydrogen, threshold branching ratios γ, R_c and R_n together with the experimental data

	$V_{\bar{K}N-\pi\Sigma}^{1,SIDD}$	$V_{\bar{K}N-\pi\Sigma}^{2,SIDD}$	$V_{\bar{K}N-\pi\Sigma-\pi\Lambda}^{Chiral}$	Experiment
$\Delta E_{1s}^{K^-p}$	-313	-308	-313	$-283 \pm 36 \pm 6$ [18]
$\Gamma_{1s}^{K^-p}$	597	602	561	$541 \pm 89 \pm 22$ [18]
γ	2.36	2.36	2.35	2.36 ± 0.04 [32,33]
R_c	–	–	0.663	0.664 ± 0.011 [32,33]
R_n	–	–	0.191	0.189 ± 0.015 [32,33]
$R_{\pi\Sigma}$	0.709	0.709	–	0.709 ± 0.011 Eq. (7)
a_{K^-p}	$-0.76 + i\,0.89$	$-0.74 + i\,0.90$	$-0.77 + i\,0.84$	
z_1	$1426 - i\,48$	$1414 - i\,58$	$1417 - i\,33$	
z_2	–	$1386 - i\,104$	$1406 - i\,89$	

The additional $R_{\pi\Sigma}$ ratio, see Eq. (6), with its "experimental" value is shown as well. Scattering length of the K^-p system a_{K^-p} (fm) and pole(s) z_1, z_2 (MeV) forming the $\Lambda(1405)$ resonance are also demonstrated

The $\Lambda(1405)$ resonance can manifest itself as a bump in elastic $\pi\Sigma$ cross-sections or in K^-p amplitudes. In the last case, the real part of the amplitude crosses zero, while the imaginary part has a maximum near the resonance position. It is demonstrated in [21,23] that the elastic $\pi\Sigma$ cross-sections, provided by the three potentials, have a bump near the PDG value [28] for the mass of the $\Lambda(1405)$ resonance with appropriate width.

The physical characteristics of the three antikaon–nucleon potentials are shown in Table 1. In addition, the K^-p scattering length a_{K^-p} and the pole(s) forming the $\Lambda(1405)$ resonance, given by the potentials, are demonstrated. The elastic and inelastic K^-p cross-sections, provided by the three potentials, are plotted in Fig. 1 together with the experimental data. It is seen form the Table 1 and Fig. 1 that the one- and two-pole phenomenological potentials and the chirally motivated potential describe all the experimental data with equal high accuracy. Therefore, it is not possible to choose one of the three models of the $\bar{K}N$ interaction looking at the two-body system.

Approximate Versions of the Coupled-Channel Potential

In order to check approximations used in other theoretical works, two approximate versions of the coupled-channel potentials Eq. (11), which have only one $\bar{K}N$ channel, were also used. They are: an exact optical potential and a simple complex one.

The exact optical one-channel potential, corresponding to a two-channel one, is given by Eq. (11) with $\bar{\alpha}, \bar{\beta} = 1$ and the strength parameter defined as

$$\lambda_I^{11,Opt} = \lambda_I^{11} + \frac{(\lambda_I^{12})^2 \left\langle g_I^2 \right| G_0^{(2)} \left(z^{(2)}\right) \left| g_I^2 \right\rangle}{1 - \lambda_I^{22} \left\langle g_I^2 \right| G_0^{(2)} \left(z^{(2)}\right) \left| g_I^2 \right\rangle}. \tag{18}$$

Here $\lambda_I^{\bar{\alpha},\bar{\beta}}$ are the strength parameters of the two-channel potential, and $|g_I^2\rangle$ is the form-factor of the second channel. Since a two-body free Green's function $G_0^{(2)}$ depends on the corresponding two-body energy, the parameter $\lambda_I^{11,Opt}$ of the exact optical potential is an energy-dependent complex function. The exact optical potential has exactly the same elastic amplitudes of the $\bar{K}N$ scattering as the elastic part of the full potential with coupled channels.

A simple complex potential is quite often miscalled "an optical" one, however, it is principally different. The strength parameter $\lambda_I^{11,Complex}$ of a simple complex potential is a complex constant, therefore, the simple complex potential is energy independent. The strength parameter of a simple complex potential is usually chosen in such a way, that the potential reproduces only some characteristics of the interaction. The simple complex potential as well as the exact optical one take into account flux losses into inelastic channels through imaginary parts of the strength parameters.

2.3 Nucleon–Nucleon and $\Sigma N(-\Lambda N)$ Potentials

NN Interaction

Different NN potentials, in particular, TSA-A, TSA-B and PEST, were used in order to investigate dependence of the three-body results on the nucleon–nucleon interaction models.

A two-term separable NN potential [41], called TSA, reproduces Argonne $V18$ [42] phase shifts and, therefore, is repulsive at short distances. Two versions of the potential (TSA-A and TSA-B) with slightly different form-factors

$$g^{A,NN}_{(m)}(k) = \sum_{n=1}^{2} \frac{\gamma^A_{(m)n}}{\left(\beta^A_{(m)n}\right)^2 + k^2}, \quad \text{for } (m) = 1, 2$$

$$g^{B,NN}_{(1)}(k) = \sum_{n=1}^{3} \frac{\gamma^B_{(1)n}}{\left(\beta^B_{(1)n}\right)^2 + k^2}, \quad g^{B,NN}_{(2)}(k) = \sum_{n=1}^{2} \frac{\gamma^B_{(2)n}}{\left(\beta^B_{(2)n}\right)^2 + k^2} \tag{19}$$

were used. TSA-A and TSA-B potentials properly reproduce the NN scattering lengths and effective radii, they also give correct binding energy of the deuteron in the $^3S_1$ state. For more details see Ref. [17].

A separabelization of the Paris model of the NN interaction, called PEST potential [43], was also used. The strength parameter of the one-term PEST is equal to -1, while the form-factor is defined by

$$g^{NN}_I(k) = \frac{1}{2\sqrt{\pi}} \sum_{n=1}^{6} \frac{c^{NN}_{n,I}}{k^2 + \left(\beta^{NN}_{n,I}\right)^2} \tag{20}$$

with $c^{NN}_{n,I}$ and $\beta^{NN}_{n,I}$ being the parameters. PEST is equivalent to the Paris potential on and off energy shell up to $E_{\text{lab}} \sim 50\,\text{MeV}$. It also reproduces the deuteron binding energy in the $^3S_1$ state, as well as the triplet and singlet NN scattering lengths.

The quality of reproducing the $^3S_1$ phase shifts by the three NN potentials is shown in Fig. 8 of [17], were they are compared with those given by the Argonne $V18$ model. The two-term TSA-A and TSA-B potentials are very good at reproducing of the Argonne $V18$ phase shifts. They cross the real axis, which is a consequence of the NN repulsion at short distances. The one-term PEST potential does not have such a property, but its phase shifts are also close to the "standard" ones at lower energies.

Only isospin-singlet NN potential enters the AGS equations for the K^-d system and only isospin-triplet one enters the equations describing the K^-pp system after antisymmetrization.

ΣN Interaction

A spin dependent V^{Sdep} and an independent of spin V^{Sind} versions of the ΣN potential were constructed in [17] in such a way, that they reproduce the experimental ΣN and ΛN cross-sections [44–48]. The one-term separable potentials with Yamaguchi form-factors were used for the two possible isospin states, but with different number of the channels.

Parameters of the one-channel $I = \frac{3}{2}$ state were fitted to the $\Sigma^+ p \to \Sigma^+ p$ cross-sections. The ΣN system in isospin one half state is connected to the ΛN channel, therefore, a coupled-channel potential of the $I = \frac{1}{2}$ $\Sigma N - \Lambda N$ interaction was constructed first. The coupled-channel $I = \frac{1}{2}$ potential together with the one-channel $I = \frac{3}{2}$ potential reproduce the $\Sigma^- p \to \Sigma^- p$, $\Sigma^- p \to \Sigma^0 n$, $\Sigma^- p \to \Lambda n$, and $\Lambda p \to \Lambda p$ cross-sections. It is seen in Fig. 9 of [17] that both V^{Sdep} and V^{Sind} versions of the $I = 3/2$ ΣN and $I = 1/2$ $\Sigma N - \Lambda N$ potentials reproduce the experimental data perfectly. Parameters of the potentials and the scattering lengths $a^{\Sigma N}_{\frac{1}{2}}$, $a^{\Sigma N}_{\frac{3}{2}}$, and $a^{\Lambda N}_{\frac{1}{2}}$, given by them, are shown in Table 5 of [17].

For the three-body $\bar{K}NN$ calculations, where a channel containing Λ is not included directly, not a coupled-channel, but one-channel ΣN models of the interaction in the $I = \frac{1}{2}$ state were used. They are the exact optical $V^{\Sigma N,\text{Opt}}$ potential and a simple complex $V^{\Sigma N,\text{Complex}}$ one, corresponding to the $I = \frac{1}{2}$ $\Sigma N - \Lambda N$ model with coupled channels. The exact optical potential has an energy dependent strength parameter defined by Eq. (18), it reproduces the elastic ΣN amplitude of the corresponding two-channel potential exactly. The simple complex potential gives the same scattering lengths, as the two-channel potential.

2.4 Antikaon–Antikaon Interaction

Lack of an experimental information on the $\bar{K}\bar{K}$ interaction means that it is not possible to construct the $\bar{K}\bar{K}$ potential in the same way as the $\bar{K}N$ or ΣN ones. Due to this, theoretical results of a modified model describing the $\pi\pi - K\bar{K}$ system developed by the Jülich group [49,50] were used. The original model yields a good description of the $\pi\pi$ phase shifts up to partial waves with total angular momentum $J = 2$ and for energies up to $z_{\pi\pi} \approx 1.4$ GeV. In addition, the $f_0(980)$ and $a_0(980)$ mesons result as dynamically generated states.

Based on the underlying SU(3) flavor symmetry, the interaction in the $\bar{K}\bar{K}$ system was directly deduced from the KK interaction without any further assumptions. The $\bar{K}\bar{K}$ scattering length predicted by the modified Jülich model is $a_{\bar{K}\bar{K},I=1} = -0.186$ fm, therefore, it is a repulsive interaction. This version of the $\bar{K}\bar{K}$ interaction was called "Original".

Recent results for the KK scattering length from lattice QCD simulations suggest values of $a_{\bar{K}\bar{K},I=1} = (-0.141 \pm 0.006)$ fm [51] and $a_{\bar{K}\bar{K},I=1} = (-0.124 \pm 0.006 \pm 0.013)$ fm [52]. Those absolute values are noticeably smaller than the one predicted by the Original Jülich meson-exchange model and, accordingly, imply a somewhat less repulsive $\bar{K}\bar{K}$ interaction. Due to this, another version of the interaction that is in line with the lattice QCD results was also constructed. It yields scattering length $a_{\bar{K}\bar{K},I=1} = -0.142$ fm. This version of the $\bar{K}\bar{K}$ interaction was called "Lattice motivated".

However, the models of the $\bar{K}\bar{K}$ interaction described above cannot be directly used in the AGS equations. Due to this, the $\bar{K}\bar{K}$ interaction was represented in a form of the one-term separable potential with form factors given by

$$g(k) = \frac{1}{\beta_1^2 + k^2} + \frac{\gamma}{\beta_2^2 + k^2}. \tag{21}$$

The strength parameters λ, γ and range parameters β were fixed by fits to the $\bar{K}\bar{K}$ phase shifts and scattering lengths of the "Original" and the "Lattice motivated" models of the antikaon–antikaon interaction.

3 AGS Equations for Coupled $\bar{K}NN - \pi\Sigma N$ and $\bar{K}\bar{K}N - \bar{K}\pi\Sigma$ Channels

The three-body Faddeev equations in the Alt–Grassberger–Sandhas (AGS) form [53] were used for the most of the three-body calculations. The equations were extended in order to take the $\pi\Sigma$ channel, coupled to the $\bar{K}N$ subsystem, directly. In practice it means that all operators entering the system

$$U_{ij}^{\alpha\beta} = \delta_{\alpha\beta}(1 - \delta_{ij})\left(G_0^\alpha\right)^{-1} + \sum_{k,\gamma=1}^{3}(1 - \delta_{ik})T_k^{\alpha\gamma}G_0^\gamma U_{kj}^{\gamma\beta}, \tag{22}$$

namely, transition operators U_{ij}, two-body T-matrices T_i and the free Green function G_0,—have additional channel indices $\alpha, \beta = 1, 2, 3$ in addition to the Faddeev partition indices $i, j = 1, 2, 3$. The additional $\pi\Sigma N$ ($\alpha = 2$) and $\pi N\Sigma$ ($\alpha = 3$) channels were added to the $\bar{K}NN$ system, while the $\bar{K}\bar{K}N$ system was extended to the $\bar{K}\pi\Sigma$ ($\alpha = 2$) and $\pi\bar{K}\Sigma$ ($\alpha = 3$) channels. A Faddeev index i, as usual, defines a particle and the remained pair, now in the particular particle channel α. The combinations of the (i, α) indices with possible two-body isospin values can be found in [4] for the $\bar{K}NN$ system and in [27] for the $\bar{K}\bar{K}N$ systems, respectively.

Since the separable potentials Eq. (1), leading to the separable T-matrices Eq. (2), were used as an input, the system Eq. (22) turned into the new system of operator equations

$$X_{ij,I_iI_j}^{\alpha\beta} = \delta_{\alpha\beta}Z_{ij,I_iI_j}^\alpha + \sum_{k=1}^{3}\sum_{\gamma=1}^{3}\sum_{I_k}Z_{ik,I_iI_k}^\alpha \tau_{k,I_k}^{\alpha\gamma}X_{kj,I_kI_j}^{\gamma\beta} \tag{23}$$

with $X_{ij,I_iI_j}^{\alpha\beta}$ and $Z_{ij,I_iI_j}^{\alpha\beta}$ being new transition and kernel operators respectively

$$X_{ij,I_iI_j}^{\alpha\beta} = \left\langle g_{i,I_i}^\alpha | G_0^\alpha U_{ij,I_iI_j}^{\alpha\beta} G_0^\beta | g_{j,I_j}^\beta \right\rangle, \tag{24}$$

$$Z_{ij,I_iI_j}^{\alpha\beta} = \delta_{\alpha\beta}Z_{ij,I_iI_j}^\alpha = \delta_{\alpha\beta}(1 - \delta_{ij})\left\langle g_{i,I_i}^\alpha | G_0^\alpha | g_{j,I_j}^\alpha \right\rangle. \tag{25}$$

$\bar{K}NN - \pi\Sigma N$ System

The two states of the strangeness $S = -1$ $\bar{K}NN$ system were considered. The $K^- pp$ and $K^- d$ systems differ from each another by the total spin value, which leads to different symmetry of the operators describing the system containing two identical baryons, NN. This fact is taken into account when the three-body coupled-channel equations are antisymmetrized.

All calculations were performed under or slightly above the $\bar{K}NN$ threshold, so that orbital angular momentum of all two-body interactions was set to zero and, therefore, the total orbital angular momentum is also $L = 0$. In particular, the main $\bar{K}N$ potential was constructed with orbital angular momentum $l = 0$ since the interaction is dominated by the s-wave $\Lambda(1405)$ resonance. The interaction of π-meson with a nucleon is weaker than the other interactions, therefore, it was omitted in the equations. An experimental information about the ΣN interaction is very poor, and there is no reason to assume significant effect of higher partial waves. Finally, the NN interaction was also taken in $l = 0$ state only, since physical reasons for sufficient effect of higher partial waves in the present calculation are not seen.

Spin of the $\bar{K}NN$ system is given by spin of the two baryons, which also defines the NN isospin due to the symmetry properties. Looking for the quasi-bound state in $\bar{K}NN$, the isospin $I = 1/2$ and spin zero state, usually denoted as $K^- pp$, was chosen due to its connection to experiment. Another possible configuration with the same isospin and spin one, which is $K^- d$, was also studied. As for the $\bar{K}NN$ state with isospin $I = 3/2$, it is governed by the isospin $I_i = 1$ $\bar{K}N$ interaction, which is much weaker attractive than the one in the $I_i = 0$ state or even repulsive. Therefore, no quasi-bound state is expected there.

The nucleons, entering the highest $\bar{K}NN$ channel, require antisymmetrization of the operators entering the system of Eq. (23). Two identical baryons with symmetric spatial components ($L_i = 0$) has antisymmetric ($S_i = 0$) spin components for the pp state of the NN subsystem or symmetric ($S_i = 1$) ones for the d state. The operator $X_{1,1}^1$ has symmetric NN isospin components, therefore, it has the correct symmetry properties for the $K^- pp$ system (here and in what follows the right-hand indices of X are omitted: $X_{ij,I_i I_j}^{\alpha\beta} \rightarrow X_{i,I_i}^{\alpha}$). Another operator, $X_{1,0}^1$, has antisymmetric NN isospin components, so it drops out the equations for the $K^- pp$ system, but remains in the equations describing the $K^- d$ system. All the remaining operators form symmetric and antisymmetric pairs. At the end there is a system of 9 (with PEST NN potential) or 10 (with TSA nucleon–nucleon model) coupled operator equations, which has the required symmetry properties.

The system of operator Eq. (23) written in momentum space turns into a system of integral equations. To search for a quasi-bound state in a three-body system means to look for a solution of the homogeneous system corresponding to Eq. (23). Calculation of three-body scattering amplitudes require solution of the inhomogeneous system. In the both cases the integral equations are transformed into algebraic ones. The methods of solution are different for the quasi-bound state and scattering problems, so they are discussed in the corresponding sections.

More details on the three-body equations with coupled $\bar{K}NN - \pi\Sigma N$ channels can be found in [4] for the $K^- pp$ and in [17] for the $K^- d$ systems.

$\bar{K}\bar{K}N - \bar{K}\pi\Sigma$ System

As for the strangeness $S = -2$ $\bar{K}\bar{K}N$ system, its total spin is equal to one half since an antikaon is a pseudoscalar meson. Since the two-body interactions, namely the $\bar{K}N - \pi\Sigma$ and $\bar{K}\bar{K}$ potentials, were chosen to have zero orbital angular momentum, the total angular momentum is also equal to 1/2. As in the case of the $\bar{K}NN$ system, the state of the $\bar{K}\bar{K}N$ system with the lowest possible value of the isospin $I = 1/2$ was considered.

Two identical antikaons should have a symmetric way function, therefore, the $\bar{K}\bar{K}$ pair in s-wave can be in isospin one state only. Accordingly, the three-body operators entering the AGS system for the $\bar{K}\bar{K}N$ system were symmetrized. Similarly to the $\bar{K}NN$ case, the transition operator $X_{1,0}^1$, entering the equations describing $\bar{K}\bar{K}N$, already has the proper symmetry properties. The remaining operators form pairs with proper symmetry properties.

It is necessary to note that while Coulomb potential was directly included in the two-body equations, used for fitting the antikaon–nucleon potentials, the three-body calculations were performed without it (except the case of kaonic deuterium calculations, of cause). The reason is that the expected effect of its inclusion is small. In addition, the isospin averaged masses were used in all three-body calculations in contrast to the two-body $\bar{K}N$ case. Accuracy of this approximation was checked in [22?], and it turned out to be quite high.

4 Quasi-Bound States

It was shown in Sect. 2.2 that the phenomenological $\bar{K}N$ potentials with one- and two-pole structure of the $\Lambda(1405)$ resonance and the chirally motivated antikaon–nucleon potential can reproduce near-threshold experimental data on K^-p scattering and kaonic hydrogen with equal accuracy. Therefore, it is not possible to choose one of these models looking at the two-body system only. Due to this, the three-body calculations were performed using all three models of the antikaon–nucleon interaction.

The quasi-bound state in the K^-pp system was the phenomenon, which attracted the present interest to the antikaon-nucleus systems. Additionally to being an interesting exotic object, the state could clarify still unanswered questions on the antikaon–nucleon interaction, in particular, the nature of the $\Lambda(1405)$ resonance.

$\bar{K}\bar{K}N$ is one more possible candidate for a strange three-body system with the quasi-bound state in it. However, the strangeness $S = -2$ system contains $\bar{K}\bar{K}$ interaction, which is repulsive. The question was, whether the repulsion is strong enough to overtake $\bar{K}N$ attraction and by this exclude the possibility of the quasi bound state formation.

4.1 Two Ways of a Quasi-Bound State Evaluation

The quasi-bound state, which is a bound state with a non-zero width, for the higher $\bar{K}NN$ (or $\bar{K}\bar{K}N$) channel, is at the same time a resonance for the lower $\pi\Sigma N$ ($\bar{K}\pi\Sigma$) channel. Therefore, the corresponding pole should be situated between the $\bar{K}NN$ ($\bar{K}\bar{K}N$) and $\pi\Sigma N$ ($\bar{K}\pi\Sigma$)) thresholds on the physical energy sheet of the higher channel and on an unphysical sheet of the lower channel. Two methods of searching of the complex pole position by solving the homogeneous system of equations were used.

The first one is the direct pole search with contour rotation. The correct analytical continuation of the equations from the physical energy sheet to the proper unphysical one is achieved by moving the momentum integration into the complex plane. Namely, the integration was performed along a ray in the fourth quadrant of the complex plane with some condition on the momentum variable. After that the position z_0 of a quasi-bound state was found by solving the equation $\text{Det}(z_0) = 0$, where $\text{Det}(z)$ is the determinant of the linear system, obtained after discretization of the integral equations, corresponding to Eq. (23).

Another way of a quasi-bound state searching, which avoids integration in the complex plane, was suggested and used in [24]. The idea is that every isolated and quite narrow resonance should manifest itself at real energies. Namely, resonances are usually seen in cross-sections of some reactions. The function $1/\text{Det}(z)$ enters all possible amplitudes, described by a system of three-body integral equations. Therefore, the function $1/|\text{Det}(z)|^2$, entering all possible cross-sections, can be calculated instead of some cross-sections. The function is universal, it does not contain additional information about the particular processes in the three-body system. The corresponding bump of the $1/|\text{Det}(z)|^2$ function, calculated at real energies, can be fitted by a Breit–Wigner curve with arbitrary background. In this way an information on the resonance position and width can be obtained.

It is clear that the second method can work only if the resonance bump is isolated and not too wide. The bump corresponding to the K^-pp quasi-bound state satisfies these conditions [24], as is seen in Fig. 2. The calculated $1/|\text{Det}(z)|^2$ functions of the AGS system of equations are shown there as symbols while the corresponding Breit–Wigner fitting curves are drown in lines. The results obtained with the three $\bar{K}N$ potentials, described in Sect. 2.2, are shown in the figure.

Since direct search of the complex root is a non-trivial task, the Breit–Wigner values of the $1/|\text{Det}(z)|^2$ function can give a good starting point for it. On the other hand, the $1/|\text{Det}(z)|^2$ method can be used as a test of the directly found pole position, which is free of the possible uncertainty of the proper choice of the Riemann sheet. However, the $1/|\text{Det}(z)|^2$ method is not easier than the direct search, since the calculation of the determinant, which is almost equal to the solving of a scattering problem, should be performed.

4.2 K^-pp Quasi-Bound State: Results

The first dynamically exact calculation of the quasi-bound state in the K^-pp system was published in [3], while the extended version of the results appeared in [4]. Existence of the $I = 1/2$, $J^\pi = 0^-$ three-body quasi-bound state in the $\bar{K}NN$ system, predicted in [1,2], was confirmed there, but the evaluated binding energy and width were strongly different. However, the $\bar{K}N - \pi\Sigma$ potentials used in [3,4] do not reproduce the experimental

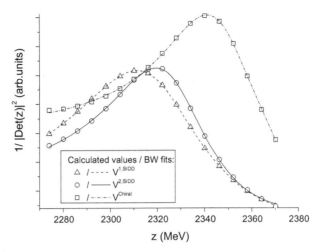

Fig. 2 Calculated $1/|\text{Det}(z)|^2$ functions of the AGS system of equations for the $K^- pp$ system (*symbols*) and the corresponding Breit–Wigner fits of the obtained curves (*lines*) [24] for the one-pole $V_{\bar{K}N-\pi\Sigma}^{1,\text{SIDD}}$ (*triangles* and *dashed line*), the two-pole $V_{\bar{K}N-\pi\Sigma}^{2,\text{SIDD}}$ (*circles* and *solid line*) phenomenological potentials, and the chirally motivated $V_{\bar{K}N-\pi\Sigma-\pi\Lambda}^{Chiral}$ (*squares* and *dash-dotted line*) potential

Table 2 Binding energy $B_{K^- pp}$ (MeV) and width $\Gamma_{K^- pp}$ (MeV) of the quasi-bound state in the $K^- pp$ system [24]: the results of the direct pole search and of the Breit–Wigner fit of the $1/|\text{Det}(z)|^2$ function at real energy axis

	Direct pole search		BW fit of $1/|\text{Det}(z)|^2$	
	$B_{K^- pp}$	$\Gamma_{K^- pp}$	$B_{K^- pp}$	$\Gamma_{K^- pp}$
$V_{\bar{K}N-\pi\Sigma}^{1,SIDD}$	53.3	64.8	54.0	66.6
$V_{\bar{K}N-\pi\Sigma}^{2,SIDD}$	47.4	49.8	46.2	51.8
$V_{\bar{K}N-\pi\Sigma-\pi\Lambda}^{Chiral}$	32.2	48.6	30.3	46.6

The AGS calculations were performed with the one-pole $V_{\bar{K}N-\pi\Sigma}^{1,SIDD}$, two-pole $V_{\bar{K}N-\pi\Sigma}^{2,SIDD}$ phenomenological potentials from [21] and the chirally motivated $V_{\bar{K}N-\pi\Sigma-\pi\Lambda}^{Chiral}$ potential from [23]

data on the $K^- p$ system as accurately, as those described in Sect. 2.2. Due to this, the calculations devoted to the $K^- pp$ system were repeated in [24]. The one-pole $V_{\bar{K}N-\pi\Sigma}^{1,SIDD}$, two-pole $V_{\bar{K}N-\pi\Sigma}^{2,SIDD}$ phenomenological potentials from [21] and the chirally motivated $V_{\bar{K}N-\pi\Sigma-\pi\Lambda}^{Chiral}$ potential from [23], described in Sect. 2.2, were used as an input. The other two potentials were the two-term TSA-B NN potential [41] together with the spin-independent exact optical ΣN potential in isospin $I = 1/2$ state and the one-channel $V_{\Sigma N}$ in $I = 3/2$, see Sect. 2.3.

The results of the last calculations [24] of the $K^- pp$ quasi-bound state are shown in Table 2. First of all, comparison of the results obtained using the direct pole search and the $1/|\text{Det}(z)|^2$ method demonstrates that they are very close each to other for the every given $\bar{K}N$ potential. Therefore, the suggested $1/|\text{Det}(z)|^2$ method of finding mass and width of a subthreshold resonance is efficient for the $K^- pp$ system, and the two methods supplement each another.

Another fact, seen from the results in Table 2, is strong dependence of the binding energy $B_{K^- pp}$ of the quasi-bound state and its width $\Gamma_{K^- pp}$ on the $\bar{K}N$ interaction models. It was already observed in [3,4], when older phenomenological antikaon–nucleon potentials were used. In particular, it is seen from Table 2 that the quasi-bound states resulting from the phenomenological potentials lie about 15–20 MeV deeper than those of the chirally motivated one. This probably is due to the energy dependence of the chirally motivated model of the interaction. Really, all three potentials were fitted to the experimental data near the $\bar{K}N$ threshold. When the $K^- pp$ quasi-bound state is calculated at lower energies, the strengths of the phenomenological models of the $\bar{K}N$ interaction are unchanged. As for the chirally motivated potential, its energy dependence reduces the attraction at the lower energies in the $\bar{K}NN$ quasi-bound state region, thus producing the states with less binding.

The widths of the three quasi-bound states are also different: those of the two-pole models of the $\bar{K}N$ interaction are almost coinciding, while the width evaluated using the one-pole $V_{\bar{K}N-\pi\Sigma}^{1,SIDD}$ potential is much larger. It is seen from Table 1 that the potentials with the two-pole $\Lambda(1405)$ structure have very close positions of the higher poles, while the pole of the one-pole potential is different. Therefore, the difference in widths might be connected with the different pole structure of the corresponding $\bar{K}N$ interaction models.

Importance of the proper inclusion of the second $\pi\Sigma N$ channel in the calculations was first demonstrated in [4]. A simple complex version of the $\bar{K}N - \pi\Sigma$ potentials, described in Sect. 2.2, was used in [4] together with the full version with coupled channels. This allowed to check importance of the proper inclusion of the second channel. Comparison of the result of the one-channel complex calculation ($B_{K^-pp}^{1\,complex}$, $\Gamma_{K^-pp}^{1\,complex}$) with the coupled-channel one ($B_{K^-pp}^{2\,coupled}$, $\Gamma_{K^-pp}^{2\,coupled}$)

$$B_{K^-pp}^{2\,coupled} = 55.1 \text{ MeV} \qquad \Gamma_{K^-pp}^{2\,coupled} = 101.8 \text{ MeV} \tag{26}$$

$$B_{K^-pp}^{1\,complex} = 40.2 \text{ MeV} \qquad \Gamma_{K^-pp}^{1\,complex} = 77.4 \text{ MeV} \tag{27}$$

shows that the quasi-bound state obtained in the full calculation with coupled channels is much deeper and broader than the approximate one-channel one (the values for the binding energy and width in Eq. (26) differ from those in Table 2 since another $\bar{K}N$ potential was used in [4]). This means that the $\pi\Sigma$ channel plays an important dynamical role in forming the three-body quasi-bound state, over its obvious role of absorbing flux from the $\bar{K}N$ channel. Thus, proper inclusion of the second $\pi\Sigma$ channel is crucial for the $\bar{K}NN$ system.

It was found later in [24] that use of the exact optical $\bar{K}N$ potential can serve an alternate way of direct inclusion of the $\pi\Sigma$ channel. An accuracy of use of the exact optical $\bar{K}N$ potential, which gives exactly the same on- and off-shell elastic $\bar{K}N$ amplitude as the original potential with coupled channels, was checked in one-channel AGS calculations for the three actual $\bar{K}N$ potentials. The "exact optical" binding energies differ only slightly from the full coupled-channel results from Table 2, while the widths gain more visible error:

$$B_{K^-pp}^{1,SIDD,Opt} = 54.2 \text{ MeV} \qquad \Gamma_{K^-pp}^{1,SIDD,Opt} = 61.0 \text{ MeV} \tag{28}$$

$$B_{K^-pp}^{2,SIDD,Opt} = 47.4 \text{ MeV} \qquad \Gamma_{K^-pp}^{2,SIDD,Opt} = 46.0 \text{ MeV} \tag{29}$$

$$B_{K^-pp}^{Chiral,Opt} = 32.9; \text{ MeV} \qquad \Gamma_{K^-pp}^{Chiral,Opt} = 48.8 \text{ MeV}. \tag{30}$$

However, the difference in widths is not dramatic, so the one-channel Faddeev calculation with the exact optical $\bar{K}N$ potential could be quite satisfactory approximation to the full calculation with coupled channels.

4.3 K^-pp Quasi-Bound State: Comparison to Other Results

The three binding energy B_{K^-pp} and width Γ_{K^-pp} values of the K^-pp quasi-bound state, shown in Table 2, can be compared with worth mentioning other theoretical results. Those are: the original prediction of the deep and narrow quasi-bound K^-pp state [1,2], the results obtained in the earlier Faddeev calculation [4], the most recent results of alternative calculation using the same equations with different input [54] and two variational results [55,56]. Only calculations presented in [54] together with the earlier ones [3,4] were performed with directly included $\pi\Sigma N$ channel. All others take it into account approximately. The second problem is that none of the $\bar{K}N$ potentials, used in all other K^-pp calculations, reproduce data on near-threshold K^-p scattering with the same level of accuracy as those described in Sect. 2.2. In addition, none of them reproduce the $1s$ level shift of kaonic hydrogen directly.

The binding energy of the quasi-bound K^-pp state and its width were obtained in [1,2] from a G-matrix calculation, which is a many-body technics. The one-channel simple complex $\bar{K}N$ potential used in those calculations does not reproduce the actual K^-p experimental data. Finally, the authors of [1,2] take into account only the $\bar{K}NN$ channel. In a funny way all the defects of the calculation presented in [1,2] led to the binding energy (48 MeV) and width (61 MeV), which are quite close to the exact results from Table 2 obtained with the two-pole and the one-pole phenomenological potentials, respectively.

The earlier result for the binding energy $B_{K^-pp} = 55.1$ MeV [4] is very close to the actual one from Table 2 calculated with the one-pole phenomenological $\bar{K}N - \pi\Sigma$ potential. In fact, in both cases the same three-body equations with coupled $\bar{K}NN$ and $\pi\Sigma N$ channels were solved. In addition, the same model of the antikaon–nucleon interaction was used, but with different sets of parameters. This difference influenced the width of the quasi-bound state: the older $\Gamma_{K^-pp} = 100.2$ MeV is much larger.

Coupled-channel AGS equations were also solved in [54] with chirally motivated energy dependent and independent $\bar{K}N$ potentials. Therefore, in principle, those results obtained with the energy dependent version of the $\bar{K}N$ potential $V_{\bar{K}N}^{E-dep}$ should give a result, which is close to those from Table 2 with chirally motivated model of interaction $V_{\bar{K}N-\pi\Sigma-\pi\Lambda}^{Chiral}$. However, only those width (34–46 MeV) is comparable to the $\Gamma_{K^-pp}^{Chiral}$, while the binding energy obtained in [54] (9–16 MeV) is much smaller than the actual one from Table 2. The situation is opposite when the actual results are compared with those obtained in [54] using the energy independent antikaon–nucleon potential. Namely, binding energies reported in that paper (44–58 MeV) are comparable with those from Table 2 evaluated using phenomenological $V_{\bar{K}N}^{1,SIDD}$ and $V_{\bar{K}N}^{2,SIDD}$ potentials, however, their widths 34–40 MeV are much smaller.

The authors of [54] neglect the ΣN interaction in their calculations. It was shown in [3,4] that dependence of the three-body K^-pp pole position on ΣN is weak. However, when the interaction is switched off completely, like in the case of [54], some visible effect manifests itself.

An approximation used in the chirally motivated models used in [54] is more serious reason of the difference. Namely, the energy-dependent square root factors, responsible for the correct normalization of the $\bar{K}N$ amplitudes, are replaced by constant masses. This can be reasonable for the highest $\bar{K}N$ channel, however, it is certainly a poor approximation for the lower $\pi\Sigma$ and $\pi\Lambda$ channels. The role of this approximation in the AGS calculations was checked in [24], the obtained approximate binding energy 25 MeV is really much smaller than the original one 32 MeV, presented in Table 2. The remaining difference between the results from Table 2 and those from [54] could be explained by the lower accuracy of reproducing experimental K^-p data by the $\bar{K}N$ potentials from [54].

Finally, no second pole in the K^-pp system reported in [54] was found in [24]. The search was performed with all three $\bar{K}N$ potentials in the corresponding energy region (binding energy 67–89 MeV and width 244–320 MeV).

Variational calculations, performed by two groups, reported the results which are very close to those obtained in [54] with the energy-dependent potential: $B_{K^-pp} = 17$–23 MeV, $\Gamma_{K^-pp} = 40$–70 MeV in [55] and $B_{K^-pp} = 15.7$ MeV, $\Gamma_{K^-pp} = 41.2$ MeV in [56]. However, there are a few problematic points in [55, 56]. First of all, the variational calculations were performed solely in the $\bar{K}NN$ channel. The authors of the variational calculations used a one-channel $\bar{K}N$ potential, derived from a chirally motivated model of interaction with many couped channels. However, the potential is not the "exact optical" one. In fact, it is not clear, how this one-channel potential is connected to the original one and whether it still reproduces some experimental K^-p data.

Moreover, the position of the K^-pp quasi-bound state was determined in [55,56] using only the real part of this complex $\bar{K}N$ potential, as a real bound state. The width was estimated as the expectation value of the imaginary part of the potential. This, essentially perturbative, treatment of the inelasticity might be justified for quite narrow resonances, but the K^-pp quasi-bound state is certainly not of this type.

Another serious problem of the variational calculations is their method of treatment of the energy dependence of the $\bar{K}N$ potential in the few-body calculations. It was already shown in the previous subsection that the energy dependence of the chirally motivated model of the $\bar{K}N$ interaction is very important for the K^-pp quasi-bound state position. While momentum space Faddeev integral equations allow the exact treatment of this energy dependence, variational calculations in coordinate space can use only energy independent interactions. Due to this the energy of the $\bar{K}N$ potential was fixed in [55,56] at a "self-consistent" value $z_{\bar{K}N}$.

A series of calculations using the exact AGS equations was performed in [24] with differently fixed two-particle $\bar{K}N$ energies $z_{\bar{K}N}$ in the couplings of the chirally motivated interaction. The conclusion was, that it is not possible to define an "averaged" $z_{\bar{K}N}$, for which the fixed-energy chirally motivated interaction, even in the correct three-body calculation, can yield a correct K^-pp quasi-bound state position. First, the calculations of [24] show, that a real $z_{\bar{K}N}$ has absolutely no chance to reproduce or reasonably approximate the exact quasi-bound state position, even with correct treatment of the imaginary part of the interaction, unlike in [55,56]. Second, the way, how the "self-consistent" value of (generally complex) $z_{\bar{K}N}$ is defined in the papers does not seem to guarantee, that the correct value will be reached or at least approximated. In view of the above considerations, the results of [55,56] can be considered as rough estimates of what a really energy dependent $\bar{K}N$ interaction will produce in the K^-pp system.

After publication of the exact results [24] one more paper on the K^-pp system appeared [57]. Hyper-spherical harmonics in the momentum representation and Faddeev equations in configuration space were used there. However, the authors collected all defects of other approximate calculations. In particular, they used simple complex antikaon–nucleon potentials and, therefore, neglected proper inclusion of the $\pi \Sigma N$ channel, which is crucial for the system. In addition, the $\bar{K}N$ potentials are those from [2,55], which have problems with reproducing of the experimental K^-p data. Keeping all this in mind, the results of [57] hardly can be reliable.

4.4 K^-d Quasi-Bound State

The strongly attractive isospin-zero part of the $\bar{K}N$ potential plays less important role in the spin-one K^-d state of the $\bar{K}NN$ system than in the spin-zero K^-pp. Therefore, if a quasi-bound state exists in K^-d, it should have smaller binding energy than in K^-pp. The Faddeev calculations of the K^-d scattering length a_{K^-d}, described in Sect. 5, gave some evidence that such a state could exists [23]. A simple analytical continuation of the effective range formula below the K^-d threshold suggests a K^-d quasi-bound state with binding energy 14.6–19.6 MeV (the energy is measured from the K^-d threshold) and width 15.6–22.0 MeV for the three antikaon–nucleon potentials from Sect. 2.2.

However, a systematic search for these states, performed in [23] with the same two-body input as for the K^-pp system, did not find the corresponding poles in the complex energy plane between the $\pi \Sigma N$ and K^-d thresholds. The reason of discrepancy between the effective range estimations and the direct calculations must be the validity of the effective range formula, which is limited to the vicinity of the corresponding threshold. Since the K^-d state is expected to have, similarly to K^-pp, rather large width, it is definitely out of this region.

It was demonstrated in [23] that increasing of the attraction in isospin-zero $\bar{K}N$ subsystem by hands (in the phenomenological antikaon–nucleon potentials only) leads to appearing of K^-d quasi-bound states. Therefore, the isospin-zero attraction in the $\bar{K}N$ system is not strong enough to bind antikaon to deuteron. It is necessary to note that the K^-d system with strong two-body interactions only is considered here. An atomic state caused by Coulomb interaction, kaonic deuterium, exists and will be considered later.

4.5 $\bar{K}\bar{K}N$ System: Results

The calculations of the quasi-bound state in the $\bar{K}\bar{K}N$ system were performed with the two $\bar{K}\bar{K}$ interactions described in Sect. 2.4 (Original $V_{\bar{K}\bar{K}}^{Orig}$ and Lattice-motivated $V_{\bar{K}\bar{K}}^{Latt}$) and three $\bar{K}N$ potentials from Sect. 2.2: the phenomenological one-pole $V_{\bar{K}N-\pi\Sigma}^{1,SIDD}$ and two-pole $V_{\bar{K}N-\pi\Sigma}^{2,SIDD}$ phenomenological potentials together with the chirally-motivated potential $V_{\bar{K}N-\pi\Sigma-\pi\Lambda}^{Chiral}$. The results are presented in Table 3. It turned out that all combinations of the two-body interactions lead to a quasi-bound state in the three-body $\bar{K}\bar{K}N$ system. The quasi-bound state exists in the strangeness $S = -2$ system in spite of the repulsive character of the $\bar{K}\bar{K}$ interaction Comparison with the K^-pp characteristics from Table 2 shows that the quasi-bound state in the strangeness $S = -2$ $\bar{K}\bar{K}N$ system is much shallower and broader than the one in the $S = -1$ K^-pp system for the given $\bar{K}N$ potential.

Two methods of the quasi-bound state evaluation were used: the direct search method and the Breit–Wigner fit of the inverse determinant. It is seen from the Table 3 that the accuracy of the inverse determinant method is much lower for the phenomenological $\bar{K}N$ interactions than for the chirally motivated one (and for the K^-pp system too). The reason is the larger widths of the "phenomenological" $\bar{K}\bar{K}N$ states, which means that the corresponding bumps are less pronounced, so they hardly can be fitted reliably by Breit–Wigner curves.

The found $\bar{K}\bar{K}N$ quasi-bound state has the same quantum numbers as a Ξ baryon with $J^P = (1/2)^+$. The available experimental information on the Ξ spectrum is rather limited, see PDG [28]. There is a $\Xi(1950)$ listed by the PDG, but its quantum numbers J^P are not determined, and it is unclear whether it should be identified with the quark-model state. It is possible that there are more than one resonance in this region. However, in spite of the fact, that the $\Xi(1950)$ state is situated above the $\bar{K}\bar{K}N$ threshold, four of the experimental values would be roughly consistent with the quasi-bound state found in the calculation [27]. Specifically, the experiment reported in Ref. [58] yielded a mass 1894 ± 18 MeV and a width 98 ± 23 MeV that is compatible with the range of values for the evaluated pole position.

An investigation on the $\bar{K}\bar{K}N$ system was also performed in [59], but several uncontrolled approximations were done there. In particular, energy-independent as well as energy-dependent potentials were used, but the

Table 3 Binding energy $B_{\bar{K}\bar{K}N}$ (MeV) and width $\Gamma_{\bar{K}\bar{K}N}$ (MeV) of the quasi-bound state in the $\bar{K}\bar{K}N$ system [27]: the results of the direct pole search and of the Breit–Wigner fit of the $1/|\mathrm{Det}(z)|^2$ function at real energy axis

	Direct pole search		BW fit of $1/|\mathrm{Det}(z)|^2$	
	$B_{\bar{K}\bar{K}N}$	$\Gamma_{\bar{K}\bar{K}N}$	$B_{\bar{K}\bar{K}N}$	$\Gamma_{\bar{K}\bar{K}N}$
$V_{\bar{K}\bar{K}}^{Orig}$ and				
$V_{\bar{K}N-\pi\Sigma}^{1,SIDD}$	11.9	102.2	17.1	110.8
$V_{\bar{K}N-\pi\Sigma}^{2,SIDD}$	23.1	91.4	23.7	77.6
$V_{\bar{K}N-\pi\Sigma-\pi\Lambda}^{Chiral}$	15.5	63.5	15.9	57.4
$V_{\bar{K}\bar{K}}^{Lattice}$ and				
$V_{\bar{K}N-\pi\Sigma}^{1,SIDD}$	19.5	102.0	23.7	103.7
$V_{\bar{K}N-\pi\Sigma}^{2,SIDD}$	25.9	84.6	26.4	76.8
$V_{\bar{K}N-\pi\Sigma-\pi\Lambda}^{Chiral}$	16.1	61.3	15.9	60.0

The AGS calculations were performed with the one-pole $V_{\bar{K}N-\pi\Sigma}^{1,SIDD}$, two-pole $V_{\bar{K}N-\pi\Sigma}^{2,SIDD}$ phenomenological potentials from [21] and the chirally motivated $V_{\bar{K}N-\pi\Sigma-\pi\Lambda}^{Chiral}$ potential from [23]. $V_{\bar{K}\bar{K}}^{Orig}$ and $V_{\bar{K}\bar{K}}^{Lattice}$ models of the antikaon–antikaon interaction were used

two-body energy of the latter was fixed arbitrarily. Moreover, the imaginary parts of all complex potentials were completely ignored in the variational calculations in [59], the widths of the state were estimated separately. As a result, the binding energies are compared to the exact ones from [27], but the widths of the $\bar{K}\bar{K}N$ state are strongly underestimated.

5 Near-Threshold K^-d Scattering

5.1 Methods and Exact Results

The K^-pp quasi-bound state is a very interesting exotic object. However, it is not clear whether the accuracy of experimental results will be enough to draw some conclusions from comparison of the data with theoretical predictions. No strong quasi-bound state was found in the $\bar{K}NN$ system with other quantum numbers K^-d [23], but an atomic state, kaonic deuterium, exists, and its energy levels can be accurately measured. In addition, scattering of an antikaon on a deuteron can be studied.

Exact calculations of the near-threshold K^-d scattering were performed in [21,23] using the three antikaon–nucleon potentials, described in Sect. 2.2, and different versions of the ΣN and NN potentials, described in Sect. 2.3. Namely, the exact optical and a simple complex versions of the spin-dependent and spin-independent $\Sigma N - \Lambda N$ potentials were used. The calculations were performed with TSA-A, TSA-B, and PEST models of the NN interaction.

The inhomogeneous system of the integral AGS equations, corresponding to Eq. (23) and describing the K^-d scattering, was transformed into the system of algebraic equations. It is known, that the original, one-channel, integral Faddeev equations have moving logarithmic singularities in the kernels when scattering above a three-body breakup threshold is considered. The K^-d amplitudes were calculated from zero up to the three-body breakup $\bar{K}NN$ threshold, so, in principle, the equations could be free of the singularities. However, the lower $\pi\Sigma N$ channel is opened when the K^-d scattering is considered, which causes appearance of the logarithmic singularities even below the three-body breakup $\bar{K}NN$ threshold. The problem was solved by interpolating of the unknown solutions in the singular region by certain polynomials and subsequent analytical integrating of the singular part of the kernels.

The K^-d scattering lengths a_{K^-d} obtained with the one- $V_{\bar{K}N-\pi\Sigma}^{1,SIDD}$ and two-pole $V_{\bar{K}N-\pi\Sigma}^{2,SIDD}$ versions of the phenomenological $\bar{K}N$ potential in [21] together with the chirally-motivated potential $V_{\bar{K}N-\pi\Sigma-\pi\Lambda}^{Chiral}$ in [23] are shown in Table 4. It is seen, that the chirally motivated potential leads to slightly larger absolute value of the real and the imaginary part of the scattering length than the phenomenological ones. However, the difference is small, so the three different models of the $\bar{K}N$ interaction, which reproduce the low-energy data on the K^-p

Table 4 Scattering length a_{K^-d} (fm) and effective range $r^{eff}_{K^-d}$ (fm) of K^-d system obtained from AGS calculations with the one-pole $V^{1,SIDD}_{\bar{K}N-\pi\Sigma}$, two-pole $V^{2,SIDD}_{\bar{K}N-\pi\Sigma}$ phenomenological potentials and the chirally-motivated $V^{Chiral}_{\bar{K}N-\pi\Sigma-\pi\Lambda}$ potential

$\bar{K}N$ potential used	a_{K^-d}	$r^{eff}_{K^-d}$	$\Delta E^{K^-d}_{1s}$	$\Gamma^{K^-d}_{1s}$
$V^{1,SIDD}_{\bar{K}N}$	$-1.49 + i\,1.24$	$0.69 - i\,1.31$	-785	1018
$V^{2,SIDD}_{\bar{K}N}$	$-1.51 + i\,1.25$	$0.69 - i\,1.34$	-797	1025
$V^{Chiral}_{\bar{K}N}$	$-1.59 + i\,1.32$	$0.50 - i\,1.17$	-828	1055

Approximate results for the $1s$ level shift $\Delta E^{K^-d}_{1s}$ (eV) and width $\Gamma^{K^-d}_{1s}$ (eV) of kaonic deuterium, corresponding to the AGS results on the near-threshold elastic amplitudes, are also shown

scattering and kaonic hydrogen with the same level of accuracy, give quite similar results for low-energy K^-d scattering. It means that it is not possible to solve the question of the number of the poles forming the $\Lambda(1405)$ resonance from the results on the near-threshold elastic K^-d scattering.

The small difference between the "phenomenological" and "chiral" results of the a_{K^-d} calculations is opposite to the results obtained for the K^-pp system (see Sect. 4.2), where the binding energy and width of the K^-pp quasi-bound state were calculated using the same equations (the homogeneous ones with properly changed quantum numbers, of cause) and input. In that case the three-body observables obtained with the three $\bar{K}N$ potentials turned out to be very different each from the other. The reason of this difference between the results for the near-threshold scattering and the quasi-bound state calculations could be the fact, that while the a_{K^-d} values were calculated near the $\bar{K}NN$ threshold, the K^-pp pole positions are situated far below it.

The amplitudes of the elastic K^-d scattering for kinetic energy from 0 to E_{deu}, calculated using the three versions of the $\bar{K}N$ potential, are shown in Fig. 3 of [21] and in Fig 5 of [23] in a form of $k\cot\delta(k)$ function. The chosen representation demonstrates that the elastic near-threshold K^-d amplitudes can be approximated by the effective range expansion rather accurately since the lines are almost straight. The calculated effective ranges $r^{\text{eff}}_{K^-d}$ of the K^-d scattering, evaluated using the obtained K^-d amplitudes, are shown in Table 4.

The dependence of the full coupled-channel results on the NN and $\Sigma N(-\Lambda N)$ interaction models was investigated in [17]. The antikaon–nucleon phenomenological potentials, used there, reproduce the earlier KEK data on kaonic hydrogen and not the actual ones by SIDDHARTA. However, the results, obtained with those phenomenological $\bar{K}N$ potentials, are relative, so they must be valid for the actual potentials as well. In order to investigate dependence of the three-body results on the NN model of interaction, TSA-A, TSA-B, and PEST nucleon–nucleon potentials were used. It turned out that the difference for the K^-d scattering length is very small even for the potentials with and without repulsion at short distances (TSA and PEST, respectively). Therefore, the s-wave NN interaction plays a minor role in the calculations. Most likely, it is caused by the relative weakness of the NN interaction as compared to the $\bar{K}N$ one. Indeed, the quasi bound state in the latter system (which is the $\Lambda(1405)$ resonance with $E_{\bar{K}N} \approx -23$ MeV) is much deeper than the deuteron bound state ($E_{\text{deu}} \approx 2$ MeV). Due to this, some visible effect from higher partial waves in NN is also not expected.

The dependence of a_{K^-d} on the $\Sigma N(-\Lambda N)$ interaction was also investigated in [17]. The K^-d scattering lengths were calculated with the exact optical and the simple complex versions of the spin dependent V^{Sdep} and spin independent V^{Sind} potentials. The results obtained with the two versions of the $\Sigma N(-\Lambda N)$ potential V^{Sdep} and V^{Sind} in exact optical form are very close, while their simple complex versions are slightly different. However, the largest error does not exceed 3%, therefore, the dependence of the K^-d scattering length a_{K^-p} on the $\Sigma N - (\Lambda N)$ interaction is also weak.

5.2 Approximate Calculations and Comparison to Other Results

It is hard to make a comparison with other theoretical results due to different methods and inputs used there. Due to this, several approximate calculations, in particular, one-channel $\bar{K}NN$ calculations with a complex and the exact optical $\bar{K}N$ potentials, were performed in [17]. In addition, a so-called FCA method was tested there.

In order to investigate the importance of the direct inclusion of the $\pi\Sigma N$ channel, the one-channel AGS calculations were performed in [17]. It means that Eq. (23) with $\alpha = \beta = 1$ were solved, therefore, only the $\bar{K}N$ and NN T-matrices enter the equations. The exact optical and two simple complex one-channel $\bar{K}N(-\pi\Sigma)$ potentials approximating the full coupled-channel one- and two-pole phenomenological models of the interaction were used. As written in Sect. 2.2, the exact optical potential V^{Opt} provides exactly the same elastic $\bar{K}N$ amplitude as the coupled-channel model of the interaction. Its energy-dependent strength parameters are defined by Eq. (18) with $\bar{\alpha}, \bar{\beta} = 1, 2$ stands for the $\bar{K}N$ and $\pi \Sigma$ channels, respectively.

The complex constants of the simple complex potentials were obtained in two ways. The first version of the simple complex $\bar{K}N$ potential reproduces the $K^- p$ scattering length $a_{K^- p}$ and the pole position z_1 of the corresponding coupled-channel version of the potentials. The second one gives the same isospin $I_i = 0$ and $I_i = 1$ $\bar{K}N$ scattering lengths as the full $\bar{K}N - \pi \Sigma$ potential.

It was found in [17] that the one-channel AGS calculation with the exact optical $\bar{K}N$ potential, giving exactly the same elastic $\bar{K}N$ amplitude as the corresponding coupled-channel phenomenological potential, is the best approximation. Its error does not exceed 2% (the same is true for the results obtained with the chirally motivated $\bar{K}N$ potential, see [23]). On the contrary, the both simple complex $\bar{K}N$ potentials led to very inaccurate three-body results. Therefore, the one-channel Faddeev-type calculation with a simple complex antikaon–nucleon potential is not a good approximation for the low-energy elastic $K^- d$ scattering.

One more approximate method, used for the $a_{K^- d}$ calculations, is a so-called "Fixed center approximation to Faddeev equations" (FCA), introduced in [60]. In fact, it is a variant of FSA or a two-center formula. The fixed-scatterer approximation (FSA) or a two-center problem assumes, that the scattering of a projectile particle takes place on two much heavier target particles, separated by a fixed distance. The motion of the heavy particles is subsequently taken into account by averaging of the obtained projectile-target amplitude over the bound state wave function of the target. The approximation is well known and works properly in atomic physics, where an electron is really much lighter than a nucleon or an ion. Since the antikaon mass is just a half of the mass of a nucleon, it was expected, that FSA hardly can be a good approximation for the $K^- d$ scattering length calculation.

The derivations of the FCA formula from Faddeev equations presented in [60] already rises questions, while the proper derivations of the FSA formula was done in [61]. The accuracy of the FCA was checked in [17] using the same input as in the AGS equations in order to make the comparison as adequate as possible.

First of all, the $\bar{K}N$ scattering lengths provided by the coupled-channel $\bar{K}N - \pi \Sigma$ potentials together with the deuteron wave function, corresponding to the TSA-B NN potential, were used in the FCA formula. Second, all $\bar{K}^0 n$ parts were removed from the formula because they drop off the AGS system of equations after the antisymmetrization. Finally, the fact, that the FCA formula was obtained for a local $\bar{K}N$ potential, while the separable $\bar{K}N - \pi \Sigma$ potentials were used in the Faddeev equations, was took into account, and the corresponding changes in the FCA formula were made.

The results of using of the FCA formula without "isospin-breaking effects" stay far away from the full calculation. While the errors for the imaginary part are not so large, the absolute value of the real part is underestimated by about 30%. Therefore, the calculations performed in [17] show that FCA is a poor approximation for the $K^- d$ scattering length calculation. It is also seen from the figure that the accuracy is lower for the two-pole model of the $\bar{K}N$ interaction.

Therefore, among the approximate results the FCA was demonstrated to be the least accurate approximation, especially in reproducing of the real part of the $K^- d$ scattering length. On the contrary, the one-channel AGS calculation with the exact optical $\bar{K}N(-\pi \Sigma)$ potential gives the best approximation to the full coupled-channel result. All approximations are less accurate for the two-pole phenomenological model of the $\bar{K}N - \pi \Sigma$ interaction.

Calculations of the $K^- d$ scattering length were performed by other authors using Faddeev equations in [62–65], while the FCA method was used in [60,66]. The result of the very recent calculation with coupled channels [62] has real part of $a_{K^- d}$, which almost coincides with the result for chirally motivated potential shown in Table 4. The imaginary part of the $K^- d$ scattering length from [62] is slightly larger. It might be caused by the fact that the model of the $\bar{K}N$ interaction, used there, was not fitted to the kaonic hydrogen data directly, but through the $K^- p$ scattering length and the Deser-type approximate formula, which has larger error for the imaginary part of the level shift.

The two old $a_{K^- d}$ values [63,64], obtained within coupled-channel Faddeev approach, significantly underestimate the imaginary part of the $K^- d$ scattering length, while their real parts are rather close to those in Table 4.

One more result of a Faddeev calculation [65] lies far away from all the others with very small absolute value of the real part of $a_{K^- d}$. One of the reasons is that the $K^- d$ scattering length was obtained in [65] from one-channel Faddeev equations with a complex $\bar{K}N$ potential. However, the underestimation of the absolute value of its real part in comparison to other Faddeev calculations is so large, that it cannot be explained by the method only. The additional reason of the difference must be the $\bar{K}N$ potential, used in the paper. It gives so high position of the K^-p quasi bound state (1439 MeV), that it is situated above the K^-p threshold.

The a_{K-d} values of [60] obtained using FCA method differ significantly from all other results. The absolute value of the real part of a_{K-d} from [60] and its imaginary part are too large, which is caused by two factors. The first one is the FCA formula itself, which was shown to be inaccurate for the present system. The second reason are too large $\bar{K}N$ scattering lengths, used as the inputs.

The result of [66] was obtained by simple applying of two approximate formulas: FCA and the corrected Deser formula, used for calculation of the $\bar{K}N$ scattering lengths, entering the FCA. The values of [66] suffer not only from the cumulative errors from the two approximations, but from using of the DEAR results on kaonic hydrogen $1s$ level shift and width as well. Indeed, it was already written that the error of the corrected Deser formula makes about 10% for two-body case, the accuracy of the FCA was shown to be poor. As for the problems with DEAR experimental data, they were demonstrated in [14] and in other theoretical works.

6 $1s$ Level Shift of Kaonic Deuterium

The shift of the $1s$ level in the kaonic deuterium (which, strictly speaking, is the "antikaonic" deuterium) and its width are caused by the presence of the strong interactions in addition to the Coulomb one. It is a directly measurable value, which is free of a few uncertainties connected with an experiment on the K^-pp quasi-bound state. However, from theoretical point of view it is harder task due to necessity to take Coulomb potential into account directly together with the strong ones.

There are two ways to solve three-body problems accurately: solution of Faddeev equations or use of variational methods. However, for the case of an hadronic atom both methods face serious difficulties. The problem of the long range Coulomb force exists in the Faddeev approach, while variational methods suffer from the presence of two very different distance scales, which both are relevant for the calculations.

Due to this, at the first step the $1s$ level energy of the kaonic deuterium was calculated approximately using a two-body model of the atom. At the next step a method for simultaneous treatment of a short range plus Coulomb forces in three-body problems based on Faddeev equations [26] was used, and the lowest level of kaonic deuterium was calculated dynamically exactly.

6.1 Approximate Calculation of Kaonic Deuterium $1s$ Level

The approximate calculation of the kaonic deuterium was performed assuming that the atom can be considered as a two-body system consisting of a point-like deuteron, interacting with an antikaon through a complex strong $K^- - d$ potential and Coulomb. By this the size of a deuteron was taken into account only effectively through the strong potential, which reproduces the elastic three-body K^-d amplitudes, evaluated before. Keeping in mind the relative values of a deuteron and Bohr radius of the kaonic deuterium, the approximation seemed well grounded.

The complex two-body $K^- - d$ potential, constructed and used for investigation of the kaonic deuterium by Lippmann–Schwinger equation, is a two-term separable potential

$$V_{K-d}\left(\mathbf{k}, \mathbf{k}'\right) = \lambda_{1,K-d}\, g_1\left(\mathbf{k}\right) g_1\left(\mathbf{k}'\right) + \lambda_{2,K-d}\, g_2\left(\mathbf{k}\right) g_2\left(\mathbf{k}'\right) \tag{31}$$

with Yamaguchi form-factors

$$g_i(k) = \frac{1}{\beta_{i,K-d}^2 + k^2}, \qquad i = 1, 2. \tag{32}$$

The complex strength parameters $\lambda_{1,K-d}$ and $\lambda_{2,K-d}$ were fixed by the conditions, that the V_{K-d} potential reproduces the K^-d scattering length a_{K-d} and the effective range r_{K-d}^{eff}, obtained with one of the $\bar{K}N - \pi\Sigma$ potentials and presented in Table 4. A variation of the real $\beta_{1,K-d}$ and $\beta_{2,K-d}$ parameters allowed to reproduce the full near-threshold K^-d amplitudes from [21,23] more accurately. As a result, the near-threshold amplitudes obtained from the three-body calculations $f_{K-d}^{(3)}$ are reproduced by the two-body $K^- - d$ potentials through the interval $[0, E_{\mathrm{deu}}]$ with such accuracy, that the two-body functions $k \cot \delta^{(2)}(k)$ are indistinguishable from the three-body $k \cot \delta^{(3)}(k)$.

The parameters of the potentials are shown in Table 3 of [21] and in Eqs. (18,19) of [23]. Both $\beta_{1,K-d}$ and $\beta_{2,K-d}$ parameters for every $K^- - d$ potential are much smaller than the corresponding $\beta^{\bar{K}N}$ parameter of the

⧬ Springer

$\bar{K}N$ potential. The constructed two-body complex potentials V_{K-d} were used in the Lippmann–Schwinger equation. The calculations of the binding energy of a two-body system, described by the Hamiltonian with the strong and Coulomb interactions were performed in the same way as those of the K^-p system, see Sect. 2.2.

The shifts $\Delta E_{1s}^{K^-d}$ and widths $\Gamma_{1s}^{K^-d}$ of the 1s level of kaonic deuterium, corresponding to the three models of the $\bar{K}N$ interaction, described in Sect. 2.2, are shown in Table 4. It is seen that the "chirally motivated" absolute values of the level shift $\Delta E_{1s}^{K^-d}$ and the width $\Gamma_{1s}^{K^-d}$ are both slightly larger than those obtained using the phenomenological $\bar{K}N - \pi\Sigma$ potentials. However, all three results do not differ one from the other more than several percents. It is similar to the case of the K^-d scattering length calculations, which turned out to be very close for the three $\bar{K}N$ potentials. The important point here is the fact that all three $\bar{K}N$ potentials reproduce the low-energy experimental data on K^-p scattering and kaonic hydrogen with the same level of accuracy. It is also important that the 1s level of kaonic deuterium is situated not far from the $\bar{K}NN$ threshold.

The closeness of the results for kaonic deuterium means that comparison of the theoretical predictions with eventual experimental results hardly could choose one of the models of the $\bar{K}N$ interaction, especially taking into account the large widths $\Delta E_{1s}^{K^-d}$. Therefore, it could not be possible to say, whether the potential of the antikaon–nucleon interaction should have one- or two-pole structure of the $\Lambda(1405)$ resonance and whether the potential should be energy dependent or not. It is seen from Tables 1 and 4 that there is no correlation between the pole or poles of the $\Lambda(1405)$ resonance given by a $\bar{K}N$ potential and the three-body K^-d elastic scattering or kaonic deuterium characteristics obtained using the potential.

Inaccuracy of the corrected Deser formula Eq. (10) was already shown for the two-body K^-p system, but some authors use it for the kaonic deuterium as well. Due to this, an accuracy of the formula was checked for this three-body system. The results were obtained using the a_{K-d} values from Table 4. Being compared to the ΔE_{K-d} and Γ_{K-d} from the same table, the "corrected Deser" results show large error for all three versions of the antikaon–nucleon interaction. While difference for the shift is not so drastic, the width of the 1s level of the kaonic deuterium is underestimated by the corrected Deser formula by $\sim 30\%$.

The 1s level shift and width presented in Table 4 are not exact, they were evaluated using the two-body approximation, which, however, is well-grounded. Information on the three-body strong part is taken into account indirectly through the $K^- - d$ potential, reproducing the exact elastic K^-d amplitudes. On the contrary, the corrected Deser formula contains no three-body information at all since the only input is a K^-d scattering length, which is a complex number. Moreover, the formula relies on further approximations, which are absent in the accurate approximate calculations.

6.2 Exact Calculation of Kaonic Deuterium: Faddeev Equations with Coulomb Interaction

Exact calculations of the kaonic deuterium were performed using a method [26] for simultaneous treatment of short range plus Coulomb forces in three-body problems, based on Faddeev equations. The method was successfully applied for purely Coulomb systems with attraction and repulsion and for the short range plus repulsive Coulomb forces. The case of an hadronic atom with three strongly interacting particles and Coulomb attraction between certain pairs was not considered before.

The basic idea of the method is to transform the Faddeev integral equations into a matrix form using a special discrete and complete set of Coulomb Sturmian functions as a basis. Written in coordinate space the Coulomb Sturmian functions are orthogonal with the weight function $1/r$. So that they form a bi-orthogonal and complete set with their counter-parts. The most remarkable feature of this particular set is, that in this representation the matrix of the two-body $(z - h^c)$ operator, where z is an energy and h^c is the pure two-body Coulomb Hamiltonian, is tridiagonal. When this property is used for evaluation of the matrix elements of the two-body Coulomb Green's function g^c, an infinite tridiagonal set of equations, which can be solved exactly, is obtained. The same holds for the matrix elements of the free two-body Green's function g^0.

The system of equations with the Coulomb and strong interactions was solved in [25] for kaonic deuterium. This calculation is different from all other three-body calculations, described before. Already the initial form of the Faddeev equations for the kaonic deuterium differs from those for the pure strong interactions, described in Sect. 3. First, the equations should be written in coordinate space, while the AGS equations were written in momentum space. Second, since the Coulomb interaction acts between K^- and the proton, the particle basis was used and not the isospin one. Finally, the Faddeev equations with Coulomb do not define the transition operators, as e.g. those in Eq. (22), but the wave functions.

The equations are written in the Noble form [67], when the Coulomb interaction appears in the Green's functions. As usual for Faddeev-type equations, there are three partition channels $\alpha = (pn, K^-)$, (pK^-, n),

(nK^-, p) and three sets of Jacobi coordinates. The system of homogeneous equations to be solved contains the matrix elements of the overlap between the basis functions from different Jacobi coordinate sets and of the strong potentials. They all can be calculated directly. The remaining parts of the kernel are matrix elements of the three-body partition Green's functions G_α. They are the basic quantities of the method, and their calculation depends on the partition channel.

The partition Green function $G_{(pK^-,n)}$ of the (pK^-, n) channel contains Coulomb interaction in its "natural" coordinate. It describes the (pK^-) subsystem and the neutron, which do not interact between themselves. Due to this, $G_{(pK^-,n)}$ can be calculated taking a convolution integral along a suitable contour in the complex energy plane over two two-body Green functions. As for the matrix elements of the two-body Green functions, they can be calculated using the properties of the Coulomb Sturmian basis and solving a resolvent equation.

The situation with the remaining G_α functions is more complicated. In the case of the $\alpha = (pn, K^-)$ and (nK^-, p) channels the Coulomb interaction is written not in its "natural" coordinates. Due to this, it should be rewritten as a sum of the Coulomb potential in the natural coordinates plus a short range potential U_α, which is a "polarization potential". The three-body Green function G_α^{ch} containing Coulomb potential in natural for the channel coordinates is called the "channel Green function", and it is evaluated similarly to the previous $\alpha = (pK^-, n)$ case. Namely, since the function describes a two-body subsystem and the non-interacting with it third particle, the $G_{(pn,K^-)}^{ch}$ and $G_{(nK^-,p)}^{ch}$ functions can be found by taking a convolution integral with two two-body Green's functions. At the last step the G_α function is found from the equation, containing the obtained channel Green function G_α^{ch} and the polarization potential U_α

$$G_\alpha(z) = G_\alpha^{ch}(z) + G_\alpha^{ch}(z)U_\alpha G_\alpha(z). \tag{33}$$

For the kaonic deuterium calculations it was necessary to take the isospin dependence of the $\bar{K}N$ interaction into account. In particle representation it means that the strong $V_{pK^-}^s$ potential is a 2×2 matrix, containing V_{pK^-,pK^-}^s, $V_{pK^-,n\bar{K}^0}^s$ and $V_{n\bar{K}^0,n\bar{K}^0}^s$ elements. Due to this, the final equations for the kaonic deuterium have four Faddeev components, including the additional one in the $(n\bar{K}^0, n)$ channel.

The solution of the Faddeev-type equations with Coulomb gave the full energy of the $1s$ level. Since the aim of [25] was evaluation of the $1s$ level shift of kaonic deuterium caused by the strong interactions between the antikaon and the nucleons, it was necessary to define the energy, from which the real part of the shift is measured. It can be the lowest eigenvalue of the channel Green function or of the "original" Green function of the (pn, K^-) channel. The first one corresponds to a deuteron and an antikaon feeling a Coulomb force from the center of mass of the deuteron. The second reflects the fact that the antikaon interacts via Coulomb force not with the center of the deuteron, but with the proton. In principle, the correct one should be the second variant, however, all approximate approaches use an analogy of the first one as the basic point, due to this it was used in [25] as well. In any case, the difference between both versions is small.

6.3 Exact Calculation of Kaonic Deuterium: Results

The calculation performed in [25] was considered as a first test of the method for the description of three-body hadronic atoms. Due to this, the second three-body particle channel $\pi\Sigma N$ was not directly included and no energy dependent potentials (exact optical or chirally motivated one) were used. The $\bar{K}N$ and NN interactions were described using one-term separable complex potentials with Yamaguchi form factors. Four versions of the $\bar{K}N$ potential V_I, V_{II}, V_{III} and V_{IV}, used in the calculations, give the $1s$ level shift of the kaonic hydrogen within or close to the SIDDHARTA data and a reasonable fit to the elastic $K^-p \rightarrow K^-p$ and charge exchange $K^-p \rightarrow \bar{K}^0n$ cross-sections. Parameters of the potentials can be found in Table I of [25]. The nucleon–nucleon potential reproduces the NN scattering lengths, low-energy phase shifts and the deuteron binding energy in the np state.

The results of the dynamically exact calculations of the kaonic deuterium are presented in Table 5. The absolute values of the $1s$ level shift were found in the region 641–736 eV, while the width variates between 826–980 eV. Both observables are smaller than the accurate results from [23], shown in Table 4. However, it is necessary to remember that both calculations differ not only by the three-body methods, but also by the two-body input. To make the comparison reasonable, the two-body approximate calculation, described in Sect. 6.1, was repeated with the simple complex $\bar{K}N$ potentials V_I, V_{II}, V_{III}, and V_{IV}. The corrected Deser formula was also checked for these potentials. The approximate results are shown in Table 5.

Table 5 Exact $1s$ level shifts ΔE (eV) and widths Γ (eV) of the kaonic deuterium for the four complex $\bar{K}N$ potentials V_I, V_{II}, V_{III}, and V_{IV}

	Corrected Deser		Complex V_{K^--d}		Exact Faddeev	
	$\Delta_{1s}^{K^-d}$	$\Gamma_{1s}^{K^-d}$	$\Delta_{1s}^{K^-d}$	$\Gamma_{1s}^{K^-d}$	$\Delta_{1s}^{K^-d}$	$\Gamma_{1s}^{K^-d}$
V_I	-675	702	-650	868	-641	856
V_{II}	-694	740	-658	920	-646	888
V_{III}	-795	780	-747	1034	-732	980
V_{IV}	-750	620	-740	844	-736	826

The approximate results obtained using the corrected Deser formula and the complex $K^- - d$ potential are also shown

It is seen that the two-body approximate calculation, described in Sect. 6.1, makes $\leq 2\%$ error for the shift and $\leq 5\%$ for the width, so it is quite accurate. It is an expected result keeping in mind the relative values of deuteron and Bohr radius of kaonic deuterium. The corrected Deser formula Eq. (10) leads to 2–8% error in the shift, and strongly, up to 25%, underestimates the width.

It is also possible to compare the approximate results obtained in [25] with the four complex $\bar{K}N$ potentials V_I, V_{II}, V_{III}, and V_{IV} and in [23] with the coupled-channel models of the antikaon–nucleon interaction (the phenomenological $V_{\bar{K}N-\pi\Sigma}^{1,SIDD}$ and $V_{\bar{K}N-\pi\Sigma}^{2,SIDD}$ with one- and two-pole structure of the $\Lambda(1405)$ resonance respectively, and the chirally motivated $V_{\bar{K}N-\pi\Sigma-\pi\Lambda}^{Chiral}$). It is seen that the "complex one-channel" absolute values of the $1s$ level shift and width shown in Table 5 are smaller than the "coupled-channel" ones presented in Table 4. The similar situation was observed with the exactly evaluated characteristics of the strong pole in the K^-pp system, while a one-channel simple complex antikaon - nucleon potential led to more narrow and less bound quasi-bound state than the coupled-channel version [see Eqs. (26, 27)]. But the differences for the kaonic deuterium are smaller than those for the K^-pp quasi-bound state.

The very recent calculations [68] of the kaonic deuterium $1s$ level shift were performed using the same Faddeev-type equations with Coulomb interaction as in [25], but with energy-dependent $\bar{K}N$ potentials. Namely, the exact optical versions of the one- and two-pole phenomenological $\bar{K}N - \pi\Sigma$ potentials and of the chirally motivated $\bar{K}N - \pi\Sigma - \pi\Lambda$ interaction model were used. The predicted $1s$ level shifts (800 ± 30 eV) and widths (960 ± 40 eV) are larger by absolute value than the exact ones from Table 5 evaluated with the simple complex antikaon nucleon potentials.

Keeping in mind good accuracy of the results obtained with the exact optical $\bar{K}N$ potentials for all three-body $\bar{K}NN$ observables, demonstrated in the present paper, the predictions of [68] for the kaonic deuterium must be the most accurate ones up to date. The two-body approximation used in [21,23], being compared to the more accurate approach of [68], gives very accurate value of the $1s$ level shift (the error is $\leq 2\%$), while the error for the width is larger ($\leq 9\%$).

7 Summary

The three-body antikaon nucleon systems could provide an important information about the antikaon nucleon interaction. It is quite useful since the two-body $\bar{K}N$ potentials of different type can reproduce all low-energy experimental data with the same level of accuracy. This fact was demonstrated on the example of the phenomenological $\bar{K}N - \pi\Sigma$ potentials with one and two-pole structure of the $\Lambda(1405)$ resonance together with the chirally motivated $\bar{K}N - \pi\Sigma - \pi\Lambda$ potential. Being used in the three-body calculations, the three $\bar{K}N$ potentials allowed to investigate the influence of the $\bar{K}N$ model on the results.

It was found that while the quasi-bound state position in the K^-pp and K^-K^-p systems strongly depends on the model of the $\bar{K}N$ interaction, the near-threshold observables (K^-d scattering length, elastic near-threshold K^-d amplitudes, $1s$ level shift and width of kaonic deuterium) are almost insensitive to it. Therefore, some conclusions on the number of poles of the $\Lambda(1405)$ resonance could be done only if a hight accuracy measurement of K^-pp binding energy and width will be done. Probably, one of the existing experiments: by HADES [9], LEPS [10] collaborations or in J-PARC [11,12], hopefully will clarify the situation with - will do it. While dependence of the three-body results on the $\bar{K}N$ potentials is different for the different systems and processes, dependence on NN and ΣN interactions is weak in all cases.

Comparison of the exact results with some approximate ones revealed the most accurate approximations. In particular, the one-channel Faddeev calculations give results, which are very close to the coupled-channel calculations if the exact optical $\bar{K}N$ potential is used. This fact gives a hope for four-body calculations, which are already very complicated without additional coupled-channel structure. It is necessary to note here that the "exact optical" potential is defined as an energy dependent potential, which exactly reproduces the elastic amplitudes of the corresponding potential with coupled channels.

As for the kaonic deuterium, influenced mainly by Coulomb interaction, the shift of its $1s$ level caused by the strong interactions is described quite accurately in the two-body approximation. The $K^- - d$ complex potential should herewith reproduce the exact elastic three-body K^-d amplitudes, and the Lippmann–Schwinger equation must be solved exactly with Coulomb plus the strong potentials. Of cause, the exact calculation is more precise and, therefore, is preferable. The predicted by the exact calculations $1s$ level energy could be checked by SIDDHARTA-2 collaboration [69].

The suggested $1/|\mathrm{Det}(z)|^2$ method of theoretical evaluation of an underthreshold resonance is quite accurate for rather narrow and well pronounced resonances. It could supplement the direct search of the pole providing the first estimation and working as a control. The method is free from the uncertainties connected with the calculations on the complex plane, but it has the logarithmic singularities in the kernels of the integral equations.

The next step in the field of the few-body systems consisting of antikaons and nucleons should be done toward the four body systems. They could give more possibilities, but theoretical investigations of them are much more complicated.

References

1. Y. Akaishi, T. Yamazaki, Nuclear $\bar{K}$ bound states in light nuclei. Phys. Rev. C **65**, 044005 (2002)
2. T. Yamazaki, Y. Akaishi, (K^-, π^-) production of nuclear $\bar{K}$ bound states in proton-rich systems via Λ^* doorways. Phys. Lett. B **535**, 70 (2002)
3. N.V. Shevchenko, A. Gal, J. Mareš, Faddeev calculation of a K^-pp quasi-bound state. Phys. Rev. Lett. **98**, 082301 (2007)
4. N.V. Shevchenko, A. Gal, J. Mareš, J. Révai, $\bar{K}NN$ quasi-bound state and the $\bar{K}N$ interaction: coupled-channel Faddeev calculations of the $\bar{K}NN - \pi\Sigma N$ system. Phys. Rev. C **76**, 044004 (2007)
5. Y. Ikeda, T. Sato, Strange dibaryon resonance in the $\bar{K}NN - \pi YN$ system. Phys. Rev. C **76**, 035203 (2007)
6. M. Agnello et al., Evidence for a kaon-bound State K^-pp produced in K^- absorption reactions at rest. Phys. Rev. Lett. **94**, 212303 (2005)
7. G. Bendiscioli et al., Search for signals of bound $\bar{K}$ nuclear states in antiproton $-^4$He annihilations at rest. Nucl. Phys. A **789**, 222 (2007)
8. T. Yamazaki et al., Indication of a deeply bound and compact K^-pp state formed in the $pp \rightarrow p\Lambda K^+$ reaction at 2.85 GeV. Phys. Rev. Lett. **104**, 132502 (2010)
9. L. Fabbietti et al., $p\Lambda K^+$ final state: towards the extraction of the ppK^- contribution. Nucl. Phys. A **914**, 60 (2013)
10. A.O. Tokiyasu et al., Search for the K^-pp bound state via $\gamma d \rightarrow K^+\pi^- X$ reaction at $E_\gamma = 1.5-2.4$ GeV. Phys. Lett. B **728**, 616 (2014)
11. S. Ajimura et al., A search for deeply-bound kaonic nuclear state at the J-PARC E15 experiment. Nucl. Phys. A **914**, 315 (2013)
12. Y. Ichikawa et al., J-PARC E27 experiment to search for a nuclear kaon bound state K^-pp. Few Body Syst. **54**, 1191 (2013)
13. J. Révai, N.V. Shevchenko, On extracting information about hadron-nuclear interaction from hadronic atom level shifts. Few Body Syst. **42**, 83 (2008)
14. J. Révai, N.V. Shevchenko, Isospin mixing effects in low-energy $\bar{K}N - \pi\Sigma$ interaction. Phys. Rev. C **79**, 035202 (2009)
15. A. Cieplý, J. Smejkal, Chirally motivated $\bar{K}N$ amplitudes for in-medium applications. Nucl. Phys. A **881**, 115 (2012)
16. J.A. Oller, U.-G. Meißner, Chiral dynamics in the presence of bound states: kaon–nucleon interactions revisited. Phys. Lett. B **500**, 263 (2001)
17. N.V. Shevchenko, One- versus two-pole $\bar{K}N - \pi\Sigma$ potential: K^-d scattering length. Phys. Rev. C **85**, 034001 (2012)
18. M. Bazzi et al. (SIDDHARTA Collaboration), A new measurement of kaonic hydrogen X-rays. Phys. Lett. B **704**, 113 (2011)
19. G. Beer et al., Measurement of the kaonic hydrogen X-ray spectrum. Phys. Rev. Lett. **94**, 212302 (2005)
20. M. Iwasaki et al., Observation of kaonic hydrogen K_α X rays. Phys. Rev. Lett. **78**, 3067 (1997)
21. N.V. Shevchenko, Near-threshold K^-d scattering and properties of kaonic deuterium. Nucl. Phys. A **890–891**, 50–61 (2012)
22. J. Révai, Signature of the $\Lambda(1405)$ resonance in neutron spectra from the $K^- + d$ reaction. Few Body Syst. **54**, 1865–1876 (2013)
23. N.V. Shevchenko, J. Révai, Faddeev calculations of the $\bar{K}NN$ system with chirally-motivated $\bar{K}N$ interaction. I. Low-energy K^-d scattering and antikaonic deuterium. Phys. Rev. C **90**, 034003 (2014)
24. J. Révai, N.V. Shevchenko, Faddeev calculations of the $\bar{K}NN$ system with chirally-motivated $\bar{K}N$ interaction. II. The K^-pp quasi-bound state. Phys. Rev. C **90**, 034004 (2014)
25. P. Doleschall, J. Révai, N.V. Shevchenko, Three-body calculation of the $1s$ level shift in kaonic deuterium. Phys. Lett. B **744**, 105–108 (2015)
26. Z. Papp, W. Plessas, Coulomb-Sturmian separable expansion approach: three-body Faddeev calculations or Coulomb-like interactions. Phys. Rev. C **54**, 50 (1996)
27. N.V. Shevchenko, J. Haidenbauer, Exact calculations of a quasibound state in the $\bar{K}\bar{K}N$ system. Phys. Rev. C **92**, 044001 (2015)

28. K.A. Olive et al. (Particle Data Group), The review of particle physics (2015). Chin. Phys. C. **38**, 090001 (2014) (**and 2015 update**)
29. S. Ohnishi, Y. Ikeda, T. Hyodo, E. Hiyama, W. Weise, $K^-d \to \pi \Sigma n$ reactions and structure of the (1405). J. Phys. Conf. Ser. **569**(1), 012077 (2014). arXiv:1408.0118 [nucl-th]
30. S. Ohnishi, Y. Ikeda, T. Hyodo, W. Weise, Structure of the (1405) and the $K^-d \to \pi \Sigma n$ reaction. Phys. Rev. C **93**, 025207 (2016)
31. K. Moriya (for the CLAS Collaboration), Measurement of the $\Sigma \pi$ photoproduction line shapes near the $\Lambda(1405)$. Phys. Rev. C **87**, 035206 (2013)
32. D.N. Tovee et al., Some properties of the charged Σ hyperons. Nucl. Phys. B **33**, 493 (1971)
33. R.J. Nowak et al., Charged Σ hyperon production by K^- meson interactions at rest. Nucl. Phys. B **139**, 61 (1978)
34. U.-G. Meißner, U. Raha, A. Rusetsky, Spectrum and decays of kaonic hydrogen. Eur. Phys. J. C **35**, 349 (2004)
35. M. Sakitt et al., Low-energy K^--meson interactions in hydrogen. Phys. Rev. B **139**, 719 (1965)
36. J.K. Kim, Low-energy K^--p interaction and interpretation of the 1405-MeV Y_0^* resonance as a $\bar{K}N$ bound state. Phys. Rev. Lett. **14**, 29 (1965)
37. J.K. Kim, Multichannel phase-shift analysis of $\bar{K}N$ interaction in the region 0 to 550 MeV/c. Phys. Rev. Lett. **19**, 1074 (1967)
38. W. Kittel, G. Otter, I. Wacek, The K^- proton charge exchange interactions at low energies and scattering lengths determination. Phys. Lett. **21**, 349 (1966)
39. J. Ciborowski et al., Kaon scattering and charged Sigma hyperon production in K^-p interactions below 300 MeV/c. J. Phys. G **8**, 13 (1982)
40. D. Evans et al., Charge-exchange scattering in K^-p interactions below 300 MeV/c. J. Phys. G **9**, 885 (1983)
41. P. Doleschall, *Private communication*
42. R.B. Wiringa, V.G.J. Stoks, R. Schiavilla, Accurate nucleon–nucleon potential with charge-independence breaking. Phys. Rev. C **51**, 38 (1995)
43. H. Zankel, W. Plessas, J. Haidenbauer, Sensitivity of N-d polarization observables on the off-shell behavior of the N–N interaction. Phys. Rev. C **28**, 538 (1983)
44. G. Alexander et al., Study of the $\Lambda - N$ system in low-energy $\Lambda - p$ elastic scattering. Phys. Rev. **173**, 1452 (1968)
45. B. Sechi-Zorn, B. Kehoe, J. Twitty, R.A. Burnstein, Low-energy Λ-proton elastic scattering. Phys. Rev. **175**, 1735 (1968)
46. F. Eisele et al., Elastic $\Sigma^\pm p$ scattering at low energies. Phys. Lett. B **37**, 204 (1971)
47. R. Engelmann, H. Filthuth, V. Hepp, E. Kluge, Inelastic $\Sigma^- p$-interactions at low momenta. Phys. Lett. **21**, 587 (1966)
48. V. Hepp, M. Schleich, A new determination of the capture ratio $r_c = \frac{\Sigma^- p \to \Sigma^0 n}{(\Sigma^- p \to \Sigma^0 n)+(\Sigma^- p \to \Lambda^0 n)}$, the Λ^0-lifetime and the $\Sigma^- - \Lambda^0$ mass difference. Z. Phys. **214**, 71 (1968)
49. D. Lohse, J.W. Durso, K. Holinde, J. Speth, Meson exchange model for pseudoscalar meson–meson scattering. Nucl. Phys. A **516**, 513 (1990)
50. G. Janssen, B.C. Pearce, K. Holinde, J. Speth, Structure of the scalar mesons $f_0(980)$ and $a_0(980)$. Phys. Rev. D **52**, 2690 (1995)
51. S.R. Beane et al. (NPLQCD Collaboration), The K^+K^+ scattering length from lattice QCD. Phys. Rev. D **77**, 094507 (2008)
52. K. Sasaki, N. Ishizuka, M. Oka, T. Yamazaki (PACS-CS Collaboration), Scattering lengths for two pseudoscalar meson systems. Phys. Rev. D **89**, 054502 (2014)
53. E.O. Alt, P. Grassberger, W. Sandhas, Reduction of the three-particle collision problem to multi-channel two-particle Lippmann–Schwinger equations. Nucl. Phys. B **2**, 167 (1967)
54. Y. Ikeda, T. Sato, Energy dependence of $\bar{K}NN$ interactions and resonance pole of strange dibaryons. Prog. Theor. Phys. **124**, 533 (2010)
55. A. Doté, T. Hyodo, W. Weise, Variational calculation of the ppK^- system based on chiral SU(3) dynamics. Phys. Rev. C **79**, 014003 (2009)
56. N. Barnea, A. Gal, E.Z. Liverts, Realistic calculations of $\bar{K}NN$, $\bar{K}NNN$, and $\bar{K}\bar{K}NN$ quasibound states. Phys. Lett. B **712**, 132 (2012)
57. R. Ya. Kezerashvili et al., Three-body calculations for the K^-pp system within potential models. J. Phys. G: Nucl. Part. Phys. **43**, 065104 (2016)
58. P.M. Dauber, J.P. Berge, J.R. Hubbard, D.W. Merrill, R.A. Muller, Production and decay of cascade hyperons. Phys. Rev. **179**, 1262 (1969)
59. Y. Kanada-En'yo, D. Jido, $\bar{K}\bar{K}N$ molecular state in three-body calculation. Phys. Rev. C **78**, 025212 (2008)
60. S.S. Kamalov, E. Oset, A. Ramos, Chiral unitary approach to the K^--deuteron scattering length. Nucl. Phys. A **690**, 494 (2001)
61. V.V. Peresypkin, Consideration of the recoil effect under a light particle scatteriing by two heavy particles. Ukr. Fiz. Zh. **23**, 1256 (1978)
62. T. Mizutani, C. Fayard, B. Saghai, K. Tsushima, Faddeev-chiral unitary approach to the K^-d scattering length. Phys. Rev. C **87**, 035201 (2013)
63. G. Toker, A. Gal, J.M. Eisenberg, The YN interactions and K^- reactions on deuterium at low energies. Nucl. Phys. A **362**, 405 (1981)
64. M. Torres, R.H. Dalitz, A. Deloff, K^- absorption reactions from rest in deuterium. Phys. Lett. B **174**, 213 (1986)
65. A. Deloff, $\eta - d$ and $K^- - d$ zero-energy scattering: a Faddeev approach. Phys. Rev. C **61**, 024004 (2000)
66. U.-G. Meissner, U. Raha, A. Rusetsky, Kaon–nucleon scattering lengths from kaonic deuterium experiments. Eur. Phys. J. C **47**, 473 (2006)
67. J.V. Noble, Three-body problem with charged particles. Phys. Rev. **161**, 945 (1967)
68. J. Révai, Three-body caculation of the $1s$ level shift in kaonic deuterium with realistic potentials. arXiv:1608.01802 [nucl-th]
69. C. Curceanu et al., Unlocking the secrets of the kaonnucleon/nuclei interactions at low-energies: the SIDDHARTA(-2) and the AMADEUS experiments at the DAΦNE collider. Nucl. Phys. A **914**, 251 (2013)

Few-Body Syst (2017) 58:1–12
DOI 10.1007/s00601-016-1168-z

Craig D. Roberts

Perspective on the Origin of Hadron Masses

Received: 11 June 2016 / Accepted: 13 October 2016 / Published online: 9 December 2016
© Springer-Verlag Wien (Outside the USA) 2016

Abstract The energy–momentum tensor in chiral QCD, $T_{\mu\nu}$, exhibits an anomaly, viz. $\Theta_0 := T_{\mu\mu} \neq 0$. Measured in the proton, this anomaly yields m_p^2, where m_p is the proton's mass; but, at the same time, when computed in the pion, the answer is $m_\pi^2 = 0$. Any attempt to understand the origin and nature of mass, and identify observable expressions thereof, must explain and unify these two apparently contradictory results, which are fundamental to the nature of our Universe. Given the importance of Poincaré-invariance in modern physics, the utility of a frame-dependent approach to this problem seems limited. That is especially true of any approach tied to a rest-frame decomposition of $T_{\mu\nu}$ because a massless particle does not possess a rest-frame. On the other hand, the dynamical chiral symmetry breaking paradigm, connected with a Poincaré-covariant treatment of the continuum bound-state problem, provides a straightforward, simultaneous explanation of both these identities, and also a diverse array of predictions, testable at existing and proposed facilities. From this perspective, $\langle\pi|\Theta_0|\pi\rangle = 0$ owing to exact, symmetry-driven cancellations which occur between one-body dressing effects and two-body-irreducible binding interactions in any well-defined computation of the forward scattering amplitude that defines this expectation value in the pseudoscalar meson. The cancellation is incomplete in any other hadronic bound state, with a remainder whose scale is set by the size of one-body dressing effects.

1 Introduction

Classical chromodynamics (CCD) is a non-Abelian local gauge field theory. As with all such theories formulated in four spacetime dimensions, no mass-scale exists in the absence of Lagrangian masses for the fermions. There is no dynamics in a scale-invariant theory, only kinematics: the theory looks the same at all length-scales and hence there can be no clumps of anything. Bound-states are therefore impossible and, accordingly, our Universe cannot exist. Spontaneous symmetry breaking, as realised via the Higgs mechanism, does not solve this problem because normal matter is constituted from light-quarks, u and d, and the masses of the neutron and proton, the kernels of all visible matter, are roughly 100-times larger than anything the Higgs can produce in connection with u- and d-quarks. Consequently, the question of how did the Universe come into being is inseparable from the questions of how does a mass-scale appear and why does it have the value we observe?

Modern quantum field theories are not built simply on Lorentz invariance. The effect of space-time translations must also be considered and thus enters the group of Poincaré transformations. In this connection, consider the energy–momentum tensor in CCD, $T_{\mu\nu}$, which can always be made symmetric [1]. Conservation of energy and momentum in a quantum field theory is a consequence of spacetime translational invariance,

---

This article belongs to the special issue "30th anniversary of Few-Body Systems".

Craig D. Roberts (✉)
Physics Division, Argonne National Laboratory, Argonne, IL 60439, USA
E-mail: cdroberts@anl.gov

one of the family of Poincaré transformations. Consequently,

$$\partial_\mu T_{\mu\nu} = 0. \tag{1}$$

Consider now a global scale transformation in the Lagrangian of the classical theory:

$$x \to x' = e^{-\sigma} x, \tag{2a}$$

$$A_\mu^a(x) \to A_\mu^{a\prime}(x') = e^{-\sigma} A_\mu^a(e^{-\sigma} x), \tag{2b}$$

$$q(x) \to q'(x') = e^{-(3/2)\sigma} q(e^{-\sigma} x), \tag{2c}$$

where $A_\mu^a(x)$, $q(x)$ are the gluon and quark fields. The Noether current associated with this transformation is the dilation current

$$\mathcal{D}_\mu = T_{\mu\nu} x_\nu. \tag{3}$$

In the absence of fermion masses, the classical action is invariant under Eq. (2), i.e. the theory is scale invariant, and hence

$$\partial_\mu \mathcal{D}_\mu = 0 = [\partial_\mu T_{\mu\nu}] x_\nu + T_{\mu\nu} \delta_{\mu\nu} = T_{\mu\mu}, \tag{4}$$

where the last equality follows from Eq. (1). Plainly, the energy–momentum tensor is traceless in a scale invariant theory.[1]

Massless CCD is not a meaningful framework in modern physics for many reasons; amongst them the fact that strong interactions in the Standard Model are empirically known to be characterised by a large mass-scale, characterised, e.g. by the value of $\Lambda_{\text{QCD}} \approx 0.23$ GeV [4]. In quantising the theory, regularisation and renormalisation of (ultraviolet) divergences introduces a mass-scale. This is "dimensional transmutation": mass-dimensionless quantities become dependent on a mass-scale, and this entails the violation of Eq. (4), i.e. the appearance of the chiral-limit "trace anomaly":

$$T_{\mu\mu} = \beta(\alpha(\zeta)) \tfrac{1}{4} G_{\mu\nu}^a G_{\mu\nu}^a =: \Theta_0, \tag{5}$$

where $\beta(\alpha)$ is the QCD β-function, ζ is the renormalisation scale, $G_{\mu\nu}^a$ is the gluon field-strength tensor, and this expression assumes the chiral limit for all current-quarks.

There is a simple, nonperturbative derivation of this identity, which eliminates any need for a diagrammatic or perturbative analysis [5]. Namely, under the scale transformation in Eq. (2a), the mass-scale $\zeta \to e^\sigma \zeta$. Considering infinitesimal transformations of this type, it is straightforward to show:[2]

$$\alpha \to \sigma\, \alpha\beta(\alpha), \ \mathcal{L} \to \sigma\, \alpha\beta(\alpha) \frac{\delta\mathcal{L}}{\delta\alpha} \Rightarrow \partial_\mu \mathcal{D}_\mu = \frac{\delta\mathcal{L}}{\delta\sigma} = \alpha\beta(\alpha) \frac{\delta\mathcal{L}}{\delta\alpha}. \tag{6}$$

In order to compute the final product, one may first absorb the gauge coupling into the gluon field, i.e. express the action in terms of $\tilde{A}_\mu^a = g A_\mu^a$, in which case the running coupling appears only as an inverse multiplicative factor connected with the pure-gauge term:

$$\mathcal{L}(\alpha) = -\frac{1}{4\pi\alpha} \frac{1}{4} \tilde{G}_{\mu\nu}^a \tilde{G}_{\mu\nu}^a + \alpha\text{-independent terms}, \tag{7}$$

where $\tilde{G}_{\mu\nu}^a$ is the field-strength tensor expressed using $\tilde{A}_\mu^a$. It then follows that

$$T_{\mu\mu} = \partial_\mu \mathcal{D}_\mu = \alpha\beta(\alpha) \frac{\delta\mathcal{L}}{\delta\alpha} = \alpha\beta(\alpha) \frac{1}{4\pi\alpha^2} \frac{1}{4} \tilde{G}_{\mu\nu}^a \tilde{G}_{\mu\nu}^a = \beta(\alpha) \tfrac{1}{4} G_{\mu\nu}^a G_{\mu\nu}^a, \tag{8}$$

viz. Eq. (5) is recovered.

It is worth emphasising here that the appearance of a trace anomaly has nothing to do with the non-Abelian nature of QCD. Indeed, it will be apparent from the derivation just sketched that quantum electrodynamics

---

[1] It is possible for a quantum field theory to be scale invariant but not conformally-invariant; but examples are rare. In fact, there is no known example of a scale invariant but non-conformal field theory in four dimensions under a small set of seemingly reasonable assumptions. For this reason, the terms are often used interchangeably in the context of quantum field theory, even though the scale symmetry group is smaller. (See, e.g. Refs. [2,3].)

[2] For this discussion the Euclidean action expressed in the form of Eq. (2.1.8) in Ref. [6] is used, but an overall negative-sign is included when defining the Euclidean Lagrangian so that the derivation more closely resembles that in Minkowski space: $S := -\int d^4x\, L$.

(QED) must also possess a trace anomaly. However, QED is nonperturbatively undefined: four-fermion operators become relevant in strong-coupling QED and must be included in order to obtain a well-defined (albeit trivial) continuum limit (see, e.g. Refs. [7–9]). As a consequence, QED does not have a chiral limit. The QED trace anomaly is only meaningful in perturbation theory and its scale is determined by the Higgs mechanism.

In the presence of nonzero current-quark masses, Eq. (5) becomes

$$\Theta := T_{\mu\mu} = \tfrac{1}{4}\beta(\alpha(\zeta))G_{\mu\nu}^a G_{\mu\nu}^a + [1 + \gamma(\alpha(\zeta))]\sum_f m_f^\zeta \bar{q}_f q_f, \tag{9}$$

where m_f^ζ are the current-quark masses and $[1 + \gamma(\alpha)]$ is the analogue for the dressed-quark running-mass of $\beta(\alpha)$ for the running coupling. [In fact, $\gamma(\alpha)$ is the anomalous dimension of the current-quark mass in QCD.] It is notable that in the massive-case the trace anomaly is not homogeneous in the running coupling, $\alpha(\zeta)$. Consequently, renormalisation-group-invariance does not entail form invariance of the right-hand-side (rhs) [10]. This is important because discussions typically assume (perhaps implicitly) that all operators and identities are expressed in a partonic basis, viz. using simple field operators that can be renormalised perturbatively, in which case the hadronic state-vector represents an extremely complicated wave function. That perspective is not valid at renormalisation scales $\zeta \lesssim m_p$, where m_p is the proton mass; and this is where a metamorphosis from parton-basis to quasiparticle-basis may occur: under reductions in resolving scale, ζ, light partons evolve into heavy dressed-partons, corresponding to complex superpositions of partonic operators; and using these dressed-parton operators, the wave functions can be expressed in a relatively simple form. (A relevant illustration is discussed in Refs. [11,12].)

2 Magnitude of the Scale Anomaly: Mass and Masslessness

2.1 A Binary Problem

Simply knowing that a trace anomaly exists does not deliver a great deal: it only indicates that there is a mass-scale. The crucial issue is whether or not one can compute and/or understand the magnitude of that scale.

One can certainly measure the size of the scale anomaly, for consider the expectation value of the energy–momentum tensor in the proton (e.g., Ref. [13]):

$$\langle p(P)|T_{\mu\nu}|p(P)\rangle = -P_\mu P_\nu, \tag{10}$$

where the rhs follows from the equations-of-motion for an asymptotic one-particle proton state. At this point it is clear that, in the chiral limit,

$$\langle p(P)|T_{\mu\mu}|p(P)\rangle = -P^2 = m_p^2 \tag{11a}$$

$$= \langle p(P)|\Theta_0|p(P)\rangle; \tag{11b}$$

namely, there is a clear sense in which it is possible to say that the entirety of the proton mass is produced by gluons. The trace anomaly is measurably large; and that property must logically owe to gluon self-interactions, which are also responsible for asymptotic freedom.

This is a valid conclusion. After all, what else could be responsible for a mass-scale in QCD? QCD is all about gluon self-interactions; and it's gluon self-interactions that (potentially) enable one to rigorously (nonperturbatively) define the expectation value in Eq. (11). On the other hand, it's only a sensible conclusion when the operator and the wave function are defined at a resolving-scale $\zeta \gg m_p$. It will be necessary to return to this point.

There is also another issue, which can be exposed by returning to Eq. (10) and replacing the proton by the pion:

$$\langle \pi(q)|T_{\mu\nu}|\pi(q)\rangle = -q_\mu q_\nu. \tag{12}$$

Then, in the chiral limit:

$$\langle \pi(q)|\Theta_0|\pi(q)\rangle = 0 \tag{13}$$

because the pion is a massless Nambu–Goldstone mode. Equation (12) could mean that the scale anomaly vanishes trivially in the pion state, viz. that gluons and their self-interactions have no impact within a pion because each term in the expression of the operator vanishes when evaluated in the pion. However, that is a difficult way to achieve Eq. (13). It is easier, perhaps, to imagine that Eq. (13) owes to cancellations between different operator-component contributions. Of course, such precise cancellation should not be an accident. It could only arise naturally because of some symmetry and/or symmetry-breaking pattern; and, as will be argued below, that is precisely the manner by which Eq. (13) is realised.

Equations (11) and (13) present a quandary, which highlights that no understanding of the origin of the proton's mass can be complete unless it simultaneously explains the meaning of Eq. (13). Given that a massless particle doesn't have a rest-frame, any approach based on a rest-frame decomposition of the energy–momentum tensor (e.g. Refs. [14,15]) cannot readily be useful in this dual connection.

2.2 Frame Independent Resolution

Both Eqs. (11) and (13) are Poincaré invariant statements. Quantum field theories are the only known realisation of the Poincaré algebra in quantum mechanics with a particle interpretation. This entails that asymptotic one-particle states, even those which are composite, are characterised by just two invariants [16], i.e. the eigenvalues of M^2 and W^2, where the former is the mass-squared operator and W_μ is the Pauli-Lubanski four-vector, both constituted for the entire (composite) system. The eigenvalues of the mass-squared operator, m^2, need no further explanation.

Concerning the Pauli-Lubanski four-vector, note that, by definition, W_μ contains no information about orbital angular momentum: it is sensitive only to the total-spin of the system. For massive particles, the eigenvalues of W^2 are the products $m^2 j(j+1)$, where $j(j+1)$ is the rest-frame eigenvalue of $\mathbf{J} \cdot \mathbf{J}$, with $\mathbf{J}$ being the total angular momentum operator, viz. j is the particle's spin.

In the massless case, the eigenvalues of W^2 vanish, the Pauli-Lubanski four-vector is proportional to the four-momentum, and the constant of proportionality is the particle's helicity. Consequently, massless one-particle states are labelled by their helicity which, for fermions, is related to their chirality. Massless fermions are either left-handed or right-handed and no Poincaré transformation can alter the assignment.

These statements entail that the only unambiguous labels that can be attached to a composite hadron state are its mass and spin := total-angular-momentum; and different observers, characterised by distinct reference frames, will only necessarily agree on these two quantities. Hence, e.g. no two observers need necessarily agree on a massive hadron's polarisation. Furthermore, in a quantum field theory, no separation of the total angular momentum into a sum $L + S$ can be Poincaré invariant. Such a separation is frame-dependent, and that includes frames made distinct by boosts. Consequently, no two observers (or calculators) need agree on the values of $\langle L \rangle$ and $\langle S \rangle$, even though they will agree on $\langle W^2 \sim (L + S)^2 \rangle$. This fact lies at the heart of the so-called "spin-crisis", which could therefore have been avoided. Finally, even using light-front quantisation, both the nature and size of contributions from constituents to any observable property of the composite hadron itself change with resolving scale, ζ. It seems, therefore, that a simultaneous elucidation of the meaning and consequences of Eqs. (11) and (13) cannot satisfactorily be achieved by focusing on a particular reference frame and that the picture one finds most satisfactory will likely depend on the scale at which the resolution is presented.

It is also worth recalling that CCD is still a non-Abelian local gauge theory. Consequently, the concept of local gauge invariance persists. However, without a mass-scale there is no confinement. For example, three quarks can be prepared in a colour-singlet combination and colour rotations will keep the three-body system colour neutral; but the quarks involved need not have any proximity to one another. Indeed, proximity is meaningless because all lengths are equivalent in a scale invariant theory. Hence, the question of "Whence mass"? is equivalent to "Whence a mass-scale?", which is equivalent to "Whence a confinement scale?". Thus, understanding the origin, Eq. (11), and absence, Eq. (13), of mass in QCD is quite likely inseparable from the task of understanding confinement; and existence, alone, of a scale anomaly answers neither question.

2.3 Value of a Hadronic Scale

As noted above, the energy–momentum tensor is typically considered in connection with partonic operators, which are simple and can be computed perturbatively. In this approach, however, the wave functions are extremely complicated; and they have never been computed in four dimensions, even approximately.[3] The partonic perspective is valid for $\zeta \gg \zeta_2 := 2\,\text{GeV}$. It might be valid at smaller scales, too; but it cannot be used for $\zeta \lesssim \zeta_c \approx 0.5\,\text{GeV}$, which corresponds to the mass-scale at which coloured two-point functions in QCD exhibit an inflection point. The value ζ_c is known from continuum- and lattice-QCD analyses (see, e.g. Ref. [18] and citations therein).

Suppose one is working at renormalisation scales $\zeta \geq \zeta_2$. If one begins to speak about a wave function with a quantum mechanical interpretation, then it is necessary to employ light-front quantisation. In this case, a hadron wave function, Ψ, is independent of the hadron's four-momentum and, in principle, has a meaningful Fock-space decomposition [19]. On the other hand, it is not independent of the renormalisation scale: $\Psi = \Psi(\zeta)$, such that the relative strength of each Fock-space element changes with ζ. These changes are described by QCD evolution equations (e.g. Refs. [20–26]), which are known to some low/finite-order in perturbative QCD. A method for evolving the light-front Hamiltonian to scales $\zeta \lesssim \zeta_c$ has long been sought, e.g. Ref. [27]. In this case, as already noted, the operators become complicated, describing strongly-dressed quasi-particles; but the wave functions become simple, expressed in terms of a few quasi-particle degrees of freedom. The light-front holographic approach to QCD [28] and Dyson-Schwinger equation (DSE) analyses [29–31] may be viewed as modern attempts to achieve the goal identified in Ref. [27].

It is evident now that the question: "How does one understand a hadron's mass in terms of the contributions from its constituents and their interaction dynamics?" is ill-posed because the meaning of the question depends on the energy-scale being used to probe the hadron and the reference frame to which the question is addressed.

The DSEs are useful here because they provide a symmetry-preserving (and hence Poincaré covariant) framework with a traceable connection to the Lagrangian of QCD. The known limitation of this approach is the need to employ a truncation in order to define a tractable continuum bound-state problem. That truncation might also involve an *Ansatz* for the infrared behaviour of one or more coloured Schwinger functions, although that need is passing [32]. Concerning truncation, much has been learnt in the past twenty years, so that one may now separate DSE predictions into three classes: (*A*) model-independent statements about QCD; (*B*) illustrations of such statements using well-constrained model elements and possessing a traceable connection to QCD; (*C*) analyses that can fairly be described as QCD-based but whose elements have not been computed using a truncation that preserves a systematically-improvable connection with QCD.

The dressed-quark mass-function depicted in Fig. 1 is a Class-A prediction combined with a Class-B illustration: owing to gluon self-interactions in QCD, massless quarks acquire a momentum-dependent mass-function, which is large in the infrared. This figure is a clean demonstration of the scale anomaly at work: zero parton-mass becomes a mass-function, whose value depends on the scale at which the subsystem is probed.

3 Confinement and Dynamical Chiral Symmetry Breaking

3.1 Millennium Prize

The Clay Mathematics Institute has published a Millennium Prize Problem [36], whose solution will contain a proof of confinement in pure Yang-Mills theory. The opening statement of the challenge reads: "Prove that for any compact simple gauge group G, a non-trivial Yang-Mills theory exists on $\mathbb{R}^4$ and has a mass gap $\Delta > 0$." There is strong evidence supporting the existence of a mass gap, found especially in the fact that numerical simulations of lattice-regularised QCD (lQCD) predict $\Delta \gtrsim 1.5\,\text{GeV}$ [37]. However, even allowing for the Higgs mechanism, with this value of the mass gap, one computes $\Delta^2/m_\pi^2 \gtrsim 100$; and, therefore, a conundrum presents itself: can the mass-gap in pure Yang-Mills theory really play any role in understanding confinement when DCSB, driven by the same dynamics, ensures the existence of an almost-massless strongly-interacting excitation in our Universe? If the answer is not *no*, then it must at least be that one cannot claim to provide a pertinent understanding of confinement without simultaneously explaining its connection with DCSB. The pion must play a critical role in any explanation of confinement in the Standard Model; and any discussion that omits reference to the pion's role is *practically irrelevant*.

From this perspective, the potential between infinitely-heavy quarks measured in simulations of quenched lQCD—the so-called static potential [38]—is disconnected from the question of confinement in our Universe. This is because light-particle creation and annihilation effects are essentially nonperturbative in QCD, so it is impossible in principle to compute a quantum mechanical potential between two light quarks [39–41]. It

---

[3] The problem of QCD in two-dimensions is discussed in Ref. [17]. However, two-dimensional theories have little in common with their four-dimensional counterparts, so the analysis in Ref. [17] has not yet led to much progress in four-dimensional QCD.

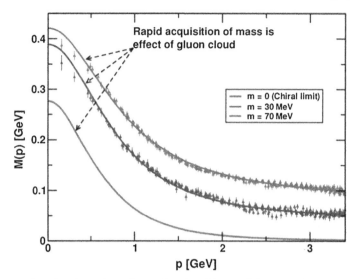

Fig. 1 Renormalisation-group-invariant dressed-quark mass function, $M(p)$: *solid curves* DSE results, explained in Refs. [33,34], "data" numerical simulations of lattice-regularised QCD [35]. (*N.B.* $m = 70$ MeV is the uppermost curve and current-quark mass decreases from *top* to *bottom*.) The current-quark of perturbative QCD evolves into a constituent-quark as its momentum becomes smaller. The constituent-quark mass arises from a cloud of low-momentum gluons attaching themselves to the current-quark. This is dynamical chiral symmetry breaking (DCSB): an essentially nonperturbative effect that generates a quark *mass from nothing*; namely, it occurs even in the chiral limit. The size of $M(0)$ is a measure of the magnitude of the QCD scale anomaly in $n = 1$-point Schwinger functions

follows that there is no flux tube in a Universe with light quarks and consequently that the flux tube is not the correct paradigm for confinement.[4]

DCSB is the key here. It ensures the existence of (pseudo-)Nambu–Goldstone modes; and in the presence of these modes, it is unlikely that any flux tube between a static colour source and sink can have a measurable existence. To explain this statement, consider such a tube being stretched between a source and sink. The potential energy accumulated within the tube may increase only until it reaches that required to produce a particle-antiparticle pair of the theory's pseudo-Nambu–Goldstone modes. Simulations of lQCD show [39,40] that the flux tube then disappears instantaneously along its entire length, leaving two isolated colour-singlet systems. The length-scale associated with this effect in QCD is $r_{\sigma} \simeq (1/3)$ fm and hence if any such string forms, it would dissolve well within a hadron's interior.

An alternative realisation associates confinement with dramatic, dynamically-driven changes in the analytic structure of QCD's coloured propagators and vertices. That leads these coloured n-point functions to violate the axiom of reflection positivity and hence forces elimination of the associated excitations from the Hilbert space associated with asymptotic states [44]. This is a sufficient condition for confinement [29,45–48]. It should be noted, however, that the appearance of such alterations when analysing some truncation of a given theory does not mean that the theory itself is truly confining: unusual spectral properties can be introduced by approximations, leading to a truncated version of a theory which is confining even though the complete theory is not, e.g. Refs. [49,50]. Notwithstanding exceptions like these, a computed violation of reflection positivity by coloured functions in a veracious treatment of QCD does express confinement. Moreover, via this mechanism, confinement is achieved as the result of an essentially dynamical process.

Figure 1 highlights that quarks acquire a running mass distribution in QCD; and this is also true of gluons (see, e.g. Refs. [18,32,51–55]). The generation of these masses leads to the emergence of a length-scale $\varsigma \approx 0.5$ fm, whose existence and magnitude is evident in all existing studies of dressed-gluon and -quark propagators, and which characterises the dramatic change in their analytic structure that has just been described. In models based on such features [56], once a gluon or quark is produced, it begins to propagate in spacetime; but after each "step" of length ς, on average, an interaction occurs so that the parton loses its

---

[4] It is sometimes argued that hadron bound-states lie on linear Regge trajectories and there must therefore be a flux tube. However, empirical evidence for the existence of such towers of states tied to parallel, linear trajectories is poor, e.g. Refs. [42,43]; the potential required to produce such trajectories is dependent both on the frame and the quantisation-scheme employed, i.e. their appearance and nature is strongly dependent on the model used; and no approach whose parameters can rigorously be connected with real-world QCD has ever produced such trajectories.

identity, sharing it with others. Finally a cloud of partons is produced, which coalesces into colour-singlet final states. This picture of parton propagation, hadronisation and confinement can be tested in experiments at modern and planned facilities [57–59].

3.2 Nambu–Goldstone modes

Returning to Eq. (13), suppose now that the Higgs mechanism is active, in which case

$$\langle \pi(q)|\Theta|\pi(q)\rangle = (m_u^\zeta + m_d^\zeta)\frac{\rho_\pi^\zeta}{f_\pi} \tag{14}$$

where [60,61] f_π is the pseudovector projection of pion's Bethe-Salpeter wave function onto the origin in configuration space (pion's leptonic decay constant) and ρ_π^ζ is the analogous pseudoscalar projection.[5]

Plainly,"heavy"-quarks play no role in generating m_π via the trace anomaly, so they're probably not going to play a large part in m_p. In fact, Class-B and -C DSE analyses indicate that, with physical values of the Higgs-generated current-masses, explicit mass terms for u- and d-quarks contribute $f_{u+d} = 6\%$ of the proton mass and that of the s-quark, just $f_s = 2\%$ [63–65]. These predictions are consistent with contemporary lQCD results [66,67]. There is no concrete reason to expect a material contribution from quarks with larger current-masses and arguments to suggest that they are small. For example, simple scaling within the frameworks used to produce the preceding fractions suggests that c-quarks contribute $\lesssim 0.1\%$ of m_p, viz.

$$f_c = f_s \frac{f_K^2}{f_D^2} \frac{m_K^2}{m_D^2} \frac{M_c}{M_s} \approx 0.1, \tag{15}$$

where $f_{K,D}$ are meson leptonic decay constants, $m_{K,D}$ are meson masses, and $M_{s,c}$ are constituent-like quark masses [68,69].

The rhs of Eq. (14) only involves light-quark current-masses. However, this does *not* mean that gluons contribute nothing to the pion's mass because both f_π and ρ_π^ζ are order parameters for DCSB, confinement leads to DCSB, gluons generate confinement, and consequently gluons are at least responsible for the strength of ρ_π^ζ/f_π and hence the rate at which m_π increases with current-quark mass. Gluons actually play a far greater role than this, as will be discussed further below.

In addressing the pion mass and its connection with DCSB, it is crucial to grasp some basic identities in QCD. To that end, consider the pion's Bethe-Salpeter amplitude:

$$\Gamma_\pi(k; P) = i\gamma_5\left[E_\pi(k; P) + \gamma \cdot P\, F_\pi(k; P) + \gamma \cdot k\, k \cdot P\, G_\pi(k; P) + \sigma_{\mu\nu}k_\mu P_\nu H_\pi(k; P)\right]. \tag{16}$$

In the chiral limit the axial-vector Ward–Green–Takahashi identity entails the following array of Goldberger–Treiman relations [60,70]:

$$f_\pi^0 E_\pi^0(y, w = 0; P^2 = 0) = B_0(y), \tag{17a}$$

$$F_A^0(y, w = 0; P^2 = 0) + 2f_\pi^0 F_\pi^0(y, w = 0; P^2 = 0) = A_0(y), \tag{17b}$$

$$G_A^0(y, w = 0; P^2 = 0) + 2f_\pi^0 G_\pi^0(y, w = 0; P^2 = 0) = 2A_0'(y), \tag{17c}$$

$$H_A^0(y, w = 0; P^2 = 0) + 2f_\pi^0 H_\pi^0(y, w = 0; P^2 = 0) = 0, \tag{17d}$$

where $y = k^2$, $w = k \cdot P$, F_A^0, G_A^0, H_A^0 are regular functions that appear in the axial-vector vertex, and the chiral-limit dressed-quark propagator is

$$S_0(p) = 1/[i\gamma \cdot p\, A_0(p^2, \zeta^2) + B_0(p^2, \zeta^2)] = Z_0(p^2, \zeta^2)/[i\gamma \cdot p + M_0(p^2)]. \tag{18}$$

It follows that in the chiral limit, DCSB is a sufficient and necessary condition for the appearance of a massless pseudoscalar bound-state that dominates the axial-vector vertex on $P^2 \simeq 0$ and whose constituents are described by a momentum-dependent mass-function, which may be arbitrarily large. Furthermore, the appearance of a dynamically-generated nonzero mass-function in the solution of QCD's chiral-limit one-quark problem entails, through Eq. (17) in general, and Eq. (17a) in particular, that the isospin-nonzero pseudoscalar two-body problem is solved, well-nigh completely and without additional effort, once the solution to the one-

---

[5] The extension to charge-neutral pseudoscalar mesons is readily accomplished following Ref. [62].

body dressed-quark problem is known; and, moreover, that the quark-level Goldberger–Treiman relation in Eq. (17a) is the most basic expression of Goldstone's theorem in QCD, viz.

Goldstone's theorem is fundamentally an expression of equivalence between the one-body problem and the two-body problem in QCD's colour-singlet pseudoscalar channel.

Equations (17) are a Class-A prediction. They are model-independent, gauge-independent and scheme-independent: any continuum approach to bound-states in QCD that faithfully expresses and preserves chiral symmetry and the pattern by which it is broken will generate these identities. There are numerous Class-B illustrations (see, e.g. Ref. [71] and references thereto).

It follows from Eq. (17) that pion properties are an almost direct measure of the mass function depicted in Fig. 1. Moreover, these identities are the reason the pion is massless in the chiral limit and indirectly the explanation for a proton mass of around 1 GeV. Thus, enigmatically, properties of the nearly-massless pion are the cleanest expression of the mechanism that is responsible for almost all the visible mass in the Universe.[6]

Given its importance, one might ask whether the dressed-quark mass function is observable; and this issue is now readily addressed. As just noted, Eqs. (17) entail that in the neighbourhood of the chiral limit, the dressed-quark mass function (almost) completely determines the pion's Bethe-Salpeter wave-function. Thus, like a wave function in any field of physics, the dressed-quark mass function is not strictly observable. On the other hand, no one can reasonably doubt the enormous value of possessing (nearly) complete knowledge of a bound-state's wave function. Moreover, the pion's Poincaré-covariant Bethe-Salpeter wave function can be projected onto the light-front. The object thus obtained is strictly a probability amplitude and the moments of a probability measure are truly observable. Consequently, there is a mathematically strict sense in which moments of the dressed-quark mass function are observable. One should note in addition that generalised parton distributions can rigorously be defined as an overlap of light-front wave functions [73–76]. Practically, therefore, the dressed-quark mass function can be "measured" because it influences and determines a vast array of experimental observables and there is at least one tractable framework, the DSEs, through which to relate those observables to QCD. In this sense, $M(p^2)$ is as readily observable as, e.g. the parton distribution amplitudes and functions which are the focus of a wide variety of extant and proposed experiments.

3.3 Positive and Negative Contributions from the Trace Anomaly

Return again now to Eq. (13). The pion's Poincaré-invariant mass and Poincaré-covariant wave function are obtained by solving a Bethe-Salpeter equation. This is a scattering problem. In the chiral limit, two massless fermions interact via exchange of massless gluons, i.e. the initial system is massless; and it remains massless at every order in perturbation theory. The complete calculation of the scattering process, however, involves an enumerable infinity of dressings and scatterings, as illustrated in Fig. 2. This can be represented by a coupled set of gap- and Bethe-Salpeter equations. At $\zeta = \zeta_2$, it is practical to build the kernels using a dressed-parton basis, viz. from valence-quarks with a momentum-dependent running mass produced by self-interacting gluons, which have given themselves a running mass.

In the chiral limit one can prove algebraically [77–80] that, at any finite order in a symmetry-preserving construction of the kernels for the gap (quark dressing) and Bethe-Salpeter (bound-state) equations, there is a precise cancellation between the mass-generating effect of dressing the valence-quarks and the attraction introduced by the scattering events. This cancellation guarantees that the simple system, which began massless, becomes a complex system, with a nontrivial bound-state wave function that is attached to a pole in the scattering matrix, which remains at $P^2 = 0$, i.e. the bound-state is also, therefore, massless.

The precise statement is that in the pseudoscalar channel, the dynamically generated mass of the two fermions is precisely cancelled by the attractive interactions between them, if and only if, Eqs. (17) are satisfied. It can be expressed as follows:

$$\langle \pi(q)|\theta_0|\pi(q)\rangle \overset{\zeta \gg \zeta_2}{=} \langle \pi(q)|\tfrac{1}{4}\beta(\alpha(\zeta))G^a_{\mu\nu}G^a_{\mu\nu}|\pi(q)\rangle \overset{\zeta \simeq \zeta_2}{\to} \langle \pi(q)|\mathcal{D}_1 + \mathcal{I}_2|\pi(q)\rangle, \tag{19a}$$

$$\mathcal{D}_1 = \sum_{f=u,d} M_f(\zeta)\,\bar{\mathcal{Q}}_f(\zeta)\mathcal{Q}_f(\zeta), \tag{19b}$$

$$\mathcal{I}_2 = \tfrac{1}{4}[\beta(\alpha(\zeta))\mathcal{G}^a_{\mu\nu}\mathcal{G}^a_{\mu\nu}]_{2\mathrm{PI}}, \tag{19c}$$

---

[6] Additional notable consequences of DCSB are described, e.g. in Ref. [72], which explains some of the results that follow from the fact that the leptonic decay constant of every radially-excited pseudoscalar meson must vanish in the chiral limit.

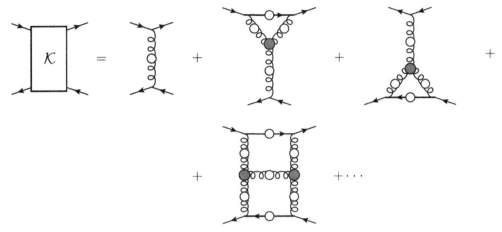

Fig. 2 Some of the contributions to the quark-antiquark scattering kernel, $\mathcal{K}$: solid line with open circle, dressed-quark propagator; open-circle spring, dressed-gluon propagator, and shaded (*blue*) circle at the junction of three gluon lines, dressed three-gluon vertex. The contribution drawn in the second row is an example of an H-diagram. It is two-particle-irreducible and cannot be expressed as a correction to either vertex in a ladder kernel nor as a member of the class of crossed-box diagrams. (Additional details are provided in Ref. [77]) (colour figure online)

which describes the metamorphosis of the parton-basis chiral-limit expression of the expectation-value of the trace anomaly into a new expression, written in terms of a nonperturbatively-dressed quasi-particle basis, with dressed-quarks denoted by $\mathcal{Q}$ and the dressed-gluon field strength tensor by $\mathcal{G}$. Here, the first term is positive, expressing the one-body-dressing content of the trace anomaly. Plainly, a massless valence-quark (antiquark) acquiring a large mass through interactions with its own gluon field is an expression of the trace-anomaly in what might be termed the one-quasiparticle subsector of a complete pion wave function. The second term is negative, expressing the two-particle-irreducible (2PI) scattering-event content of the forward scattering process represented by this expectation-value of the scale-anomaly. This term acquires a scale because the couplings, and the gluon- and quark-propagators in the 2PI processes have all acquired a mass-scale. Away from the chiral limit, and in other channels, such as the proton, the cancellation is incomplete.[7]

It is worth reiterating that these statements can be verified algebraically. Exact cancellation for the pion in the chiral limit was first demonstrated in Ref. [82]; and in connection with the gap and pseudoscalar-channel Bethe-Salpeter equations, Ref. [83] provides the simplest realisation. The same framework yields algebraic formulae for the masses of the nucleon and Δ-baryon, showing that these states possess masses on the order of the DCSB mass-scale in the chiral limit [84] despite the pion's masslessness. The most sophisticated continuum analyses of the pion and proton masses may be traced from Refs. [80,85,86].

The fact that a Poincaré-invariant analysis of the simultaneous impact of the trace anomaly in the pion and proton cannot yield a sum of terms that is each individually positive precludes a useful "pie diagram" breakdown of the distribution of mass within a hadron.

4 Conclusion

Explanations of the origin of a hadron's mass and its distribution within that state depend on the observer's preferred frame of reference and resolving scale. At a scale typical of contemporary continuum- and lattice-QCD ($\zeta = \zeta_2 := 2\,\text{GeV}$), the dynamical chiral symmetry breaking (DCSB) paradigm provides an excellent and intuitive method to explicate and understand the associated, emergent phenomena. This perspective leads to numerous predictions that can be tested at contemporary hadron physics facilities, related e.g. to hadron elastic and transition form factors, and a diverse array of parton distribution amplitudes and functions.

At this point it is reasonable to enquire after the possible impact of new facilities, such as an electron ion collider (EIC), on our understanding of the emergence of mass in the Standard Model. In response, one should first note that an EIC will enable access to a wide array of phenomena within the valence-quark domain,

---

[7] Notably, if dealing with exotic or hybrid bound-states possessing a valence-gluon degree-of-freedom, then the $\mathcal{D}_1$ term in Eq. (19) would also include an explicit and positive gauge-invariant contribution associated with the dynamical generation of a dressed-gluon mass. Moreover, such cancellations are a typical feature of bound-state problems, apparent already in both rudimentary and sophisticated uses of quantum mechanics [81].

and, consequently, the predictions made for contemporary facilities are equally relevant for an EIC and might there be better be addressed. In addition, hadron tomography will be a major focus of an EIC; and the DCSB paradigm has a great deal to say about that subject, providing a means for the QCD-connected computation of the pointwise behaviour of generalised and transverse-momentum-dependent parton distributions. Finally, a major focus of an EIC will be low-x. Gluons dominate in this region; but they also acquire a dynamically-generated mass. That is likely to have a significant impact, e.g. in connection with gluon saturation phenomena. The challenge for theory now, therefore, is to identify those observables which are most sensitive to the emergence of mass as described above. There is time; but also some urgency.

Acknowledgements I would like to thank Z.-E. Meziani and J.-W. Qiu for organising the Temple Workshop: *The Proton Mass: At the heart of most visible matter*, and they and the participants for hours of engaging discussions, which served to refocus my attention on some of the issues discussed herein. Constructive comments were subsequently received from I. C. Cloët, R. J. Holt, V. Mokeev, J. Papavassiliou and J. Rodríguez-Quintero, and are also gratefully acknowledged. Work supported by: the U.S. Department of Energy, Office of Science, Office of Nuclear Physics, under Contract No. DE-AC02-06CH11357.

References

1. F.J. Belinfante, On the current and the density of the electric charge, the energy, the linear momentum and the angular momentum of arbitrary fields. Physica **7**, 449–474 (1940)
2. Y. Nakayama, Scale invariance vs conformal invariance. Phys. Rept. **569**, 1–93 (2015)
3. J.D. Qualls, *Lectures on Conformal Field Theory*. arXiv:1511.04074 [hep-ph] (2015)
4. K.A. Olive et al., Review of particle physics. Chin. Phys. C **38**, 090001 (2014)
5. R.J. Gonsalves, Anomalies in quantum field theory. www.physics.buffalo.edu/gonsalves/phy522/. Lecture 8 (2009)
6. C.D. Roberts, S.M. Schmidt, Dyson–Schwinger equations: density, temperature and continuum strong QCD. Prog. Part. Nucl. Phys. **45**, S1–S103 (2000)
7. P.E.L. Rakow, Renormalization group flow in QED: an investigation of the Schwinger–Dyson equations. Nucl. Phys. B **356**, 27–45 (1991)
8. M. Reenders, On the nontriviality of Abelian gauged Nambu–Jona–Lasinio models in four dimensions. Phys. Rev. D **62**, 025001 (2000)
9. F. Akram, A. Bashir, L.X. Gutierrez-Guerrero, B. Masud, J. Rodriguez-Quintero et al., Vacuum polarization and dynamical chiral symmetry breaking: phase diagram of QED with four-fermion contact interaction. Phys. Rev. D **87**, 013011 (2013)
10. R. Tarrach, The renormalization of FF. Nucl. Phys. B **196**, 45–61 (1982)
11. J.R. Finger, J.E. Mandula, Quark pair condensation and chiral symmetry breaking in QCD. Nucl. Phys. B **199**, 168 (1982)
12. S.L. Adler, A.C. Davis, Chiral symmetry breaking in Coulomb gauge QCD. Nucl. Phys. B **244**, 469 (1984)
13. D. Kharzeev, Quarkonium interactions in QCD. Proc. Int. Sch. Phys. Fermi **130**, 105–131 (1996)
14. X.-D. Ji, A QCD analysis of the mass structure of the nucleon. Phys. Rev. Lett. **74**, 1071–1074 (1995)
15. X.-D. Ji, Breakup of hadron masses and energy–momentum tensor of QCD. Phys. Rev. D **52**, 271–281 (1995)
16. F. Coester, Null plane dynamics of particles and fields. Prog. Part. Nucl. Phys. **29**, 1–32 (1992)
17. K. Hornbostel, S.J. Brodsky, H.C. Pauli, Phys. Rev. D **41**, 3814 (1990)
18. A.C. Aguilar, D. Binosi, J. Papavassiliou, The gluon mass generation mechanism: a concise primer. Front. Phys. China **11**, 111203 (2016)
19. S.J. Brodsky, H.-C. Pauli, S.S. Pinsky, Quantum chromodynamics and other field theories on the light cone. Phys. Rept. **301**, 299–486 (1998)
20. V.N. Gribov, L.N. Lipatov, Deep inelastic electron scattering in perturbation theory. Phys. Lett. B **37**, 78–80 (1971)
21. V.N. Gribov, L.N. Lipatov, Deep inelastic ep scattering in perturbation theory. Sov. J. Nucl. Phys. **15**, 438–450 (1972)
22. Y.L. Dokshitzer, Calculation of the structure functions for deep inelastic scattering and e+ e− annihilation by perturbation theory in quantum chromodynamics. Sov. Phys. JETP **46**, 641–653 (1977). (in Russian)
23. G. Altarelli, G. Parisi, Asymptotic freedom in parton language. Nucl. Phys. B **126**, 298 (1977)
24. G.P. Lepage, S.J. Brodsky, Exclusive processes in quantum chromodynamics: evolution equations for hadronic wave functions and the form-factors of mesons. Phys. Lett. B **87**, 359–365 (1979)
25. A.V. Efremov, A.V. Radyushkin, Factorization and asymptotical behavior of pion form-factor in QCD. Phys. Lett. B **94**, 245–250 (1980)
26. G.P. Lepage, S.J. Brodsky, Exclusive processes in perturbative quantum chromodynamics. Phys. Rev. D **22**, 2157–2198 (1980)
27. K.G. Wilson, T.S. Walhout, A. Harindranath, W.-M. Zhang, R.J. Perry, S.D. Glazek, Nonperturbative QCD: a weak coupling treatment on the light front. Phys. Rev. D **49**, 6720–6766 (1994)
28. S.J. Brodsky, G.F. de Teramond, H.G. Dosch, J. Erlich, Light-front holographic QCD and emerging confinement. Phys. Rep. **584**, 1–105 (2015)
29. I.C. Cloët, C.D. Roberts, Explanation and prediction of observables using continuum strong QCD. Prog. Part. Nucl. Phys. **77**, 1–69 (2014)
30. C.D. Roberts, Three lectures on hadron physics. J. Phys. Conf. Ser. **706**, 022003 (2016)
31. T. Horn, C.D. Roberts, The pion: an enigma within the Standard Model. J. Phys. G **43**, 073001/1–073001/47 (2016)
32. D. Binosi, L. Chang, J. Papavassiliou, C.D. Roberts, Bridging a gap between continuum-QCD and ab initio predictions of hadron observables. Phys. Lett. B **742**, 183–188 (2015)

33. M. Bhagwat, M. Pichowsky, C. Roberts, P. Tandy, Analysis of a quenched lattice QCD dressed quark propagator. Phys. Rev. C **68**, 015203 (2003)
34. M.S. Bhagwat, P.C. Tandy, Analysis of full-QCD and quenched-QCD lattice propagators. AIP Conf. Proc. **842**, 225–227 (2006)
35. P.O. Bowman et al., Unquenched quark propagator in Landau gauge. Phys. Rev. D **71**, 054507 (2005)
36. A.M. Jaffe, The millennium grand challenge in mathematics. Not. Am. Math. Soc. **53**, 652–660 (2006)
37. C. McNeile, Lattice status of gluonia/glueballs. Nucl. Phys. Proc. Suppl. **186**, 264–267 (2009)
38. K.G. Wilson, Confinement of quarks. Phys. Rev. D **10**, 2445–2459 (1974)
39. G.S. Bali, H. Neff, T. Duessel, T. Lippert, K. Schilling, Observation of string breaking in QCD. Phys. Rev. D **71**, 114513 (2005)
40. Z. Prkacin et al., Anatomy of string breaking in QCD. PoS **LAT2005**, 308 (2006)
41. L. Chang, I.C. Cloët, B. El-Bennich, T. Klähn, C.D. Roberts, Exploring the light-quark interaction. Chin. Phys. C **33**, 1189–1196 (2009)
42. A. Tang, J.W. Norbury, Properties of Regge trajectories. Phys. Rev. D **62**, 016006 (2000)
43. P. Masjuan, E. Ruiz Arriola, W. Broniowski, Systematics of radial and angular-momentum Regge trajectories of light non-strange $q\bar{q}$-states. Phys. Rev. D **85**, 094006 (2012)
44. J. Glimm, A. Jaffee, *Quantum Physics. A Functional Point of View* (Springer, New York, 1981)
45. M. Stingl, Propagation properties and condensate formation of the confined Yang-Mills field. Phys. Rev. D **34**, 3863–3881 (1986). [Erratum: Phys. Rev.D36,651(1987)]
46. C.D. Roberts, A.G. Williams, G. Krein, On the implications of confinement. Int. J. Mod. Phys. A **7**, 5607–5624 (1992)
47. F.T. Hawes, C.D. Roberts, A.G. Williams, Dynamical chiral symmetry breaking and confinement with an infrared vanishing gluon propagator. Phys. Rev. D **49**, 4683–4693 (1994)
48. C.D. Roberts, A.G. Williams, Dyson–Schwinger equations and their application to hadronic physics. Prog. Part. Nucl. Phys. **33**, 477–575 (1994)
49. G. Krein, M. Nielsen, R.D. Puff, L. Wilets, Ghost poles in the nucleon propagator: vertex corrections and form-factors. Phys. Rev. C **47**, 2485–2491 (1993)
50. M.E. Bracco, A. Eiras, G. Krein, L. Wilets, Selfconsistent solution of the Schwinger–Dyson equations for the nucleon and meson propagators. Phys. Rev. C **49**, 1299–1308 (1994)
51. A. Aguilar, D. Binosi, J. Papavassiliou, Gluon and ghost propagators in the Landau gauge: deriving lattice results from Schwinger–Dyson equations. Phys. Rev. D **78**, 025010 (2008)
52. A. Aguilar, D. Binosi, J. Papavassiliou, J. Rodríguez-Quintero, Non-perturbative comparison of QCD effective charges. Phys. Rev. D **80**, 085018 (2009)
53. P. Boucaud, J.P. Leroy, A. Le-Yaouanc, J. Micheli, O. Pene, J. Rodríguez-Quintero, The infrared behaviour of the pure Yang–Mills green functions. Few Body Syst. **53**, 387–436 (2012)
54. M.R. Pennington, D.J. Wilson, Are the dressed gluon and ghost propagators in the landau gauge presently determined in the confinement regime of QCD? Phys. Rev. D **84**, 119901 (2011)
55. A. Ayala, A. Bashir, D. Binosi, M. Cristoforetti, J. Rodriguez-Quintero, Quark flavour effects on gluon and ghost propagators. Phys. Rev. D **86**, 074512 (2012)
56. M. Stingl, A systematic extended iterative solution for quantum chromodynamics. Z. Phys. A **353**, 423–445 (1996)
57. A. Accardi, F. Arleo, W.K. Brooks, D. D'Enterria, V. Muccifora, Parton propagation and fragmentation in QCD matter. Riv. Nuovo Cim. **32**, 439–553 (2010)
58. J. Dudek et al., Physics opportunities with the 12 GeV upgrade at Jefferson Lab. Eur. Phys. J. A **48**, 187 (2012)
59. A. Accardi et al., Electron ion collider: The next QCD Frontier–understanding the glue that binds us all. Eur. Phys. J. A. **52**(9), 268 (2016). doi:10.1140/epja/i2016-16268-9
60. P. Maris, C.D. Roberts, P.C. Tandy, Pion mass and decay constant. Phys. Lett. B **420**, 267–273 (1998)
61. S.J. Brodsky, C.D. Roberts, R. Shrock, P.C. Tandy, Confinement contains condensates. Phys. Rev. C **85**, 065202 (2012)
62. M.S. Bhagwat, L. Chang, Y.-X. Liu, C.D. Roberts, P.C. Tandy, Flavour symmetry breaking and meson masses. Phys. Rev. C **76**, 045203 (2007)
63. V.V. Flambaum et al., Sigma terms of light-quark hadrons. Few Body Syst. **38**, 31–51 (2006)
64. A. Höll, P. Maris, C.D. Roberts, S.V. Wright, Schwinger functions and light-quark bound states, and sigma terms. Nucl. Phys. Proc. Suppl. **161**, 87–94 (2006)
65. I.C. Cloë, C.D. Roberts, Form factors and Dyson–Schwinger equations. PoS **LC2008**, 047 (2008)
66. P.E. Shanahan, A.W. Thomas, R.D. Young, Sigma terms from an SU(3) chiral extrapolation. Phys. Rev. D **87**, 074503 (2013)
67. G.S. Bali, S. Collins, D. Richtmann, A. Schäfer, W. Sldner, A. Sternbeck, Direct determinations of the nucleon and pion σ terms at nearly physical quark masses. Phys. Rev. D **93**, 094504 (2016)
68. M.A. Ivanov, YuL Kalinovsky, C.D. Roberts, Survey of heavy-meson observables. Phys. Rev. D **60**, 034018 (1999)
69. B. El-Bennich, M.A. Paracha, C.D. Roberts, E. Rojas, Couplings between the ρ and D- and D^*-mesons (2016). arXiv:1604.01861 [nucl-th]
70. S.-X. Qin, C.D. Roberts, S.M. Schmidt, Ward–Green–Takahashi identities and the axial-vector vertex. Phys. Lett. B **733**, 202–208 (2014)
71. P. Maris, C.D. Roberts, π and K meson Bethe-Salpeter amplitudes. Phys. Rev. C **56**, 3369–3383 (1997)
72. B.L. Li, L. Chang, F. Gao, C.D. Roberts, S.M. Schmidt, H.S. Zong, Distribution amplitudes of radially-excited π and K mesons. Phys. Rev. D **93**, 114033 (2016)
73. M. Burkardt, Impact parameter dependent parton distributions and off forward parton distributions for $\zeta > 0$. Phys. Rev. D **62**, 071503 (2000)
74. M. Diehl, T. Feldmann, R. Jakob, P. Kroll, The overlap representation of skewed quark and gluon distributions. Nucl. Phys. B **596**, 33–65 (2001)
75. M. Burkardt, Impact parameter space interpretation for generalized parton distributions. Int. J. Mod. Phys. A **18**, 173–208 (2003)

76. M. Diehl, The overlap representation of skewed quark and gluon distributions. Phys. Rep. **388**, 41–277 (2003)
77. D. Binosi, L. Chang, J. Papavassiliou, S.-X. Qin, C.D. Roberts, Symmetry preserving truncations of the gap and Bethe–Salpeter equations. Phys. Rev. D **93**, 096010 (2016)
78. H.J. Munczek, Dynamical chiral symmetry breaking, Goldstone's theorem and the consistency of the Schwinger–Dyson and Bethe–Salpeter equations. Phys. Rev. D **52**, 4736–4740 (1995)
79. A. Bender, C.D. Roberts, L. von Smekal, Goldstone theorem and diquark confinement beyond rainbow-ladder approximation. Phys. Lett. B **380**, 7–12 (1996)
80. L. Chang, C.D. Roberts, Sketching the Bethe–Salpeter kernel. Phys. Rev. Lett. **103**, 081601 (2009)
81. J. Carlson, S. Gandolfi, F. Pederiva, S.C. Pieper, R. Schiavilla, K.E. Schmidt, R.B. Wiringa, Quantum Monte Carlo methods for nuclear physics. Rev. Mod. Phys. **87**, 1067 (2015)
82. Y. Nambu, G. Jona-Lasinio, Dynamical model of elementary particles based on an analogy with superconductivity. 1. Phys. Rev. **122**, 345–358 (1961)
83. H.L.L. Roberts, C.D. Roberts, A. Bashir, L.X. Gutiérrez-Guerrero, P.C. Tandy, Abelian anomaly and neutral pion production. Phys. Rev. C **82**, 065202 (2010)
84. H.L.L. Roberts, L. Chang, I.C. Cloët, C.D. Roberts, Masses of ground and excited-state hadrons. Few Body Syst. **51**, 1–25 (2011)
85. G. Eichmann, R. Alkofer, I.C. Cloët, A. Krassnigg, C.D. Roberts, Perspective on rainbow-ladder truncation. Phys. Rev. C **77**, 042202(R) (2008)
86. G. Eichmann, I.C. Cloët, R. Alkofer, A. Krassnigg, C.D. Roberts, Toward unifying the description of meson and baryon properties. Phys. Rev. C **79**, 012202(R) (2009)

⧄ Springer

Few-Body Syst (2017) 58:1–11
DOI 10.1007/s00601-016-1172-3

Bernard L. G. Bakker · Chueng-Ryong Ji

The Construction of Compton Tensors in Scalar QED

Received: 31 August 2016 / Accepted: 2 November 2016 / Published online: 9 December 2016

Abstract Current conservation is a vital condition in electrodynamics. We review the literature concerning the ways to ensure that the formalism used in calculating amplitudes for the scattering of charged particles is in compliance with current conservation. For the case of electron scattering off a scalar and a spin-1/2 target as well as Compton scattering on a scalar target, we present some novelties besides reviewing the literature.

1 Introduction

The conservation of the electric current is a corner stone of electrodynamics, both classical and quantized. Textbooks like Refs. [1] and [2], to mention only two out of a plethora of references, write down the continuity equation

$$\partial_\mu J^\mu = 0, \tag{1}$$

which in the case of classical electrodynamics expresses the fact that in any volume the time derivative of the charge inside this volume is opposite to the divergence of the current flowing out through the surface. Noether's theorem states that in general such relations as Eq. (1) point to a conserved quantity, which of course in electrodynamics is the total charge.

In this paper we review some aspects of the construction of tensorial operators like the current J^μ in QED, which lead in the case of interacting theories to the concept of form factors. Naturally, once this concept is introduced, the question arises what is the number of independent form factors in each particular case. We shall not answer this question in general, but rather illustrate some methods used in three comparatively simple cases, namely the current of a charged particle in scalar QED (sQED) and spinor QED, and the construction of the Compton tensor in sQED.

While most of this paper is concerned with the existing literature, we add some novel results, that may help to gain a new perspective on the standard approaches.

This article belongs to the special issue "30th anniversary of Few-Body Systems".

B. L. G. Bakker (✉)
Department of Physics and Astrophysics, Vrije Universiteit, Amsterdam, The Netherlands
E-mail: b.l.g.bakker@vu.nl

C.-R. Ji
Department of Physics, North Carolina State University, Raleigh, NC 27695-8202, USA
E-mail: crji@ncsu.edu

2 Electron Scattering Off a Scalar Target

We start our discussion with the electron scattering off a scalar target to introduce the idea of parametrising the amplitude in terms of a tensor built with the help of the relevant momenta. In the case of a scalar target, three momenta are involved, the momenta p and p' of the target before and after absorption of a virtual photon, and the momentum q of the latter. Because of four-momentum conservation, $p + q = p'$, only two out of three are independent. Dropping in this paper everywhere the magnitude of the charge of the target, which can easily be inserted in the photon-hadron vertices, we write the current operator as follows:

$$J^\mu = F_1 \, p^\mu + F_2 \, p'^\mu. \tag{2}$$

Implementing the transversality condition Eq. (1), we find, using $q = p' - p$

$$0 = q_\mu J^\mu = F_1 \, p \cdot (p' - p) + F_2 \, p' \cdot (p' - p). \tag{3}$$

Using the on-mass-shell condition $p'^2 = p^2 = M^2$ we find the relation $(F_1 - F_2)(p' \cdot p - M^2) = (F_1 - F_2)q^2/2 = 0$. Because the photon can only be real, i. e. $q^2 = 0$, for the special kinematics where $p' = p$, we see that $F_1 = F_2$ almost everywhere in the kinematic space. The kinematical space for this scattering process is the three-dimensional space of independent dot products p'^2, p^2, and $p' \cdot p$.

Thus the final form of the electromagnetic current of a scalar particle can be written as

$$J^\mu = F_S \, (p'^\mu + p^\mu) \equiv F_S \, \overline{P}^\mu. \tag{4}$$

We may call this the manifestly transverse form of the current, because the transversality condition is immediately seen to be satisfied by the momentum structure: $q \cdot \overline{P} = (p' - p).(p' + p) = p'^2 - p^2 = M^2 - M^2 = 0$. The same purpose is achieved if one writes the current in terms of a projector, namely in the form

$$J^\mu = \left(g^{\mu\nu} - \frac{q^\mu q^\nu}{q^2} \right)(F_S \overline{P}_\nu + F_A q_\nu) \equiv \mathcal{G}^{\mu\nu}(qq)(F_S \overline{P}_\nu + F_A q_\nu). \tag{5}$$

The projector $\mathcal{G}^{\mu\nu}(qq)$ annihilates q^μ and q^ν, which property guarantees the transversality of the current and removes the form factor F_A. Thus an equivalent form of the current operator would be

$$J^\mu = F_S \, \mathcal{G}^{\mu\nu}(qq)\overline{P}_\nu = F_S \left(\overline{P}^\mu - \frac{\overline{P} \cdot q}{q^2} q^\mu \right). \tag{6}$$

The two forms, Eqs. (4) and (6), are equivalent in the part of the kinematical space where $q^2 \neq 0$, but not in the singular subspace where q^2 vanishes. Such situations will be seen in the Compton-scattering case too. Let us note that although the vector part may vary, the number of form factors (FFs) is invariant.

If the target is not structureless, but can be excited, the argument for a single form factor does not apply anymore: there must exist an elastic form factor plus transition form factors. For the purpose of this paper we shall only consider the situation where p' and p are on the same mass shell.

We shall later use the same notation, $\overline{P} = p' + p$, when we discuss Compton scattering.

3 Electron Scattering Off a Spinor Target

Was the form of the electromagnetic current of a scalar particle almost unique, for a spin-1/2 particle it is not. In fact, one can write it in different forms. The text books usually give two forms of the current, namely one with the Dirac and Pauli FFs [3]

$$J^\mu = \gamma^\mu \, F_1 + i \frac{\sigma^{\mu\nu} q_\nu}{2M} \, F_2. \tag{7}$$

Here the normalisation of F_2 is such that $F_2(q^2 = 0)$ is equal to the anomalous magnetic moment.

Ernst et al. [4] introduced a redefinition of the FFs, namely

$$G_E = F_1 + \frac{q^2}{4M^2} F_2, \quad G_M = F_1 + F_2, \tag{8}$$

where G_E and G_M are the electric and magnetic FFs [5]. Such a redefinition is equivalent to reshuffling the operators in the expression for the current. In the derivation of the current operator, Bjorken and Drell [3] allow for three terms, namely $F_1 p'^\mu + F_2 p^\mu + F_3 \gamma^\mu$ and show that using the on-shellness of the momenta p' and p and current conservation, that the operator could be reduced to $F_1(p'^\mu + p^\mu) + F_3 \gamma^\mu$. Now a key ingredient is the Gordon decomposition [6]:

$$\bar{u}(p')\gamma^\mu u(p) = \frac{1}{2M}\bar{u}(p')((p' + p)^\mu + i\sigma^{\mu\nu}q_\nu)u(p). \tag{9}$$

Besides this classic form, one may also get the following decomposition [7]:

$$\bar{u}(p')i\sigma^{\mu\nu}q_\nu u(p) = \frac{1}{2M}\bar{u}(p')(q^2\gamma^\mu + 2i\epsilon^{\mu\nu\alpha\beta}\gamma_5\gamma_\nu p'_\alpha p_\beta)u(p). \tag{10}$$

Using both decompositions we altogether arrive at six different forms of the spin-1/2 electromagnetic current operator:

$$
\begin{aligned}
J^\mu &= \gamma^\mu F_1 + i\frac{\sigma^{\mu\nu}q_\nu}{2M} F_2, \\
&= \gamma^\mu (F_1 + F_2) - \frac{(p' + p)^\mu}{2M} F_2, \\
&= \frac{(p' + p)^\mu}{2M}\frac{4M^2 F_1 + q^2 F_2}{4M^2 - q^2} + i\epsilon^{\mu\nu\alpha\beta}\gamma_5\gamma_\nu p'_\alpha p_\beta \frac{2(F_1 + F_2)}{4M^2 - q^2}, \\
&= \frac{(p' + p)^\mu}{2M} F_1 + i\frac{\sigma^{\mu\nu}q_\nu}{2M} i(F_1 + F_2), \\
&= \gamma^\mu \left(F_1 + \frac{q^2}{4M^2} F_2\right) + i\epsilon^{\mu\nu\alpha\beta}\gamma_5\gamma_\nu p'_\alpha p_\beta \frac{F_2}{2M^2}, \\
&= i\frac{\sigma^{\mu\nu}q_\nu}{2M} \left(\frac{4M^2}{q^2} F_1 + F_2\right) - i\epsilon^{\mu\nu\alpha\beta}\gamma_5\gamma_\nu p'_\alpha p_\beta \frac{2F_1}{q^2}.
\end{aligned}
\tag{11}
$$

Several remarks must be made. First, these forms are not identical, but give identical matrix elements between spinors, which in operator theory is defined as weak identity. Secondly, neither of the two decompositions Eqs. (9) and (10) can be derived for massless spinors, as the occurrences of an M in the denominators in these equations show. Thirdly, some of these forms are manifestly transverse to q_μ, namely the third, fourth and sixth one. Transversality is trivially proven by the identity we used before, $\bar{P} \cdot q = 0$, and taking into account that $\sigma^{\mu\nu}$ as well as $\epsilon^{\mu\nu\alpha\beta}$ are antisymmetric in all indices.

4 Compton Scattering Off a Scalar Target

Real Compton scattering has been an interesting phenomenon since its interpretation by Compton [8] as well as Debye [9] played an important role in the reception of the light-quanta hypothesis [10] before quantum mechanics was established. Presently, virtual Compton scattering [11,12] has been proposed to determine the generalized-parton distributions (GPDs) of hadrons. In this interpretation, the handbag diagram, as shown in Fig. 1, plays a crucial role. In a more general setting, one considers the hadronic part of the amplitude, which

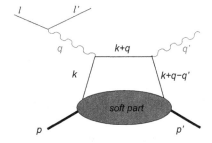

Fig. 1 Handbag diagram, including the leptonic part

describes the absorption of a virtual photon and the emission of a real photon by a hadron, without committing oneself from the beginning to the structure of the hadron except for its mass and spin, remaining oblivious to its (quark) constituents.

In this paper we shall remain on that level, discussing only the number of independent scalar functions, Compton form factors (CFFs) that can be distinguished, like the form factors in the case of the vector currents discussed in the previous sections.

4.1 Tensor Formalism

In (virtual) Compton scattering the physical invariant amplitudes are written as contractions of a tensor with the polarization vectors of the photons:

$$A(q', h'; q, h) = \epsilon_\mu^*(q'; h') T^{\mu\nu} \epsilon_\nu(q; h). \tag{12}$$

The tensor $T^{\mu\nu}$ must satisfy two transversality conditions, which are related to current conservation or gauge invariance, namely

$$q'_\mu T^{\mu\nu} = 0, \quad T^{\mu\nu} q_\nu = 0. \tag{13}$$

One may connect these relations also to the invariance of the amplitudes $A(q', h'; q, h)$ under Lorentz transformations. As discussed in Ref. [2], Ch. 8, the vector potential A^μ in QED transforms under a Lorentz transformation as a four vector only up to a gauge transformation. In the present context it means that under a Lorentz transformation the polarisation vector $\epsilon^\mu(q; h)$ changes into the formally Lorentz transformed vector $\Lambda^\mu_\nu \epsilon^\nu(q; h)$ plus a multiple of the momentum vector q^μ. Thus the invariance of the amplitude is guaranteed by the transversality of the Compton tensor.

Two problems now present themselves: how to determine the number of effective degrees of freedom, $i.e.$ the number of CFFs, and how to construct a tensor that satisfies the transversality conditions Eq. (13)? We shall sketch several ways to solve these two problems. The first one, which to the best of our knowledge was not mentioned in the physics literature, is based on a straightforward application of linear algebra. A number of constructions has been given before and we shall review the ones that are essentially different.

As a preliminary we now define our notation. To start with, we use $q(q')$ for the absorbed (emitted) photons and $p(p')$ for the target (recoiled) hadron. Because of four-momentum conservation, three out of these four momenta are independent. We shall choose the following three:

$$k_1 = \overline{P} = p' + p, \quad k_2 = q', \quad k_3 = q. \tag{14}$$

For coherent Compton scattering, where the recoiled hadron is identical to the target, one easily derives the following identities

$$
\begin{aligned}
2\overline{P} \cdot q &= 2\overline{P} \cdot q' = s - M^2 - (u - M^2), \\
(\overline{P} + q) \cdot q' &= (\overline{P} + q') \cdot q = s - M^2, \\
(\overline{P} - q) \cdot q' &= (\overline{P} - q') \cdot q = -(u - M^2), \\
2q' \cdot q &= s - M^2 + u - M^2,
\end{aligned}
\tag{15}
$$

where s and u are the usual Mandelstam variables $s = (p + q)^2 = (p' + q')^2$ and $u = (p - q')^2 = (p' - q)^2$.

Using the notation k_i, $i = 1, 2, 3$ for the momenta, we can write for the most general tensor of rank two in the following form

$$T^{\mu\nu} = t_0 g^{\mu\nu} + \sum_{i=1}^{3} \sum_{j=1}^{3} t_{ij} K_{ij}^{\mu\nu} \tag{16}$$

with

$$K_{ij}^{\mu\nu} = k_i^\mu k_j^\nu. \tag{17}$$

The ten quantities $t_0, t_{11}, \ldots t_{33}$ are scalar quantities. If no conditions are set for the tensor $T^{\mu\nu}$, we see that it has ten degrees of freedom. Thus, instead of the representation given in Eq. (16) we could, interpreting the set $g^{\mu\nu}$, $K_{11}^{\mu\nu}, \ldots, K_{33}^{\mu\nu}$ as a basis in a 10-dimensional linear space, interpret the tensor as the contraction of the vector t with scalar elements $t_0, t_{11}, \ldots, t_{33}$ with the vector of basis tensors $g^{\mu\nu}, K_{11}^{\mu\nu}, \ldots, K_{33}^{\mu\nu}$. This interpretation lends itself immediately to the application of the theory of vector spaces, which is the content of what we call the direct method.

4.1.1 Direct Method

Understanding Eq. (16) as expressing the full tensor as a linear combination of basis tensors leads to the treatment of the transversality conditions as conditions on the vector t. Let us write the contraction $q'_\mu T^{\mu\nu}$ as follows

$$k_{2\mu} T^{\mu\nu} = t_0 k_2^\nu + \sum_{i,j} t_{ij} (k_2 \cdot k_i) k_j^\nu. \tag{18}$$

Similarly, we write the other contraction as

$$T^{\mu\nu} k_{3\nu} = t_0 k_3^\mu + \sum_{i,j} t_{ij} (k_3 \cdot k_j) k_i^\mu. \tag{19}$$

The condition that these two contractions vanish, leads to two sets of three linear equations in the components of t. We shall use the notation

$$x_{ij} = k_i \cdot k_j. \tag{20}$$

From Eq. (15) we read off that $\overline{P} \cdot q' = x_{12} = x_{13} = \overline{P} \cdot q$. Taking into account that the basis vectors k_i are independent four vectors, we find the equations in the following form

$$X \cdot t = 0, \tag{21}$$

where the matrix X is given by

$$X = \begin{pmatrix} 0 & x_{13} & 0 & 0 & x_{22} & 0 & 0 & x_{23} & 0 & 0 \\ 1 & 0 & x_{13} & 0 & 0 & 0 & x_{22} & 0 & x_{23} & 0 \\ 0 & 0 & 0 & x_{13} & 0 & 0 & x_{22} & 0 & 0 & x_{23} \\ 0 & x_{13} & x_{23} & x_{33} & 0 & 0 & 0 & 0 & 0 & 0 \\ 0 & 0 & 0 & x_{13} & x_{23} & x_{23} & 0 & 0 & 0 \\ 1 & 0 & 0 & 0 & 0 & 0 & 0 & x_{13} & x_{23} & x_{33} \end{pmatrix}. \tag{22}$$

Now the two problems can be formulated as follows: what is the matrix rank of X and what is the null-space of this matrix? The answer to the first question gives the number of independent scalars, *i.e.* the number of CFFs, and the answer to the second question gives a way to write down the general form of a rank-two tensor $\tilde{T}^{\mu\nu}$ that is transverse to q' (left) and q (right).

The first question can be answered in two ways. The first one is a straightforward application of Gaussian elimination to find that the matrix rank of X is five. Indeed one finds that the following linear combination of the rows of X vanishes

$$\bar{P} \cdot q\, \text{row}_1 + q' \cdot q\, \text{row}_2 + q^2\, \text{row}_3 - \bar{P} \cdot q\, \text{row}_4 - q'^2\, \text{row}_5 - q' \cdot q\, \text{row}_6 = 0 \tag{23}$$

The second one is to contract $T^{\mu\nu}$ simultaneously with $k_{i\mu}$ and $k_{j\nu}$ and to find six contractions to vanish, namely the ones involving $k_{2\mu}$ or $k_{3\nu}$. In this set the double contraction with $k_{2\mu}$ and $k_{3\nu}$ occurs twice, leaving at most five independent ones.

Now that we know that only five scalars are involved in a general tensor that is transverse to q'_μ and q_ν, we can write such a tensor as a linear combination of a basis in the five-dimensional null-space $\mathcal{N}_X$ of the matrix X. Let us note here that while the matrix X is uniquely defined and thus so is its null-space, one may use any basis one likes in $\mathcal{N}_X$. This is the reason why there exist in the literature several forms of the Compton tensor.

A final point to consider is the issue of kinematical singularities. The coefficients in the linear combination Eq. (23) may vanish. At this point the most interesting ones are $q'^2 = 0$ and $q^2 = 0$. If both conditions apply, we are dealing with real Compton scattering (RCS), otherwise the tensor $T^{\mu\nu}$ is used to calculate the amplitudes for virtual Compton scattering (VCS) or even electro-production of vector states. Calculating the matrix rank of X one finds that only either in the case that all coefficients in Eq. (23) vanish or in the case that all but q^2 or q'^2 vanish, the ranks is lowered to either 1 or 4, respectively. Otherwise it remains five. However, we note that the simultaneous vanishing of all scalar products x_{ij} is kinematically forbidden. To see why, we write them down and calculate them explicitly:

$$\begin{aligned} x_{13} &= \overline{P} \cdot q' = \overline{P} \cdot q = 0, \\ x_{22} &= q'^2 = 0, \\ x_{23} &= q' \cdot q = 0, \\ x_{33} &= q^2 = 0. \end{aligned} \tag{24}$$

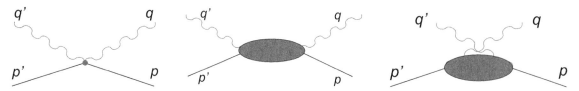

Fig. 2 Seagull, s- and u-channel amplitudes

We may thus write $q'^\mu = q'(1, \hat{q}')$, $q^\mu = q(1, \hat{q})$ and find that $x_{23} = 0$ implies that $\hat{q}' = \hat{q}$. Let us write $\overline{P}^\mu = (P^0, \boldsymbol{P})$. For x_{13} to vanish two conditions must be satisfied: (i) $\boldsymbol{P}$ is parallel to $\boldsymbol{q}'$ and (ii) P^0 must be equal to the length of $\boldsymbol{P}$. While the first condition is kinematically possible, the latter is not, because it means that $\overline{P}$ is light-like, which it is not. Substituting $p'^2 = p^2 = M^2$ and using $t = (p' - p)^2 = 2M^2 - 2p' \cdot p = (q - q')^2 = q'^2 + q^2 - 2q' \cdot q$ we find

$$\overline{P}^2 = (p' + p)^2 = 2M^2 + 2p' \cdot p = 4M^2 - t = 4M^2, \tag{25}$$

where we have used $x_{22} = x_{23} = x_{33} = 0$. Thus because $\overline{P}$ cannot be light-like, the matrix X has for all physical situations and for finite target mass M matrix rank five.

Although we are now assured that the number of CFFs is five, it does not mean that the null-spaces of X in the case that one or several dot products x_{ij} vanish must be identical. For illustrative reasons we show a basis of the null-space for the general case, substituting the dot products of physical momenta:

$$\begin{pmatrix} n^{(1)} \\ n^{(2)} \\ n^{(3)} \\ n^{(4)} \\ n^{(5)} \end{pmatrix} = \begin{pmatrix} -q'^2 & 0 & \frac{q^2}{\overline{P}\cdot q} & -\frac{q'\cdot q}{\overline{P}\cdot q} & 0 & 0 & 0 & 0 & 0 & 1 \\ -q'\cdot q & 0 & 0 & 0 & 0 & 0 & 0 & 0 & 1 & 0 \\ -\overline{P}\cdot q & -\frac{q'\cdot q}{\overline{P}\cdot q} & 1 & 0 & 0 & 0 & 0 & 1 & 0 & 0 \\ 0 & \frac{q^2 q'^2}{(\overline{P}\cdot q)^2} & 0 & -\frac{q'^2}{\overline{P}\cdot q} & -\frac{q^2}{\overline{P}\cdot q} & 0 & 1 & 0 & 0 & 0 \\ 0 & \frac{q'\cdot q\, q'^2}{(\overline{P}\cdot q)^2} & -\frac{q^2}{\overline{P}\cdot q} & 0 & -\frac{q'\cdot q}{\overline{P}\cdot q} & 1 & 0 & 0 & 0 & 0 \end{pmatrix}. \tag{26}$$

Clearly, this basis of $\mathcal{N}_X$ demonstrates that the value $\overline{P} \cdot q = 0$ is a singular point in the null-space of X. It would not help to multiply the basis vectors with a power of $\overline{P} \cdot q$ to remove the singularity, because the resulting basis, while still independent for $\overline{P} \cdot q \neq 0$, would become incomplete. In the latter situation one can determine a new basis for $\mathcal{N}_X$, that would again be five-dimensional. However, this time the basis will become singular if $q'^2 = 0$ or $q' \cdot q$ would vanish. This is an illustration of a conclusion already formulated in a CERN report by Callan and Stora [11,12], that it is not possible to construct a basis for $\mathcal{N}_X$ that spans the whole null-space for all physical values of the dot products x_{ij} and is free of kinematic singularities. We shall see that the other methods will meet the same problem.

Finally we illustrate the construction of the Compton tensor described in this section by writing it for the tree-level case and show how it is expanded in the given basis. We shall use this example to illustrate also the constructions reviewed below. The tree level $T^{\mu\nu}$ is given by

$$T_{\text{tree}}^{\mu\nu} = -2g^{\mu\nu} + \frac{(\overline{P} + q)^\mu (\overline{P} + q')^\nu}{s - M^2} + \frac{(\overline{P} - q)^\mu (\overline{P} - q')^\nu}{u - M^2}. \tag{27}$$

This tensor satisfies the transversality conditions, as it must of course. To facilitate the demonstration that $T_{\text{tree}}^{\mu\nu}$ can be expanded in the basis we found before, Eq. (26), we write the denominators in terms of the dot products of basis four vectors using Eq. (15): $s - M^2 = x_{13} + x_{23}$, and $u - M^2 = x_{13} - x_{23}$. The result for the t-vector corresponding to $T_{\text{tree}}^{\mu\nu}$ is

$$t_{\text{tree}} = -\frac{2}{(\overline{P} \cdot q)^2 - (q' \cdot q)^2} \left[-q' \cdot q\, n^{(2)} + \overline{P} \cdot q\, n^{(3)} \right], \tag{28}$$

where we have substituted the physical values for the dot products. Thus the tree-level CFFs in this formulation are given by

$$\mathcal{C}_1 = -2\frac{q'\cdot q}{(\overline{P}\cdot q)^2 - (q'\cdot q)^2} = \frac{1}{s - M^2} + \frac{1}{u - M^2}, \quad \mathcal{C}_2 = 2\frac{\overline{P}\cdot q}{(\overline{P}\cdot q)^2 - (q'\cdot q)^2} = \frac{1}{s - M^2} - \frac{1}{u - M^2}.$$
(29)

The fact that in $T_{\text{tree}}^{\mu\nu}$ only two CFFs occur is not due to kinematical singularities but to the limited dynamics involved in a tree-level amplitude.

4.1.2 Projection Methods

The relative opacity of the direct method may be the reason that one does not find it discussed in the literature. Instead, one can find several treatments of expansions of the Compton tensor as a sum of transverse basis tensors. We shall briefly discuss two proposals, one by Perrottet [13], the other by Tarrach [14] with the explicit application to the sQED case by Metz [15]. These authors treat the situation where the target hadron is a proton, but their methods include the case where the target has spin zero.

Perrottet uses the idea, presented already above for the single-photon case, to use projection operators to construct the transversal Compton tensor. In particular, he uses the projectors $\mathcal{P}$ and $\mathcal{P}'$ given by

$$\mathcal{P}^{\mu\nu} = g^{\mu\nu} - \frac{q^\mu q^\nu}{q^2}, \quad \mathcal{P}'^{\mu\nu} = g^{\mu\nu} - \frac{q'^\mu q'^\nu}{q'^2}.$$
(30)

Because these projectors annihilate q and q', respectively, they can be used to turn a general tensor $T^{\mu\nu}$ into a transversal one:

$$\tilde{T}^{\mu\nu} = \mathcal{P}'^{\mu m}\, T_{mn}\, \mathcal{P}^{n\nu},$$
(31)

Because the projectors that are used by Perrottet are singular for real photons, this method has a limited domain of application. An improvement was achieved by Tarrach, who constructs the transversal tensor $T^{\mu\nu}$ by applying a two-sided projector $\tilde{g}^{\mu\nu}(q, q')$, given by

$$\tilde{g}^{\mu\nu}(q, q') = g^{\mu\nu} - \frac{q^\mu q'^\nu}{q\cdot q'}$$
(32)

to $T^{\mu\nu}$:

$$\tilde{T}^{\mu\nu} = \tilde{g}^{\mu m}\, T_{mn}\, \tilde{g}^{n\nu}.$$
(33)

The projector $\tilde{g}^{\mu\nu}(q, q')$ is idempotent, *i.e.*, its contraction with itself is equal to $\tilde{g}^{\mu\nu}(q, q')$ again.

Because $\tilde{g}^{\mu\nu}(q, q')$ contracted from the left with $T^{\mu\nu}$ removes the terms with $q^\mu k_j^\nu$ and contracted from the right it does the same with terms of the form $k_i^\mu q'^\nu$, the result of applying $\tilde{g}^{\mu\nu}(q, q')$ simultaneously from the left and from the right to $T^{\mu\nu}$ immediately produces a transversal tensor with only five independent terms. Defining the reduced momenta, $(k = \overline{P}, q', q)$:

$$\tilde{k}_{\mathsf{L}}^\mu = \tilde{g}^{\mu\nu} k_\nu, \quad \tilde{k}_{\mathsf{R}}^\nu = k_\mu\, \tilde{g}^{\mu\nu}$$
(34)

one finds the following result for $\tilde{T}^{\mu\nu}$

$$\tilde{T}^{\mu\nu} = \mathcal{H}_0\, \tilde{g}^{\mu\nu} + \mathcal{H}_1\, \tilde{P}_{\mathsf{L}}^\mu \tilde{P}_{\mathsf{R}}^\nu + \mathcal{H}_2\, \tilde{P}_{\mathsf{L}}^\mu \tilde{q}_{\mathsf{R}}^\nu + \mathcal{H}_3\, \tilde{q}_{\mathsf{L}}'^\mu \tilde{P}_{\mathsf{R}}^\nu + \mathcal{H}_4\, \tilde{q}_{\mathsf{L}}'^\mu \tilde{q}_{\mathsf{R}}^\nu := \sum_{a=0}^{4} \mathcal{H}_a \tilde{T}_a^{\mu\nu}.$$
(35)

While we saw previously that kinematical singularities arise using the null-space basis, we see now that they also arise using the method of projectors. The difference with the one we saw explicitly in the previous subsection, namely the point $\overline{P}\cdot q = 0$, is that now the singularity is at $q'\cdot q = 0$. If we should have used another basis for the null-space $\mathcal{N}_X$ we could have removed the singularity at $\overline{P} \cdot q = 0$ but would find another one, perhaps two as in the Perrottet construction. The null-space basis corresponding to Eq. (35) is

$$
\begin{pmatrix}
n_T^{(1)} \\
n_T^{(2)} \\
n_T^{(3)} \\
n_T^{(4)} \\
n_T^{(5)}
\end{pmatrix}
=
\begin{pmatrix}
1 & 0 & 0 & 0 & 0 & 0 & 0 & 0 & -\frac{1}{q'\cdot q} & 0 \\
0 & 1 & -\frac{\overline{P}\cdot q}{q'\cdot q} & 0 & 0 & 0 & 0 & -\frac{\overline{P}\cdot q}{q'\cdot q} & \frac{(\overline{P}\cdot q)^2}{(q'\cdot q)^2} & 0 \\
0 & 0 & -\frac{q^2}{q'\cdot q} & 1 & 0 & 0 & 0 & 0 & \frac{q^2\,\overline{P}\cdot q}{(q'\cdot q)^2} & -\frac{\overline{P}\cdot q}{q'\cdot q} \\
0 & 0 & 0 & 0 & 1 & -\frac{\overline{P}\cdot q}{q'\cdot q} & 0 & -\frac{q'^2}{q'\cdot q} & \frac{q'^2\,\overline{P}\cdot q}{(q'\cdot q)^2} & 0 \\
0 & 0 & 0 & 0 & 0 & -\frac{q^2}{q'\cdot q} & 1 & 0 & \frac{q'^2 q^2}{(q'\cdot q)^2} & -\frac{q'^2}{q'\cdot q}
\end{pmatrix}.
\tag{36}
$$

This basis shows explicitly that the singularity in the projection operator also occurs in the the null-space basis.

In Tarrach's paper a method is described to remove the kinematic poles, which is a generalisation of the method given by Bardeen and Tung [16] for real Compton scattering. It consists in first taking linear combinations of the basis tensors presented in Eq. (35) to eliminate as many poles as possible. Next, if necessary, multiply any remaining double poles by $q' \cdot q$ and again take linear combinations to remove as many single poles as possible. If any single poles are left, the final step consists in multiplying them with $q' \cdot q$.

The form of the transverse tensor given in Eq. (35) is regularised by first adding a multiple of $\tilde{T}_0$ which contains only a single pole, to the four tensors with both single and double poles, multiplied by $q' \cdot q$. Finally $\tilde{T}_0$ is also multiplied by $q' \cdot q$. The result for the regularised tensors denoted with the subscript R, is

$$
\begin{aligned}
\tilde{T}_{0R}^{\mu\nu} &= q'\cdot q\,\tilde{T}_0^{\mu\nu} = q'\cdot q\,g^{\mu\nu} - q^\mu q'^\nu \\
\tilde{T}_{1R}^{\mu\nu} &= (\overline{P}\cdot q)^2\,\tilde{T}_0^{\mu\nu} + q'\cdot q\,\tilde{T}_1^{\mu\nu} = (\overline{P}\cdot q)^2\,g^{\mu\nu} + q'\cdot q\,\overline{P}^\mu \overline{P}^\nu - \overline{P}\cdot q\,(\overline{P}^\mu q'^\nu + q^\mu \overline{P}^\nu), \\
\tilde{T}_{2R}^{\mu\nu} &= \overline{P}\cdot q\,q^2\,\tilde{T}_0^{\mu\nu} + q'\cdot q\,\tilde{T}_2^{\mu\nu} = \overline{P}\cdot q\,q^2\,g^{\mu\nu} - q^2\,\overline{P}^\mu q'^\nu + q'\cdot q\,\overline{P}^\mu q^\nu - \overline{P}\cdot q\,q^\mu q^\nu, \\
\tilde{T}_{3R}^{\mu\nu} &= \overline{P}\cdot q\,q'^2\,\tilde{T}_0^{\mu\nu} + q'\cdot q\,\tilde{T}_3^{\mu\nu} = \overline{P}\cdot q\,q'^2\,g^{\mu\nu} + q'\cdot q\,q'^\mu \overline{P}^\nu - q'^2\,q^\mu \overline{P}^\nu - \overline{P}\cdot q\,q'^\mu q'^\nu, \\
\tilde{T}_{4R}^{\mu\nu} &= q'^2 q^2\,\tilde{T}_0^{\mu\nu} + q'\cdot q\,\tilde{T}_4^{\mu\nu} = q'^2 q^2\,g^{\mu\nu} - q^2\,q'^\mu q'^\nu + q'\cdot q\,q'^\mu q^\nu - q'^2\,q^\mu q^\nu.
\end{aligned}
\tag{37}
$$

When the limit $q' \cdot q \to 0$ is taken, these regularised tensors are not complete, because we can write for instance $\tilde{T}_{4R}^{\mu\nu}$ as a linear combination of three others, namely

$$
\lim_{q'\cdot q \to 0} \tilde{T}_{4R}^{\mu\nu} = -\frac{1}{\overline{P}\cdot q}\left[\frac{q'^2 q^2}{\overline{P}\cdot q}\,\tilde{T}_{1R}^{\mu\nu} - q'^2\,\tilde{T}_{2R}^{\mu\nu} + q^2\,\tilde{T}_{3R}^{\mu\nu} \right].
\tag{38}
$$

Because Metz uses in his thesis [15] a basis for $T^{\mu\nu}$ different from the one we use here, the final result obtained by him using Tarrach's algorithm differs from the result Eq. (37), namely

$$
\begin{aligned}
M_1^{\mu\nu} &= -q'\cdot q\,g^{\mu\nu} + q^\mu q'^\nu, \\
M_2^{\mu\nu} &= -(\bar{P}\cdot q)^2\,g^{\mu\nu} - q'\cdot q\,\bar{P}^\mu \bar{P}^\nu + \bar{P}\cdot q\,(\bar{P}^\mu q'^\nu + q^\mu \bar{P}^\nu), \\
M_3^{\mu\nu} &= q'^2 q^2\,g^{\mu\nu} + q'\cdot q\,q'^\mu q^\nu - q^2\,q'^\mu q'^\nu - q'^2\,q^\mu q^\nu, \\
M_4^{\mu\nu} &= \bar{P}\cdot q\,(q'^2 + q^2)\,g^{\mu\nu} - \bar{P}\cdot q\,(q'^\mu q'^\nu + q^\mu q^\nu) \\
&\quad - q^2\,\bar{P}^\mu q'^\nu - q'^2\,q^\mu \bar{P}^\nu + q'\cdot q\,(\bar{P}^\mu q^\nu + q'^\mu \bar{P}^\nu), \\
M_{19}^{\mu\nu} &= (\bar{P}\cdot q)^2\,q'^\mu q^\nu + q'^2 q^2\,\bar{P}^\mu \bar{P}^\nu - \bar{P}\cdot q\,q^2\,q'^\mu \bar{P}^\nu - \bar{P}\cdot q\,q'^2\,\bar{P}^\mu q^\nu.
\end{aligned}
\tag{39}
$$

(For historical reasons, the fifth tensor has subscript 19).

As an example we give the tree-level amplitude. In Tarrach's basic formulation of the CFFs one finds:

$$
\mathcal{H}_0^{\text{tree}} = -2, \quad \mathcal{H}_1^{\text{tree}} = \frac{1}{s - M^2} + \frac{1}{u - M^2}, \quad \mathcal{H}_2^{\text{tree}} = 0, \quad \mathcal{H}_3^{\text{tree}} = 0, \quad \mathcal{H}_4^{\text{tree}} = 0,
\tag{40}
$$

where we use the convention of Eq. (35) for the CFFs. In Metz's formulation one finds, using the symbol $\mathcal{B}$ for the CFFs:

$$
\mathcal{B}_1^{\text{tree}} = \frac{1}{s - M^2} + \frac{1}{u - M^2}, \quad \mathcal{B}_2^{\text{tree}} = -\frac{2}{(s - M^2)(u - M^2)}, \quad \mathcal{B}_3^{\text{tree}} = 0, \quad \mathcal{B}_4^{\text{tree}} = 0, \quad \mathcal{B}_{19}^{\text{tree}} = 0.
\tag{41}
$$

where we use the numbering in Eq. (39).

We have found three different results for the form of the Compton tensor, even in the simplest case, namely tree-level. This demonstrates that the choice of the basis elements used in $T^{\mu\nu}$, apart from the fact that there are only three independent four vectors to choose from, matters in identifying the CFFs. In general, one will find linear relations between the tensors used in one conventions to the ones used in another one. In general, those relations will not be free of kinematical poles, because they are obtained by solving sets of coupled linear equation, which by Cramer's rule are found as ratios of determinants.

4.1.3 A Novel Projection Method

The projection methods we discussed in the previous sub-section share the occurrence of single and double poles from the beginning, which must be removed to obtain a formulation of the Compton tensor free of kinematical singularities. Here we propose a method that is free of poles ab inito so that no regularisation is necessary. It will serve as the back bone of the Compton tensor. To this back bone, pairs of momenta are fixed by contraction, like the base pairs in DNA. So we define

$$d^{\mu\nu\alpha\beta} = g^{\mu\nu}g^{\alpha\beta} - g^{\mu\beta}g^{\nu\alpha}. \tag{42}$$

We note that $d^{\mu\nu\alpha\beta}$ is symmetric under the simultaneous interchange $\mu \leftrightarrow \nu$, $\alpha \leftrightarrow \beta$ and changes sign by the interchanges $\mu \leftrightarrow \alpha$, and $\nu \leftrightarrow \beta$. Using this back bone we construct pieces of "DNA" by contracting it with the three basis four vectors. With an obvious notation we write them as follows:

$$
\begin{aligned}
G^{\mu\nu}(q'q) &= q'_\alpha d^{\mu\nu\alpha\beta} q_\beta = q' \cdot q \, g^{\mu\nu} - q^\mu q'^\nu, \\
G^{\mu\nu}(qq) &= q_\alpha d^{\mu\nu\alpha\beta} q_\beta = q^2 \, g^{\mu\nu} - q^\mu q^\nu, \\
G^{\mu\nu}(q'q') &= q'_\alpha d^{\mu\nu\alpha\beta} q'_\beta = q'^2 \, g^{\mu\nu} - q'^\mu q'^\nu, \\
G^{\mu\nu}(\overline{P}q) &= \overline{P}_\alpha d^{\mu\nu\alpha\beta} q_\beta = \overline{P} \cdot q \, g^{\mu\nu} - q^\mu \overline{P}^\nu, \\
G^{\mu\nu}(q'\overline{P}) &= q'_\alpha d^{\mu\nu\alpha\beta} \overline{P}_\beta = \overline{P} \cdot q' \, g^{\mu\nu} - \overline{P}^\mu q'^\nu.
\end{aligned}
\tag{43}
$$

The first tensor is identical with $q' \cdot q$ times Tarrach's projector, the second and the third ones are multiples of the projectors used by Perrottet. The last two are novel. Including $\overline{P}$ in the set of building blocks of projectors, more freedom in the construction of the transverse tensor is created. These five tensors have vanishing contractions with q'_μ and q_ν and are free of kinematical singularities ab initio. The latter property obviates the necessity of the Tarrach construction to remove the single and double poles.

Given these building blocks the transverse tensor $\tilde{T}^{\mu\nu}_{DNA}$ can be written as follows

$$
\begin{aligned}
\tilde{T}^{\mu\nu}_{\mathrm{DNA}} := \sum_{i=1}^{5} \mathcal{S}_i \, \tilde{T}^{(i)\,\mu\nu}_{\mathrm{DNA}} = \; & \mathcal{S}_1 \, G^{\mu\nu}(q'q) \\
& + \mathcal{S}_2 \, G^{\mu\lambda}(q'q') \, G_\lambda{}^\nu(qq) \\
& + \mathcal{S}_3 \, G^{\mu\lambda}(q'\overline{P}) \, G_\lambda{}^\nu(\overline{P}q) \\
& + \mathcal{S}_4 \, [G^{\mu\lambda}(q'\overline{P}) \, G_\lambda{}^\nu(qq) + G^{\mu\lambda}(q'q') \, G_\lambda{}^\nu(\overline{P}q)] \\
& + \mathcal{S}_5 \, G^{\mu\lambda}(q'q') \overline{P}_\lambda \overline{P}_{\lambda'} G^{\lambda\nu}(qq).
\end{aligned}
\tag{44}
$$

By direct computation one may check that the DNA representation is simply related to Metz's as given in Eq. (39):

$$\tilde{T}^{(1)}_{\mathrm{DNA}} = -M_1, \quad \tilde{T}^{(2)}_{\mathrm{DNA}} = M_3, \quad \tilde{T}^{(3)}_{\mathrm{DNA}} = -M_2, \quad \tilde{T}^{(4)}_{\mathrm{DNA}} = M_4, \quad \tilde{T}^{(5)}_{\mathrm{DNA}} = M_{19}. \tag{45}$$

The tensor M_{19} does not fit immediately in the Bardeen-Tung construction, but was introduced in Ref. [16] as $T_{19} \equiv M_{19}/q' \cdot q$ together with two other ones that can only occur for spin-1/2 targets, in order to create more freedom to construct the Compton tensor. Metz used this tensor to replace another one in his original transverse basis. We shall not discuss this matter in more detail, but just note that in the DNA construction this tensor occurs quite naturally.

A final remark is in order here. In the literature sometimes one sees representations of the Compton tensor that are not manifestly transverse. In those cases use has been made of the equations of motion for the wave functions of the external particles, hadrons and photons. Such a representation has the disadvantage that because terms have been omitted, a check of the original equation is not possible anymore. For an illustration we take a look at the tree-level tensor $T_{\text{tree}}^{\mu\nu}$. To calculate the amplitudes, we must use Eq. (12). Using the definition $\overline{P} = p' + p$, four-momentum conservation $p' + q' = p + q$, and the property of the polarisation vectors $\epsilon_\mu^*(q'; h')q'^\mu = 0 = q^\nu \epsilon_\nu(q; h)$ we can reduce $T_{\text{tree}}^{\mu\nu}$ to a simpler form, namely

$$T_{\text{eff}}^{\mu\nu} = -2g^{\mu\nu} + 4\frac{p'^\mu p^\nu}{s - M^2} + 4\frac{p^\mu p'^\nu}{u - M^2}, \tag{46}$$

which has the same contractions with the polarisation vectors but is, however, not transverse.

5 Spin Filter

The form of the Compton tensor used by Metz or the one constructed using the DNA approach produces for any physical situation, $i.e.$ $q'^2 = 0, q' \cdot q = 0, q^2 = 0$ and $\overline{P} \cdot q' = \overline{P} \cdot q = 0$, five independent components and thus depends on five CFFs. Nevertheless, the amplitudes calculated by contracting the tensor with the photon polarisation vectors may depend on less than five CFFs. We may say that they form a kind of spin filter [17]. The reason why this works is easily understood in the Tarrach representation Eq. (35). Consider the transverse four vector $\tilde{q}_L'^\mu$:

$$\tilde{q}_L'^\mu = q'^\mu - \frac{q'^2}{q' \cdot q} q^\mu. \tag{47}$$

This four vector is a left-hand factor in $\tilde{T}_3^{\mu\nu}$ and $\tilde{T}_4^{\mu\nu}$. While $\epsilon_\mu(q'; h')$ will annihilate q'^μ these two components of the transverse tensor will survive when $q'^2 \neq 0$, but in the case that the photon in the final state is real, the second part of $\tilde{q}_L'^\mu$ vanishes, which effectively filters out the CFFs $\mathcal{H}_3$ and $\mathcal{H}_4$ from the amplitudes. A similar situation occurs when the incoming photon is real. Then $\mathcal{H}_2$ and $\mathcal{H}_4$ are filtered out. Finally, in real Compton scattering, only two CFFs are visible, namely $\mathcal{H}_0$ and $\mathcal{H}_1$.

This filter principle is reflected in the number of effective degrees of freedom in Compton scattering. In the case that both photons are virtual, there exist nine combinations of the photon helicities in the initial and final states, $h', h \in \{1, 0, -1\}$. Parity conservation gives for the amplitudes $A(-h', -h) = (-1)^{h'-h} A(h', h)$ from which it follows that there are five independent amplitudes, for instance

$$A(1, 1), \ A(1, 0), \ A(1, -1), \ A(0, 1), \text{ and } A(0, 0). \tag{48}$$

In the case that $q'^2 = 0$, the number of helicities in the final state is reduced to two: $h' = 1, -1$. Then the number of independent amplitudes is also reduced to three, for instance $A(1, 1), \ A(1, 0),$ and $A(1, -1)$, and for real Compton scattering this number is again reduced, namely to two: $A(1, 1)$ and $A(1, -1)$. This counting reflects the number of visible CFFs in these various kinematical regimes. This circumstance that the number of independent amplitudes is identical to the number of visible CFFs, provides a reason to believe that one can invert Eq. (12) to find the CFFs in terms of the helicity amplitudes. Such an extraction is indeed possible, but it is known [18] that this procedure can be very sensitive to uncertainties in the amplitudes. As in actual practice the amplitudes must be determined from experimental data, one should not expect that such an inversion will give accurate values of all visible CFFs.

6 Epilogue

It is clear now that there is a multitude of forms of the Compton tensor even for a scalar particle. How to choose a particular one is partly a matter of taste, but is sometimes motivated by arguments inspired by ideas about the physical structure of the target hadron. In deeply-virtual Compton scattering off the proton the partonic structure of the target is the main focus. Then the operator-product expansion gives a hint as to the relative importance of the GPDs. This motivation has guided the choice of the form of the Compton tensor in the literature, see for instance Ref. [19–23].

This review being devoted to the case of Compton scattering on a scalar target, does not touch the subtleties of including γ^μ as a fourth basis vector in the construction of the Compton tensor. The works of Perrottet,

Tarach, and Metz, among others, show the results of using the projector method to construct $T^{\mu\nu}$ for a spin-1/2 target. As they sometimes utilise the equations of motion of the target wave function, the Dirac equation, the transversality of the published tensors is not manifest. This point was touched also in Sect. 3, where six forms of the current operator J^{μ} were shown, some of which being manifestly transverse, but others can only be proved to be transverse after using the Dirac equation.

Possibly, the direct method and the DNA approach could be made to work for spin-1/2 targets too. This point is to our knowledge open for closer investigation.

Acknowledgements This work was supported in part by the DOE Contract No. DE-FG02-03ER41260.

References

1. J.D. Jackson, *Classical Electrodynamics*, 3rd edn. (Wiley, New York, 1998)
2. S. Weinberg, *The quantum theory of fields* (Cambridge University Press, Cambridge, 1995)
3. J.D. Bjorken, S.D. Drell, *Relativistic Quantum Mechanics* (McGraw-Hill, New York, 1964)
4. F.J. Ernst, R.G. Sachs, K.C. Wali, Electromagnetic form factors of the nucleon. Phys. Rev. **119**, 1105 (1960)
5. R.G. Sachs, High-energy behaviour of nucleon electromagnetic form factors. Phys. Rev. **126**, 2256 (1962)
6. W. Gordon, Der Strom der Diracschen Elektronentheorie. Z. Phys. **50**, 630 (1928)
7. C.-R. Ji, B.L.G. Bakker, H.-M. Choi, A. Suzuki, Ideas of four-fermion operators in electromagnetic form factor calculations. Phys. Rev. D **87**, 093004 (2013)
8. A.H. Compton, A quantum theory of the scattering of X-ray by light. Phys. Rev. **21**, 483 (1923)
9. P. Debye, Zerstreuung von Röntgenstrahlen und quantentheorie. Phys. Zeitschr. **24**, 161 (1923)
10. A. Pais, *Subtle is the Lord (The Science and Life of Albert Einstein)* (Oxford University Press, New York, 1982)
11. A.V. Belitsky, D. Müller, Refined analysis of photon leptoproduction off a spinless target. Phys. Rev. D **79**, 014017 (2009)
12. K. Kumerički, D. Müller, Deeply virtual Compton scattering ar small x_{B} and the access to the GPD H. Nucl. Phys. B **841**, 1 (2010)
13. M. Perrottet, Invariant amplitudes for Compton scattering of off-shell photons on polarised nucleons. Lett. Nuovo Cim. **7**, 915 (1973)
14. R. Tarrach, Invariant amplitudes for virtual compton scattering off polarized nucleons free from kinematic singularities, zeros and constraints. Nuovo Cim **28 A**, 409 (1975)
15. Metz, M.: *Virtuelle Comptonstreuung und die Polarisierbarkeiten des Nukleons* (in German), PhD thesis, Universität Mainz, (1997)
16. W.A. Bardeen, W.-K. Tung, Invariant amplitudes for photon processes. Phys. Rev. **173**, 1423 (1968)
17. B.L.G. Bakker, C.-R. Ji, Spin filter indeeply virtual Compton scattering amplitudes. Phys. Rev. D **83**, 091502 (2011)
18. B.L.G. Bakker, C.-R. Ji, Extraction of compton form factors in scalar QED. Few-Body Syst **56**, 275 (2015)
19. X.D. Ji, Gauge invariant decomposition of nucleon spin. Phys. Rev. Lett **78**, 610 (1997)
20. X.D. Ji, Gauge invariant decomposition of nucleon spin. Phys. Rev. D **55**, 7114 (1997)
21. A.V. Radyushkin, Scaling limit of deeply virtual Compton scattering. Phys. Lett. B **380**, 417 (1996)
22. A.V. Radyushkin, Nonforward parton distributions. Phys. Rev. D **56**, 5524 (1997)
23. D. Mueller, D. Robaschik, B. Geyer, F.M. Dittes, J. Horejsi, Wave functions, evolution equations and evolution kernels from light-ray operators of QCD. Fortsch. Phys. **42**, 101 (1994). [arXiv:hep-ph/9812448]

Few-Body Syst (2017) 58:98
DOI 10.1007/s00601-017-1267-5

J. H. Alvarenga Nogueira · T. Frederico · O. Lourenço

$B^+ \to K^-\pi^+\pi^+$: Three-Body Final State Interactions and $K\pi$ Isospin States

Received: 28 December 2016 / Accepted: 21 February 2017 / Published online: 3 March 2017
© Springer-Verlag Wien 2017

Abstract In this exploratory study, final state interactions are considered to formulate the B meson decay amplitude for the $K\pi\pi$ channel. The Faddeev decomposition of the Bethe–Salpeter equation is used in order to build a relativistic three-body model within the light-front framework. The S-wave scattering amplitude for the $K\pi$ system is considered in the 1/2 and 3/2 isospin channels with the set of inhomogeneous integral equations solved perturbatively. In comparison with previous results for the D meson decay in the same channel, one has to consider the different partonic processes, which build the source amplitudes, and the larger absorption to other decay channels appears, that are important features to be addressed. As in the D decay case, the convergence of the rescattering perturbative series is also achieved with two-loop contributions.

1 Introduction

Heavy quark decays are largely explored in the literature. Due to the large B meson mass (m_b), there are several approaches for B decays based on QCD effective field theories within heavy quark expansions [1–4]. They are based on factorization of the hadronic matrix elements and mainly consider short-distance physics. The weak effective Hamiltonian is constructed based on tools from quantum field theory, such as the operator product expansion to separate the problem in the long-distance and short-distance physics. The perturbative treatment is justified by the fact that the strong coupling constant α_s is small in high energy short-distance processes. The long-distance physics and its non-perturbative nature leads to divergent amplitudes that are complicated to deal with and requires care. The called soft final state interactions (FSI) shows to be essential in studies involving B meson decays, since it does not disappear for large m_b [5]. However, within the QCD factorization approach it was shown that type of effects are suppressed in the heavy quark limit in the case of two-body decays [1,2]. Contributions coming from long-distance inelastic rescattering is expected to be the main source of soft FSI and can be substantial in charge-parity (CP) violation distributions [6,7]. Rescattering effects can also explain the appearance of events in very suppressed decay channels. A recent experimental study of the charmless B_c decay to the $KK\pi$ channel, which within the Standard Model can only occur by weak annihilation diagrams, shows some events in the phase space of this channel [8]. This can be related with hadronic rescattering inelastic transitions to that final decay channel. QCD factorization calculations of two-body B_c decays, also suppressed, can explain that small branching ratios [9]. Contributions coming from final state interaction for the $B^+ \to J/\psi\pi^+$ decay within the QCD factorization approach was further considered in [10].

This article belongs to the Topical Collection "30th anniversary of Few-Body Systems".

J. H. A. Nogueira · T. Frederico (✉)
Instituto Tecnológico de Aeronáutica, São José dos Campos, SP 12228-900, Brazil
E-mail: tobfrederico@yahoo.com.br

O. Lourenço
Universidade Federal do Rio de Janeiro, Macaé, RJ 27930-560, Brazil

FSI play an important role in heavy meson weak decays. This interactions usually appears as suppressed non-factorizable effects in QCD factorization, but even within this approach it is shown that in the center of the Dalitz plot physical values of m_b seem not to be large enough to suppress significantly that power-corrections [11]. As a test of CP violation, FSI are essential to guarantee CPT invariance [6,7]. A practical theoretical approach was used to study these three-body charmless $B^\pm$ decays in [6]. A more general formulation, including resonances and its interferences, applied for four B decay channels is found in Ref. [7]. CP violation in the low invariant mass of the $\pi\pi$ system of the $B \to \pi\pi\pi$ decay channel is also studied in Ref. [12,13], where contributions from scalar and vector resonances are considered. The S-wave $\pi\pi$ elastic scattering in the region below the ρ mass has also the important contribution from the $f_0(600)$ resonance, as showed in Ref. [14] for a four-body semileptonic decay.

Our goal in the present work is to address the issue of three-body FSI in the specific $B^+ \to K^-\pi^+\pi^+$ decay, with emphasis in the S-wave $K^-\pi^+$ amplitude, as an exploratory first approach study. In order to proceed in such direction, we closely follow the formalism developed for the D decay in Ref. [15].

Our study is based in a relativistic model for the three-body FSI that was applied to the $D^+ \to K^-\pi^+\pi^+$ decay [15–17]. In Ref. [15], the isospin projection of the decay amplitude was performed to study different isospin state contributions to the $K^-\pi^+$ rescattering. In that model, by starting from a Bethe–Salpeter like equation and using the Faddeev decomposition, the decay amplitude was separated into a smooth term and a three-body fully interacting contribution. Moreover, the amplitude was factorized in the standard two-meson resonant amplitude times a reduced complex amplitude for the bachelor meson, that carries the effect of the three-body rescattering mechanism. The off-shell bachelor amplitude is a solution of an inhomogeneous Faddeev type integral equation, that has as input the S-wave isospin 1/2 and 3/2 $K^-\pi^+$ transition matrix. In the Faddeev formulation, the integral equation has a connected kernel, which is written in terms of the two-body amplitude. The light-front (LF) projection of the equations [18] was performed to simplify the numerical calculations, and interactions between identical charged pions were neglected. A different coupled-channel framework, considering both $\pi\pi$ and $K\pi$ empirical scattering amplitudes, was used in Ref. [19] to study the $D^+ \to K^-\pi^+\pi^+$ Dalitz plot.

Here we discuss the perturbative solutions of the LF integral equations for the bachelor amplitude in the B meson decay. To check the convergence of the series expansion, we go up to terms of third order in the two-body transition matrix. The numerical results for the $B^+ \to K^-\pi^+\pi^+$ decay with three-body FSI and $K\pi$ interactions in $I = 1/2$ and 3/2 states are presented. The S-wave $K\pi$ amplitude depends on the isospin of the system. There are two isospin states possible for this system, namely, $I = 1/2$ and $I = 3/2$. The LASS experimental data [20] shows resonances and the corresponding scattering amplitude poles only in the isospin 1/2 channel. This feature is used here to model the $K\pi$ S-matrix used in the B decay amplitude.

In the recent paper of Nakamura [19], it is discussed that the effect of the $\pi^+\pi^0$ p-wave ($I = 1$) interaction can contribute to the $D^+ \to K^-\pi^+\pi^+$ decay only through the rescattering, with this contribution being a pure coupled-channel effect. Our model is a single channel model for $B^+ \to K^-\pi^+\pi^+$. It is well known that the effect of coupled channels in single channel model, like the present one, is represented by effective absorptive interaction. In our single channel model, this will correspond to a three-body absorptive interaction. Indeed, the coupling to other channels is introduced by the ε parameter in the three-body propagator, and, as we will show, this parameter is important for the resulting three-body decay amplitude. From this point of view, other channels can have important consequences on the form of B-meson decay amplitude. The loss of flux in our three-body model corresponds to the presence of other channels. As one introduce explicitly all the coupled channels, three-body unitarity has to be satisfied. A single channel representation of the rescattering process has to include the loss of probability flux given in our case by the finite width given in the three-body propagator. Physically, this finite width has to be associated with the transition to different channels and the channel formed by the neutral kaon, neutral pion and charged pion, is particularly important, since it allows the $\pi^+\pi^0$ p-wave ($I = 1$) interaction that generates $\rho(770)$, and plays a major role in the FSI in the $D^+ \to K^-\pi^+\pi^+$ decay. The same effect is possibly present also in the $B^+ \to K^-\pi^+\pi^+$ decay. However, effectively the coupling to other channels in our approach is taken into account qualitatively by allowing to a finite width (ε) to the three-body propagator.

2 Decay Amplitude for $B^+ \to K^-\pi^+\pi^+$ Decay with FSI

2.1 S-wave $K\pi$ Scattering Amplitude

The three-body rescattering model used here to study the decay amplitude with FSI, requires a two-body transition matrix as input. In the same way we have done in the $D^+ \to K^-\pi^+\pi^+$ decay in Ref. [15], the $K\pi$

S-wave elastic scattering amplitude is introduced in the resonant $I_{K\pi} = 1/2$ and non-resonant $I_{K\pi} = 3/2$ isospin states. We use the same parametrization fitted to the LASS data [20] including two resonances above $K_0^*(1430)$, namely $K^*(1630)$ and $K_0^*(1950)$. The main reason to use the additional $K_0^*(1630)$ and $K_0^*(1950)$ resonances is the LASS data, where the whole kinematical range up to 1.89 GeV is fitted. We choose here to not introduce new resonances in the $K^-\pi^+$ $I = 1/2$ channels as it seems that no higher mass resonances are present in this channels, according to the PDG. In addition, the $I = 3/2$ channels seem to be a simple S-wave scattering parametrized by the first two terms in the effective range expansion. In order to not introduce more assumptions, we prefer to be conservative and keep what was used in the previous paper [15]. In our analysis, we also neglect the $\pi\pi$ interaction. The same approximation was also considered in the D decay case of Ref. [15].

The parametrized S-matrix ($S_{K\pi}^{1/2}$) is written as:

$$S_{K\pi}^{1/2} = \frac{k \cot\delta + i\,k}{k \cot\delta - i\,k} \prod_{r=1}^{3} \frac{M_r^2 - M_{K\pi}^2 + i\,z_r \bar{\Gamma}_r}{M_r^2 - M_{K\pi}^2 - i\,z_r \Gamma_r} \tag{1}$$

where $z_r = k\,M_r^2/(k_r\,M_{K\pi})$ and k is the c. m. momentum of each meson of the $K\pi$ pair. Following this S-matrix, the scattering amplitude reads

$$\tau_{I_{K\pi}}\left(M_{K\pi}^2\right) = 4\pi \frac{M_{K\pi}}{k}\left(S_{K\pi}^{I_{K\pi}} - 1\right). \tag{2}$$

The parameters associated to the $K_0^*(1430)$, $K_0^*(1630)$ and $K_0^*(1950)$ resonances are $(M_r, \Gamma_r, \bar{\Gamma}_r)$ given by $(1.48, 0.25, 0.25)$, $(1.67, 0.1, 0.1)$ and $(1.9, 0.2, 0.14)$, respectively [16].

The non-resonant part of the scattering amplitude is parameterized by an effective range expansion as $k \cot\delta = \frac{1}{a} + \frac{1}{2}r_0 k^2$ using $a = 1.6$ GeV^{-1} and $r_0 = 3.32$ GeV^{-1}. By using such a model, the S-wave $K\pi$ amplitude in the $I = 3/2$ state is given by $S_{K\pi}^{3/2} = \frac{k \cot\delta + i\,k}{k \cot\delta - i\,k}$, with the effective range expansion parameters $a = -1.00$ GeV^{-1} and $r_0 = -1.76$ GeV^{-1} taken from Ref. [21].

The parametrization from the three-resonance model and the $I_{K\pi} = 1/2$ S-wave phase-shift compared to the LASS data shows good agreement. The results of the parametrization for $|S_{K\pi}^{1/2} - 1|/2$ are shown and discussed in more details in Ref. [15].

2.2 Three-Body Rescattering Bethe–Salpeter Model

The full decay amplitude including the rescattering series and the $3 \to 3$ transition matrix is written as [15]:

$$
\begin{aligned}
\mathcal{A}(k_\pi, k_{\pi'}) &= B_0(k_\pi, k_{\pi'}) \\
&+ \int \frac{d^4 q_\pi d^4 q_{\pi'}}{(2\pi)^8} T(k_\pi, k_{\pi'}; q_\pi, q_{\pi'}) S_\pi(q_\pi)\, S_\pi(q_{\pi'}) S_K(K - q_{\pi'} - q_\pi) B_0(q_\pi, q_{\pi'}),
\end{aligned} \tag{3}
$$

where the momentum of the pions are k_π and $k_{\pi'}$ and K is the total momentum of the system, which is used to write the two-body invariant mass of the $K\pi$ system as $M_{K\pi}^2 = (K - k_{\pi'})^2$.

The short-distance physics resides in the $B_0(k_\pi, k_{\pi'})$ amplitude, which represents the quark level amplitude. The sum of rescattering diagrams, considered in the ladder approximation, is in the second term of Eq. (3) and composes the long range physics. This term is composed by the $3 \to 3$ transition matrix $T(k_\pi, k_{\pi'}; q_\pi, q_{\pi'})$ with the source term and the meson propagators $S_i(q_i) = i(q_i^2 - m_i^2 + i\epsilon)^{-1}$, where self-energies are neglected. The $K\pi$ transition matrix sum all $2 \to 2$ collision terms. The full transition matrix with the FSI is a solution of the Bethe–Salpeter equation, used with its Faddeev decomposition.

2.3 Decay Amplitude

The full three-body T-matrix gives the final state interactions between the mesons in the decay channel and it is a solution of the Bethe–Salpeter equation. Here we follow the formalism developed in Ref. [15], where the Faddeev decomposition including only two-body irreducible diagrams for spinless particles without self-energies is considered. Only two body interactions are considered, involving all three-particles except between the equal charged pions.

The two-body transition matrix written with a four-conservation delta factorized out reads

$$T_i\left(k'_j, k'_k; k_j, k_k\right) = (2\pi)^4 \tau_i(s_i) S_i^{-1}(k_i) \delta\left(k'_i - k_i\right), \tag{4}$$

where the Mandelstam variable $s_i = (k_j + k_k)^2$ is the only dependence considered and $\tau_i(s_i)$ is the unitary S-wave scattering amplitude of particles j and k. Using the separable form of Eq. (4) the problem is reduced to a four-dimensional integral equation in one momentum variable for the Faddeev components of the vertex function.

The full decay amplitude considering interactions between all the final states mesons reduces to

$$\mathcal{A}_0(k_i, k_j) = B_0(k_i, k_j) + \sum_\alpha \tau(s_\alpha)\xi^\alpha(k_\alpha), \tag{5}$$

where the subindex in $\mathcal{A}_0$ denotes the S-wave two-meson scattering and the bachelor amplitude $\xi(k_i)$ carries the three-body rescattering effect and is represented by the connected Faddeev-like equations

$$\xi^i(k_i) = \xi_0^i(k_i)$$
$$+ \int \frac{d^4 q_j}{(2\pi)^4} S_j(q_j) S_k(K - k_i - q_k) \tau_j(s_j) \xi^j(q_j) + \int \frac{d^4 q_k}{(2\pi)^4} S_j(K - k_i - q_k) S_k(q_k) \tau_k(s_k) \xi^k(q_k). \tag{6}$$

with $q_k = K - k_i - q_j$. In Eq. (6), both, amplitude and phase, depending on the bachelor meson on-mass-shell momentum and $\tau(s_i)$, can take into account two-meson resonances. The parameterized $K\pi$ scattering amplitude $\tau_i(M_{K\pi}^2)$ reproduces the LASS experimental [20] S-wave phase-shift in the isospin $1/2$ and $3/2$ channels.

By taking into account all the model assumptions, the decay amplitude for the $B^+ \to K^-\pi^+\pi^+$ process is given by

$$\mathcal{A}_0(k_\pi, k_{\pi'}) = B_0(k_\pi, k_{\pi'}) + \tau(M_{K\pi}^2)\xi(k_{\pi'}) + \tau(M_{K\pi'}^2)\xi(k_\pi), \tag{7}$$

where $M_{K\pi}^2 = (K - k_{\pi'})^2$, $M_{K\pi'}^2 = (K - k_\pi)^2$ and the bachelor pion on-mass-shell momentum is given by

$$|\mathbf{k}_\pi| = \left[\left(\frac{M_B^2 + m_\pi^2 - M_{K\pi'}^2}{2M_B}\right)^2 - m_\pi^2\right]^{\frac{1}{2}}. \tag{8}$$

The rescattering series comes from the solution of Eq. (9), where the second and third terms in Eq. (6) correspond to higher order loop diagrams.

The inhomogeneous integral equation for the spectator amplitude in the three-body collision process is a function only of the bachelor momentum (see [15]),

$$\xi(k) = \xi_0(k) + \int \frac{d^4 q}{(2\pi)^4} \tau\left((K - q)^2\right) S_K(K - k - q) S_\pi(q) \xi(q), \tag{9}$$

where the first term is

$$\xi_0(k) = \int \frac{d^4 q}{(2\pi)^4} S_\pi(q) S_K(K - k - q) B_0(k, q), \tag{10}$$

with the partonic decay amplitude described by $B_0(k, q)$.

The two basic contributions for the decay amplitude are the well behaved function $B_0(k_\pi, k_{\pi'})$ and three-body rescattering term $\tau\left(M_{K\pi'}^2\right)\xi(k_\pi)$. The operator τ acts on the isospin states $1/2$ and $3/2$. The complex decay amplitude can be decomposed in terms of phase and amplitude as

$$A\left(M_{K\pi'}^2\right) = \frac{1}{2}\langle K\pi\pi|B_0\rangle + \langle K\pi\pi|\tau\left(M_{K\pi'}^2\right)|\xi(k_\pi)\rangle = a_0\left(M_{K\pi'}^2\right) e^{i\Phi_0\left(M_{K\pi'}^2\right)}, \tag{11}$$

which is a function of only $M_{K\pi'}^2$ and $|K\pi\pi\rangle$ represents the state in isospin space.

84

3 FSI Light-Front Equations

The equations presented for the decay processes considering FSI effects are simplified when treated in light-front dynamics. The light-front (LF) projection of the four-dimensional coupled equations presents a three-dimensional form. Such a technique was successfully applied for the heavy meson decays presented in Ref. [15] and will also be used here to treat the $B \to K\pi\pi$ decay problem.

The light-front projection performed in the field-theoretical inhomogeneous three-body BS equation to build the integral equations used in our work, corresponds to the truncation of the light-front Fock-space to the three-meson valence component in the intermediate. The advantages of performing the LF projection is that Z-diagrams are in general suppressed [22]. In addition, the integral equations with only the valence three-meson state are covariant under seven LF kinematical transformation, namely, the ones that keep the null-plane invariant, which includes three translations, rotation around the z-direction, two other kinematical boosts, and the boost along the z-direction. The truncation of the LF Fock-space is stable under the kinematical boosts [23]. In contrast, the Fock-space truncation in the instant form has only three translations and three rotations and no-boosts.

After all manipulations, discussed in details in [15], the integral equation in terms of the LF variables reads

$$
\xi^i(y, \mathbf{k}_\perp) = \xi_0^i(y, \mathbf{k}_\perp)
$$
$$
+ \frac{i}{2(2\pi)^3} \int_0^{1-y} \frac{dx}{x(1-x-y)} \int d^2 q_\perp \left[\frac{\tau_j \left(M_{ik}^2(x, q_\perp) \right) \xi^j(x, \mathbf{q}_\perp)}{M^2 - M_0^2(x, \mathbf{q}_\perp; y, \mathbf{k}_\perp) + i\varepsilon} + (j \leftrightarrow k) \right], \quad (12)
$$

where $M^2 = K^\mu K_\mu$, $y = k_i^+/K^+$, $x = q_j^+/K^+$ or $x = q_k^+/K^+$ in the first or second integral in the right-hand side of the equation. The free three-body squared mass is

$$
M_0^2(x, \mathbf{q}_\perp; y, \mathbf{k}_\perp) = \frac{k_\perp^2 + m_i^2}{y} + \frac{q_\perp^2 + m_j^2}{x} + \frac{(\mathbf{k}_\perp + \mathbf{q}_\perp)^2 + m_k^2}{1 - x - y}. \quad (13)
$$

The argument of the two-body amplitude $\tau_j \left(M_{ik}^2(x, q_\perp) \right)$ should be understood as

$$
M_{ik}^2(x, q_\perp) = (1-x) \left(M^2 - \frac{q_\perp^2 + m_j^2}{x} \right) - q_\perp^2. \quad (14)
$$

The driven term in Eq. (12) is rewritten as

$$
\xi_0^i(y, \mathbf{k}_\perp) = \frac{i}{2(2\pi)^3} \int_0^{1-y} \frac{dx}{x(1-y-x)} \int d^2 q_\perp \frac{B_0(x, \mathbf{q}_\perp; y, \mathbf{k}_\perp)}{M^2 - M_0^2(x, \mathbf{q}_\perp; y, \mathbf{k}_\perp) + i\varepsilon} = B_0 \, \xi_0(y, k_\perp) \quad (15)
$$

where $B_0(x, \mathbf{q}_\perp; y, \mathbf{k}_\perp) = B_0$ is the short-distance amplitude, taken as a constant in this work.

Since the integral over $q_\perp$ is divergent, a regularization procedure is needed. Here we use a finite subtraction constant $\lambda(\mu^2)$, and a subtraction point within the integration kernel of Eq. (15). This method leads to the following driven term

$$
\xi_0(y, k_\perp) = \lambda(\mu^2)
$$
$$
+ \frac{i}{2} \int_0^1 \frac{dx}{x(1-x)} \int_0^{2\pi} d\theta \int_0^\infty \frac{dq_\perp q_\perp}{(2\pi)^3} \left[\frac{1}{M_{K\pi}^2(y, k_\perp) - M_{0,K\pi}^2(x, q_\perp) + i\varepsilon} - \frac{1}{\mu^2 - M_{0,K\pi}^2(x, q_\perp)} \right]
$$
$$
(16)
$$

with the $K\pi$ system free squared-mass given by $M_{0,K\pi}^2(x, q_\perp) = \frac{q_\perp^2 + m_\pi^2}{x} + \frac{q_\perp^2 + m_K^2}{1-x}$. After integration over θ and $q_\perp$, Eq. (16) is finally written as

$$
\xi_0(y, k_\perp) = \lambda(\mu^2) + \frac{i}{4} \int_0^1 \frac{dx}{(2\pi)^2} \ln \frac{(1-x) \left(x M_{K\pi}^2(y, k_\perp) - m_\pi^2 + ix\varepsilon \right) - x m_K^2}{(1-x) \left(x\mu^2 - m_\pi^2 \right) - x m_K^2}. \quad (17)
$$

4 Application in the $B^+ \to K^- \pi^+ \pi^+$ Decay

The model for the $B^+ \to K^- \pi^+ \pi^+$ decay with FSI is based on an inhomogeneous integral equation for the spectator meson, with the meson-meson scattering amplitude as input. Isospin states of the $\pi\pi$ interaction are disregarded here, unlike the $I_{K\pi} = 1/2$ and $I_{K\pi} = 3/2$ states for the $K^\mp \pi^\pm$ channel, consider in our calculations. Our parametrization for the $K\pi$ amplitude follows the experimental results of [20], where the resonant $I_{K\pi} = 1/2$ channel below $K_0^*(1430)$ dominates and the $I_{K\pi} = 3/2$ amplitude is comparable. This model is the same used in Ref. [15] to study the $D^+ \to K^- \pi^+ \pi^+$ decay. A calculation up to two loops for this same decay was performed in Ref. [17] bellow $K_0^*(1430)$.

Here the LF model is applied to the B decay and the calculations are performed up to three-loops in order to check the numerical convergence of the integrals. There are two possible total isospin states, namely, $I_T = 5/2$ and $3/2$. In our notation, the bachelor amplitude has the total isospin index and the one related with the interacting pair $\xi_{I_T, I_{K\pi}}^{I_T^z}(y, k_\perp)$, where we also consider the isospin projection index. The source amplitude written in terms of the $K\pi$ isospin state reads

$$|B_0\rangle = \sum_{I_T, I_{K\pi}} \alpha_{I_T, I_{K\pi}}^{I_T^z} |I_T, I_{K\pi}, I_T^z\rangle + \sum_{I_T, I_{K\pi'}} \alpha_{I_T, I_{K\pi'}}^{I_T^z} |I_T, I_{K\pi'}, I_T^z\rangle, \tag{18}$$

which has no dependence on the momentum variables and has an arbitrary normalization, since we are not considering explicitly short-distance processes in our calculations. For sake of simplicity we define the recoupling coefficients as $R_{I_T, I_{K\pi}, I_{K\pi'}}^{I_T^z} = \langle I_T, I_{K\pi}, I_T^z | I_T, I_{K\pi'}, I_T^z \rangle$. This allows us to write the set of isospin coupled integral equations as

$$\xi_{I_T, I_{K\pi}}^{I_T^z}(y, k_\perp) = \langle I_T, I_{K\pi}, I_T^z | B \rangle \xi_0(y, k_\perp) + \frac{i}{2} \sum_{I_{K\pi'}} R_{I_T, I_{K\pi}, I_{K\pi'}}^{I_T^z} \int_0^{1-y} \frac{dx}{x(1-y-x)} \int_0^\infty \frac{dq_\perp}{(2\pi)^3}$$
$$\times K_{I_{K\pi'}}(y, k_\perp; x, q_\perp) \xi_{I_T, I_{K\pi'}}^{I_T^z}(x, q_\perp), \tag{19}$$

where the free squared mass of the $K\pi\pi$ system is

$$M_{0, K\pi\pi}^2(x, q_\perp, y, k_\perp) = \frac{k_\perp^2 + m_\pi^2}{y} + \frac{q_\perp^2 + m_\pi^2}{x} + \frac{q_\perp^2 + k_\perp^2 + 2q_\perp k_\perp \cos\theta + m_K^2}{1-x-y}, \tag{20}$$

with the squared-mass of the virtual $K\pi$ system $M_{K\pi}^2(z, p_\perp) = (1-z)\left(M_B^2 - \frac{p_\perp^2 + m_\pi^2}{z}\right) - p_\perp^2$. The kernel carrying the $K\pi$ scattering amplitude is

$$K_{I_{K\pi'}}(y, k_\perp; x, q_\perp) = \int_0^{2\pi} d\theta \frac{q_\perp \tau_{I_{K\pi'}}\left(M_{K\pi'}^2(x, q_\perp)\right)}{M_B^2 - M_{0, K\pi\pi}^2(x, q_\perp, y, k_\perp) + i\varepsilon}. \tag{21}$$

Isospin 2 states of pion-pion interactions are not considered in the model, which will be explored as a single channel model, with the $K\pi$ S-wave interaction in the resonant $I = 1/2$, and as a coupled channel model with both $I = 1/2$ and $3/2$ $K\pi$ S-wave interactions.

The symmetrized decay amplitude with respect to the identical pions is written as

$$\mathcal{A}_0 = A_0\left(M_{K\pi'}^2\right) + A_0\left(M_{K\pi}^2\right). \tag{22}$$

The isospin projection on each term leads to

$$A_0\left(M_{K\pi'}^2\right) = \sum_{I_T, I_{K\pi'}, I_T^z} \langle K^- \pi^+ \pi^+ | I_T, I_{K\pi'}, I_T^z\rangle \left[\frac{1}{2}\langle I_T, I_{K\pi'}, I_T^z | B_0\rangle + \tau_{I_{K\pi}}\left(M_{K\pi'}^2\right)\xi_{I_T, I_{K\pi'}}^{I_T^z}(k_\pi)\right]$$
$$= a_0\left(M_{K\pi'}^2\right)e^{i\Phi_0\left(M_{K\pi'}^2\right)}. \tag{23}$$

5 Numerical Perturbative Solutions

The problem is solved by integrating the terms starting from the driving term and iterating as a perturbative series. The integrations are done up to three loops in order to check the convergence. In the coupled-channel calculations, the total isospin states $I = 3/2$ are performed coupling $I_{K\pi} = 1/2$ or $I_{K\pi} = 3/2$ states. We also consider the $I_T = 5/2$ with its single contribution in the $K\pi$ interaction for the isospin $3/2$ states.

For the single channel case we consider only $K\pi$ interaction in the resonant isospin $1/2$ states and the perturbative solution of the equation up to three-loops reduces to

$$
\begin{aligned}
\xi_{3/2,1/2}^{3/2}(y, k_\perp) = {} & \frac{1}{6}\sqrt{\frac{2}{3}}\xi_0(y, k_\perp) - \frac{i}{3}\left(\frac{1}{6}\sqrt{\frac{2}{3}}\right)\int_0^\infty \frac{dq_\perp}{(2\pi)^3}\int_0^{1-y} dx\, K_{1/2}(y, k_\perp; x, q_\perp)\xi_0(x, q_\perp) \\
& - \frac{1}{9}\left(\frac{1}{6}\sqrt{\frac{2}{3}}\right)\int_0^\infty \frac{dq_\perp}{(2\pi)^3} \\
& \times \int_0^{1-y} dx\, K_{1/2}(y, k_\perp; x, q_\perp)\int_0^\infty \frac{dq'_\perp}{(2\pi)^3}\int_0^{1-x} dx'\, K_{1/2}(x, q_\perp; x', q'_\perp)\xi_0(x', q'_\perp) \\
& + \cdots
\end{aligned}
\tag{24}
$$

where we compute driving term considering $\alpha_{3/2,1/2}^{3/2} = 1$ and the kernel $K_{1/2}$ is defined by Eq. (21). This equation has an arbitrary normalization factor, coming from the source partonic amplitude, which we assume to be constant.

The numerical integration over the radial variable is computed introducing a momentum cut-off $\Lambda = 0.8$ GeV. This is smaller than in the D decay case, but in that case the change of the cutt-off parameter from 2.0 to 0.8 GeV practically does not alter the results. In the B case, the use of $\Lambda = 2.0$ GeV is very expensive numerically, and probably this is related to the large non-physical region accessed.

The finite value of the momentum cut-off represents roughly the two-meson interaction range, that is somewhat related to the size of the mesons itself. If the $K\pi$ model interaction had a finite range, naturally a cut-off in the hadronic loop would appear and, in our case, it is brought by the momentum cut-off.

Concerning the ε parameter, the value used here was $\varepsilon = 0.5$ GeV^2, which is larger than the one used in the D decay case. In fact, since the B phase space is very large, we know that the absorption is higher comparing with the D decay. Here we mimic this effect by using a larger value for the ε parameter. We have tested different values of ε (close to $\varepsilon = 0.5$ GeV^2), obtaining a small difference in the results. The subtraction constant in the driving term is chosen to be zero.

Regarding the convergence of the loop expansion, we have studied it up to three-loops. The results concerning phase and modulus of the bachelor function is depicted in Fig. 1.

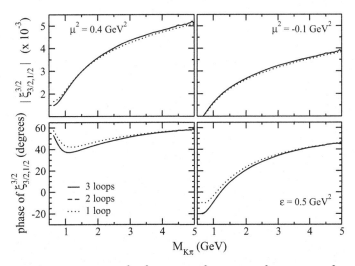

Fig. 1 Modulus and phase of $\xi_{3/2,1/2}^{3/2}$ for $\varepsilon = 0.5$ GeV^2, $\mu^2 = 0.4$ GeV^2 (*left*) and $\mu^2 = -0.1$ GeV^2 (*right*)

It was used $\mu^2 = (0.4, -0.1)$ GeV^2, in order to verify the effect of the subtraction point in the calculations. We have also used $\varepsilon = 0.5$ GeV^2. For a fixed value of μ^2 it is clear that the two-loop solution already present convergence and is enough for practical applications. This finding is similar to that observed in the D decay case, but now the results are even better concerning the convergence. The phase is always positive for $\mu^2 = 0.4$ GeV^2, but can be either positive or negative for $\mu^2 = -0.1$ GeV^2. The phase variation is large for $\mu^2 = -0.1$ GeV^2 and presents a minimum increasing again for $\mu^2 = 0.4$ GeV^2. In both cases the modulus increases for larger two-body invariant masses.

5.1 Interaction for Coupled Channels in $I_{K\pi} = 1/2$ and 3/2 States

In the coupled channels case the set of integral equations obtained from Eq. (19) reads

$$\xi_{3/2,1/2}^{3/2}(y, k_\perp) = A_w \, \xi_0(y, k_\perp)$$
$$+ \frac{i R_{3/2,1/2,1/2}^{3/2}}{2} \int_0^{1-y} \frac{dx}{x(1-y-x)} \int_0^\infty \frac{dq_\perp}{(2\pi)^3} K_{1/2}(y, k_\perp; x, q_\perp) \xi_{3/2,1/2}^{3/2}(x, q_\perp)$$
$$+ \frac{i R_{3/2,1/2,3/2}^{3/2}}{2(2\pi)^3} \int_0^{1-y} \frac{dx}{x(1-y-x)} \int_0^\infty \frac{dq_\perp}{(2\pi)^3} K_{3/2}(y, k_\perp; x, q_\perp) \xi_{3/2,3/2}^{3/2}(x, q_\perp), \qquad (25)$$

$$\xi_{3/2,3/2}^{3/2}(y, k_\perp) = B_w \, \xi_0(y, k_\perp)$$
$$+ \frac{i R_{3/2,3/2,1/2}^{3/2}}{2} \int_0^{1-y} \frac{dx}{x(1-y-x)} \int_0^\infty \frac{dq_\perp}{(2\pi)^3} K_{1/2}(y, k_\perp; x, q_\perp) \xi_{3/2,1/2}^{3/2}(x, q_\perp)$$
$$+ \frac{i R_{3/2,3/2,3/2}^{3/2}}{2} \int_0^{1-y} \frac{dx}{x(1-y-x)} \int_0^\infty \frac{dq_\perp}{(2\pi)^3} K_{3/2}(y, k_\perp; x, q_\perp) \xi_{3/2,3/2}^{3/2}(x, q_\perp). \qquad (26)$$

and for $I_T = 5/2$:

$$\xi_{5/2,3/2}^{3/2}(y, k_\perp) = C_w \, \xi_0(y, k_\perp)$$
$$+ \frac{i R_{5/2,3/2,3/2}^{3/2}}{2} \int_0^{1-y} \frac{dx}{x(1-y-x)} \int_0^\infty \frac{dq_\perp}{(2\pi)^3} K_{3/2}(y, k_\perp; x, q_\perp) \xi_{5/2,3/2}^{3/2}(x, q_\perp), \qquad (27)$$

where the isospin states related to the projection of the partonic amplitude (18) brings the weights A_w, B_w and C_w, given by $A_w = \langle I_T = 3/2, I_{K\pi} = 1/2, I_T^z = 3/2| B_0 \rangle$, $B_w = \langle 3/2, 3/2, 3/2| B_0 \rangle$ and $C_w = \langle 5/2, 3/2, 3/2| B_0 \rangle$ where the isospin coefficients are $A_w = \alpha_{3/2,1/2}^{3/2}(1 + R_{3/2,1/2,1/2}^{3/2}) + \alpha_{3/2,3/2}^{3/2} R_{3/2,1/2,3/2}^{3/2}$, $B_w = \alpha_{3/2,3/2}^{3/2}(1 + R_{3/2,3/2,3/2}^{3/2}) + \alpha_{3/2,1/2}^{3/2} R_{3/2,3/2,1/2}^{3/2}$ and $C_w = \alpha_{5/2,3/2}^{3/2}(1 + R_{5/2,3/2,3/2}^{3/2})$. The coefficients α come from the partonic decay amplitude (18) projected onto the isospin space and are defined as $\alpha_{3/2,1/2}^{3/2} = \frac{W_1}{2} C_{1/2\,1\,3/2}^{1/2\,1\,3/2} C_{1\,-1/2\,1/2}^{1\,1/2\,1/2}$, $\alpha_{3/2,3/2}^{3/2} = \frac{W_2}{2} C_{1/2\,1\,3/2}^{3/2\,1\,3/2} C_{1\,-1/2\,1/2}^{1\,1/2\,3/2}$ and $\alpha_{5/2,3/2}^{3/2} = \frac{W_3}{2} C_{1/2\,1\,3/2}^{3/2\,1\,5/2} C_{1\,-1/2\,1/2}^{1\,1/2\,3/2}$ and the Clebsch-Gordan and recoupling coefficients $C_{1/2\,1\,3/2}^{1/2\,1\,3/2} = 1$, $C_{1\,-1/2\,1/2}^{1\,1/2\,1/2} = \sqrt{2/3}$, $C_{1/2\,1\,3/2}^{3/2\,1\,3/2} = -\sqrt{2/5}$, $C_{1\,-1/2\,1/2}^{1\,1/2\,3/2} = 1/\sqrt{3}$, $C_{1/2\,1\,3/2}^{3/2\,1\,5/2} = \sqrt{3/5}$, $R_{3/2,1/2,1/2}^{3/2} = -2/3$, $R_{3/2,1/2,3/2}^{3/2} = \sqrt{5}/3$, $R_{3/2,3/2,3/2}^{3/2} = 2/3$, $R_{3/2,3/2,1/2}^{3/2} = \sqrt{5}/3$, and $R_{5/2,3/2,3/2}^{3/2} = 1$. With all these manipulations the weights A_w, B_w, and C_w reads $A_w = \sqrt{\frac{1}{54}}(W_1 - W_2)$, $B_w = \sqrt{\frac{5}{54}}(W_1 - W_2)$ and $C_w = \frac{W_3}{\sqrt{5}}$.

Also in this coupled channels case, the bachelor amplitude is computed to check the convergence. The coupled equations of Eq. (25) appear in the case $I_T = 3/2$. For $I_T = 5/2$, it is a single channel equation Eq. (27). The results are shown in Fig. 2, using $\varepsilon = 0.5$ GeV^2 and $\mu^2 = -0.1$ GeV^2, with the parameters from the expansion of the source term given by $W_1 = 1$, $W_2 = 2$ and $W_3 = 0.2$.

Again, the convergence is clear and the two-loop result is already enough for practical applications. Both, phase and modulus of the bachelor amplitudes increases with $M_{K\pi}$ and changes considerably along the large phase space available. In the channel $I_T = 3/2$, both components have similar magnitudes for the phase and are larger than that from the $I_T = 5/2$ case, same pattern observed in the D decay.

6 Results for the Phase and Amplitude in the $B^+ \to K^-\pi^+\pi^+$ Decay

Since the two-loop result presents already a good convergence for the bachelor amplitudes, we restrict our calculations hereafter to decay amplitude up to two-loops in Eq. (24). For the moment, there is no experimental

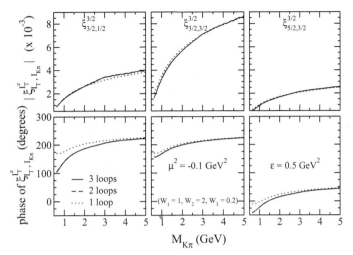

Fig. 2 Modulus and phase of $\xi^{I_T^z}_{I_T, I_{K\pi}}$ for $\varepsilon = 0.5\,\mathrm{GeV}^2$ and $\mu^2 = -0.1\,\mathrm{GeV}^2$ and the parameters $W_1 = 1$, $W_2 = 2$ and $W_3 = 0.2$

data available to perform a comparative analysis as done for the D meson decay in [15]. For the single channel calculations we consider only the S-wave $K\pi$ scattering amplitude in the isospin $1/2$ state, which is fitted to the LASS data [20]. The reduced form of the decay amplitude, that will give us both phase and modulus by means of Eq. (23), reads

$$A_0(M_{K\pi}^2) = \sqrt{\frac{2}{3}}\left[\frac{1}{12}\sqrt{\frac{2}{3}} + \tau_{1/2}\left(M_{K\pi}^2\right)\xi^{3/2}_{3/2,1/2}(k_{\pi'})\right]. \tag{28}$$

The iteration of the coupled equations (25)–(26) gives the results for the channel $I_T = 3/2$. For the the $I_T = 5/2$ state, the amplitude is given by the single expression in Eq. (27). We also consider for these calculations the results up to two loops, since the convergence is verified. The S-wave decay amplitude is

$$A_0\left(M_{K\pi}^2\right) = C_1\left[\frac{A_w}{2} + \tau_{1/2}\left(M_{K\pi}^2\right)\xi^{3/2}_{3/2,1/2}(k_{\pi'})\right] + C_2\left[\frac{B_w}{2} + \tau_{3/2}\left(M_{K\pi}^2\right)\xi^{3/2}_{3/2,3/2}(k_{\pi'})\right] + $$
$$+ C_3\left[\frac{C_w}{2} + \tau_{3/2}\left(M_{K\pi}^2\right)\xi^{3/2}_{5/2,3/2}(k_{\pi'})\right] \tag{29}$$

where the constants C_i come from the isospin projection onto the state $K\pi\pi$, Eq. (23). There are two free parameters related with the projected partonic amplitude, namely, $W_1 - W_2$ and W_3. If the first is zero and the second nonzero, only total isospin $5/2$ appears and there is no structure in the decay amplitude, as shown in Ref. [15]. This shows that it is not a good physical solution, since the isospin state contributions are not being taken into account in a reasonable way. A more detailed study of the correct weights using the LF model would be guided by experimental data, as done for the $D^+ \to K^-\pi^+\pi^+$ decay in [15]. Here we just follow that study, where the authors found a small mixture of the total isospin $5/2$ state.

In Fig. 3, we observe that the amplitude approaches a constant as $M_{K\pi}$ increases. This behavior appears because the two-body amplitude of Eq. (2) damps fast at large $M_{K\pi}$, and only the constant partonic amplitude [first term in Eq. (3)] remains. This behavior of the two-body amplitude in the dominant $I = 1/2$ channel is traced back to the inclusion of resonances below 2 GeV, as suggested by the known resonances given in PDG. Thus, by considering the current available experimental information from LASS and $D^+ \to K^-\pi^+\pi^+$ decay, we opted to be conservative, however, this may be not realistic but only further experiments can decide. Still regarding Fig. 3, we show a comparison between modulus and phase of decay amplitudes for the B^+ and D^+ mesons, both decaying to the same final state $K^-\pi^+\pi^+$. The subtraction scale is fixed in $\mu^2 = -0.1\,\mathrm{GeV}^2$, the ε parameter was chosen to be $\varepsilon = 0.5\,\mathrm{GeV}^2$, and $W_1 - W_2 = -1$ and $W_3 = 0.2$ were used. All these parameters are kept the same for both cases. In order to test the effect of the constants $W_1 - W_2$ and W_3, we have tried a second set of parameters, namely, $W_1 - W_2 = 1$ and $W_3 = 0.3$, which was used in Ref. [15] to study the experimental data for the $D^+ \to K^-\pi^+\pi^+$ decay amplitude, but the results are very similar and with only a change of sign in the phase.

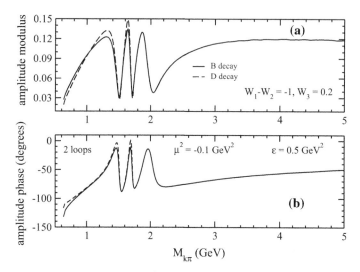

Fig. 3 Comparison of **a** modulus and **b** phase between $D^+ \to K^-\pi^+\pi^+$ and $B^+ \to K^-\pi^+\pi^+$ amplitudes for a initial state in which $W_1 - W_2 = -1$ and $W_3 = 0.2$

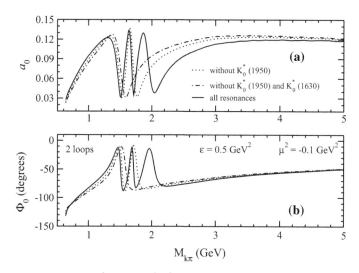

Fig. 4 Modulus (a_0) and phase (Φ_0) of the $B^+ \to K^-\pi^+\pi^+$ amplitude, in the notation of Eq. (23), comparing cases with all resonances, without $K_0^*(1950)$ and without both $K_0^*(1950)$ and $K_0^*(1630)$

In Fig. 4 we compare both modulus and phase of the $B^+ \to K^-\pi^+\pi^+$ decay amplitude with and without the resonant structure, which incorporates $K_0^*(1630)$ and $K_0^*(1950)$.

For this study, we also fix the subtraction point at $\mu^2 = -0.1$ GeV^2 and the other parameters with the same values as before. The figure shows that the inclusion of the resonances produces more bands in both modulus and phase. This is clearly related with the resonances, since the peaks are around its masses and bellow $K_0^*(1430)$ the effect is small. All the cases have the same tail when the two-body invariant mass increases. In amplitude analysis of the three-body B decay to the $K\pi\pi$ channel the $K_0^*(1630)$ resonance is usually included explicitly in the fit, insofar the $K_0^*(1950)$ is more complicated to claim that exists in the channel, therefore it appears indirectly in experimental analysis.

7 Summary and Conclusions

In this exploratory and simplified work, we have used a light-front framework to compute off-shell decay amplitudes starting from the four-dimension Bethe–Salpeter equation decomposed in the Faddeev form. The contribution of final state interactions to the $B^+ \to K^-\pi^+\pi^+$ decay is obtained. This approach can be applied for charged three-body heavy meson decays, and was used before for the D meson decay, and the calculations were compared to the experimental data expressed in terms of the modulus and phase-shift [15]. Here, we have used the same three-body rescattering model in the final state for the $B \to K\pi\pi$ decay, considering the S-wave $K\pi$ interactions in the resonant $1/2$ state, the $K_0^*(1430)$, $K_0^*(1630)$ and $K_0^*(1950)$ resonances and the non-resonant $3/2$ isospin states. The scattering matrix was parametrized and fixed with the requirement of fitting the LASS data [20], as done in the D decay case [15].

In the light-front, the inhomogeneous integral equations reduce to three-dimensional ones, solved here with a perturbative series up to three-loops and with the accuracy of the solution checked. The convergence of the series is clear and the two-loop results shows up enough for practical applications, as happened in the D meson decay case of Ref. [15]. In comparison with the decay of the lighter D meson, we needed to use a larger imaginary part for the propagators of the mesons by increasing the ε parameter. Since this parameter mimics the absorption to other decay channels, it is expected that in the B decay, ε increases due to the much larger phase space available. The momentum cut-off was chosen smaller in the B case than in the D decay, in order to have a good convergence. Such a decrease seems reasonable as B is much more massive than D. The heavier particle should have a larger number of decay channels, meaning larger absorption, and therefore the wave function of the particular decay channel at short-distances, where the absorption takes place, is suppressed. The result is that the outgoing state is more concentrated at large distances, which corresponds to the low-momentum region. The smaller cut-off in the B decay with respect to the D one, can be understood as an effective way to parametrize the physics of the larger number of open channels.

The resonant structure above the $K_0^*(1430)$ resonance is also a question that deserves a detailed analysis in face of future experimental data. While the presence of the $K_0^*(1630)$ resonance is expected, and this is in fact used in our amplitude analysis, the $K_0^*(1950)$ influence must be better understood. Other aspect that requires more study are the real weights of the three isospin components of the source amplitudes at the quark level. Three-body rescattering effects are also important because they distribute CP violation to different decay channels, since it is one of the mechanisms allowed by the CPT constraint [24]. In the near future, this light-front approach will be generalized in order to study CP violation in three-body charmless B decays, taking into account the unitarity of the S-matrix, and the CPT constraint, exactly as done in Refs. [6,7].

Acknowledgements We thank the Brazilian funding agencies Fundação de Amparo à Pesquisa do Estado de São Paulo (FAPESP) and Conselho Nacional de Desenvolvimento Científico e Tecnológico (CNPq). J.H.A.N. also acknowledges the support of Grant No. 2014/19094-8 from FAPESP.

References

1. M. Beneke, G. Buchalla, M. Neubert, C.T. Sachrajda, QCD factorization for $B \to \pi\pi$ decays: strong phases and CP violation in the heavy quark limit. Phys. Rev. Lett. **83**, 1914–1917 (1999)
2. M. Beneke, G. Buchalla, M. Neubert, C.T. Sachrajda, QCD factorization for exclusive non-leptonic B-meson decays: general arguments and the case of heavylight final states. Nucl. Phys. B **591**, 313–418 (2000)
3. Y.Y. Keum, H. Li, A.I. Sanda, Fat penguins and imaginary penguins in perturbative QCD. Phys. Lett. B **504**, 6–14 (2001)
4. C.W. Bauer, I.W. Stewart, Invariant operators in collinear effective theory. Phys. Lett. B **516**, 134–142 (2001)
5. J.F. Donoghue, E. Golowich, A.A. Petrov, J.M. Soares, Systematics of soft final-state interactions in B decays. Phys. Rev. Lett. **11**, 2178–2181 (1996)
6. I. Bediaga, T. Frederico, O. Lourenço, CP violation and CPT invariance in B$^\pm$ decays with final state interactions. Phys. Rev. D **89**, 094013 (2014)
7. J.H.A. Nogueira, I. Bediaga, A.B.R. Cavalcante, T. Frederico, O. Lourenço, CP violation: Dalitz interference, CPT, and final state interactions. Phys. Rev. D **92**, 054010 (2015)
8. R. Aaij, LHCb Collaboration et al., Study of B$_c^+$ decays to the K$^+$K$^-\pi^+$ final state and evidence for the decay B$_c^+ \to \chi_{c0}\pi^+$. Phys. Rev. D **94**, 091102(R) (2016)
9. N. Wang, Charmless B$_c \to$ PP, PV decays in the QCD factorization approach. Adv. High Energy Phys. **2016**, 6314675 (2016)
10. B. Mohammadi, H. Mehraban, Final state interaction effects on the B$^+ \to$ J/$\psi\rho^+$ decay. Adv. High Energy Phys. **2012**, 203692 (2012)
11. S. Kränkl, T. Mannel, J. Virto, Three-body non-leptonic B decays and QCD factorization. Nucl. Phys. B **899**, 247–264 (2015)
12. Z.H. Zhang, X.H. Guo, Y.D. Yang, CP violation in B$^\pm \to \pi^\pm\pi^+\pi^-$ in the region with low invariant mass of one $\pi^+\pi^-$ pair. Phys. Rev. D **87**, 076007 (2013)
13. Zhang, Z.H., Guo, X.H., Yang, Y.D., CP violation induced by the interference of scalar and vector resonances in three-body decays of bottom mesons. arXiv:1308.5242 [hep-ph]
14. X.W. Kang, B. Kubis, C. Hanhart, U.G. Meiner, B$_{l4}$ decays and the extraction of $|V_{ub}|$. Phys. Rev. D **89**, 053015 (2014)
15. K.S.F.F. Guimarães, O. Lourenço, W. de Paula, T. Frederico, A.C. dos Reis, Final state interaction in D$^+ \to$ K$^-\pi^+\pi^+$ with Kπ $I = 1/2$ and $3/2$ channels. J. High Energy Phys. **08**, 135 (2014)

16. K.S.F.F. Guimarães, I. Bediaga, A. Delfino, T. Frederico, A.C. dos Reis, L. Tomio, Three-body model of the final state interaction in heavy meson decay. Nucl. Phys. B Proc. Suppl. **199**, 341–344 (2010)
17. P.C. Magalhães, M.R. Robilotta, K.S.F.F. Guimarães, T. Frederico, W. de Paula, I. Bediaga, A.C. dos Reis, C.M. Maekawa, G.R.S. Zarnauskas, Towards three-body unitarity in $D^+ \to K^- \pi^+ \pi^+$. Phys. Rev. D **84**, 094001 (2011)
18. J.H.O. Sales, T. Frederico, B.V. Carlson, P.U. Sauer, Light front Bethe–Salpeter equation. Phys. Rev. C **61**, 044003 (2000)
19. S.X. Nakamuram, Coupled-channel analysis of $D^+ \to K^- \pi^+ \pi^+$ decay. Phys. Rev. D **93**, 014005 (2016)
20. D. Aston, N. Awaji, T. Bienz, F. Bird, J. D'Amore, W.M. Dunwoodie, R. Endorf, K. Fujii et al., A study of $K^- \pi^+$ scattering in the reaction $K^- p \to K^- \pi^+ n$ at 11 GeV/c. Nucl. Phys. B **296**, 493–716 (1988)
21. P. Estabrooks, R.K. Carnegie, A.D. Martin, W.M. Dunwoodie, T.A. Lasinski, D.W.G.S. Leith, Study of $K\pi$ scattering using the reactions $K^- + p \to K^- + \pi^+ n$ and $K^- + p \to K^- + \pi^- \Delta^{++}$ at 13 GeV/c. Nucl. Phys. B **133**, 490–524 (1978)
22. S.J. Brodsky, H.C. Pauli, S.S. Pinsky, Quantum chromodynamics and other field theories on the light-cone. Phys. Rep. **301**, 299–486 (1998)
23. R.J. Perry, A. Harindranath, K.G. Wilson, Light front Tamm–Dancoff field theory. Phys. Rev. Lett. **65**, 2959–2962 (1990)
24. J.H.A. Nogueira, I. Bediaga, T. Frederico, P.C. Magalhães, J.M. Rodriguez, Suppressed $B \to PV$ CP asymmetry: CPT constraint. Phys. Rev. D **94**, 054028 (2016)

Few-Body Syst (2017) 58:112
DOI 10.1007/s00601-017-1271-9

U. van Kolck

Unitarity and Discrete Scale Invariance

Received: 2 December 2016 / Accepted: 28 February 2017 / Published online: 16 March 2017
© Springer-Verlag Wien 2017

Abstract While the complexity of some many-body systems may stem from a profusion of distinct scales, as we approach two-body unitarity (through experimental control or as a theoretical limit) rich structures exist even though there is no more than one essential scale. I comment, from the point of view of effective field theory, on some current problems in the transition from few to many bodies in bosonic and multi-state fermion systems, where order emerges from the discrete scale invariance associated with a single, contact three-body force.

1 Introduction

It is a trivial observation, of profound consequences, that nature is not scale invariant. Tracking how physics changes with resolution is the task of the renormalization group (RG), which underlies the paradigm of effective field theories (EFTs). In this paradigm, nature is described by a succession of EFTs, each one relevant in a certain energy regime. I focus here on the scales that appear in the EFTs for non-relativistic few-body systems with short-range interactions whose two-body subsystems are near unitarity. I will argue that only one scale is essential. At its 30th anniversary, *Few-Body Syst.* continues to provide a unique forum for discussions like this, which cut across the atomic, nuclear and other communities where one can hope to understand just *how* "more" becomes "different".

Despite the obvious absence of scale invariance in nature as a whole, there have been frequent attempts to formulate one EFT or another using scale invariance as an approximate classical symmetry that is anomalous, that is, broken by quantum effects. The systems that I consider here lend themselves to such an attempt, because, by definition of "near", the two-body dynamics near unitarity is approximately characterized by no scale other than the energy. Parameters that originate in the short-distance dynamics play only a minor role, which is amenable to perturbation theory. No finite-energy two-body bound state can exist. Yet, when the bodies are identical bosons and fermions with more than two spin and/or internal states, such as nucleons with spin 1/2 and isospin 1/2, more-body bound states exist with finite energy: continuous scale invariance is broken. In EFT, this breaking is anomalous: it is *required* by the RG at a quantum level in the three-body system [1–3]. But the

---

This article belongs to the Topical Collection "30th anniversary of Few-Body Systems".

U. van Kolck (✉)
Institut de Physique Nucléaire, CNRS/IN2P3, Univ. Paris-Sud, Université Paris-Saclay, 91406 Orsay, France
E-mail: vankolck@ipno.in2p3.fr

U. van Kolck
Department of Physics, University of Arizona, Tucson, AZ 85721, USA

U. van Kolck
Kavli Institute for Theoretical Physics, University of California, Santa Barbara, CA 93016, USA

breaking preserves *discrete* scale invariance, which implies a geometric tower of few-body bound states, as first discovered by Efimov in the case of three particles [4]. The lowest states of the Efimov tower have been seen near Feshbach resonances [5,6] and in naturally occurring atomic ^{4}He [7,8], while triton is the nuclear Efimov ground state.

I offer here a *Comment* on the growing effort to study the structure of systems with more particles near unitarity. A significant motivation for this enterprise is the promise of new, similarly striking discoveries. Another is the simplicity associated with universality. Existing evidence is that a single, three-body dimensionful parameter is sufficient to describe the overall structure, with everything else, again, less important. This conclusion is not universally accepted but, if correct, it has far-reaching implications. For two-state fermions of mass m, it has long been realized that scale invariance means that energies are determined by the external constraints. For example, in the many-body system where one defines a Fermi momentum k_F, the energy can only be proportional to the free Fermi gas energy,

$$\lim_{N \to \infty} \frac{E_N}{N} \equiv \xi \frac{3k_F^2}{10m}. \tag{1}$$

ξ, known as the Bertsch parameter, is a pure, universal number that summarizes all the dynamics of this system. The sign of ξ, for example, determines the sign of the pressure. It makes for a meeting ground in the comparison of various approaches.

For bosons or multi-state fermions, I propose that similar focus be placed on the expression for N-body ground-state binding energies [9–11] that arises from the anomalous breaking of scale invariance,

$$B_N = \kappa_N B_3, \tag{2}$$

where κ_N are pure numbers ($\kappa_2 = 0$ and $\kappa_3 = 1$). As I discuss below, for small N it seems that

$$\kappa_N \approx (N - 2)^2. \tag{3}$$

I suggest that, if the growth slows down, the discrete scale invariance underlying Eq. (2) would imply, as an analog of Eq. (1), that

$$\lim_{N \to \infty} \frac{E_N}{N} = \frac{k_\rho^2}{m} \daleth \left[s_0 \ln \left(k_\rho / \sqrt{m B_3} \right) \right], \tag{4}$$

where k_ρ is the momentum associated with the density, s_0 is a constant, and $\daleth$ is a periodic function. What happens when N grows?

2 Unitarity and Scale Invariance

Non-relativistic particles interacting through a force of range R can be described at distances $r \gg R$ by a Hamiltonian containing contact interactions only. The interaction potential can be expanded in powers of R/r, starting with non-derivative contact interactions and continuing with successively more derivatives. It is important to realize that even if there is an underlying potential that is mostly two-body, its expansion at large distances will contain higher-body components, stemming from successive two-body encounters at short distances and times. In most situations these higher-body forces are relatively small, but not always. All contact interactions allowed by symmetries are non-vanishing. The problem for the theorist is to figure out which interactions are dominant.

One way to formulate the dynamics starts from the most general action consistent with the symmetries and built out of a field ψ that annihilates the particles of interest:

$$S = \int \frac{dt}{2m} \int d^3x \left\{ \psi^\dagger \left(2im \frac{\partial}{\partial t} + \nabla^2 \right) \psi - 4\pi C_0 \left(\psi^\dagger \psi \right)^2 - \frac{(4\pi)^2}{3} D_0 \left(\psi^\dagger \psi \right)^3 + \cdots \right\}, \tag{5}$$

where C_0, D_0, *etc.* are interaction strengths or, in EFT jargon, low-energy constants (LECs). The "..." include terms with more derivatives. Note that I use units where $\hbar = c = 1$. Since, with the above definitions of the LECs,[1] m enters Eq. (5) only in the combination t/m, it appears in observables together with the energy E as

---

[1] My choice here differs from the usual definition found in the literature.

$mE = k^2$. As a consequence, binding energies $B \propto 1/m$, which is at the root of the alternative units frequently used in atomic physics, where one makes $m = 1$.

In writing Eq. (5) I neglected spin projections that are important when particles have spin. For a two-state fermion, the C_0 interaction operates only between two particles in different states, and the D_0 interaction vanishes because of the exclusion principle. For more-state fermions, more than one C_0- and/or D_0-type interaction is possible, depending on the symmetries. For simplicity of notation, I will ignore such complications in the following. I assume there are sufficiently many approximate symmetries in the space of fermion states so that, as for bosons, only a single D_0-type interaction is important. This is, for example, the case for nucleons which, in addition to (presumably exact) spatial-rotation invariance, have an approximate invariance under rotations in isospin space—that is, rotations that make a nucleon a superposition of proton and neutron. In this case the isospin-breaking interaction is small and there is a single relevant D_0-type interaction, which is invariant [3] under Wigner's $SU(4)_W$ symmetry in spin–isospin space. (The approximate $SU(4)_W$ symmetry of this EFT was elaborated upon in Refs. [12,13].) The discussion can be generalized to more complicated cases is a straightforward way.

An essential aspect of this expansion around the zero-range limit is that the resulting interactions are singular. They demand an arbitrary procedure that—at intermediate steps of the calculation of observables, that is, of the S matrix—regularizes the short-range behavior of amplitudes by the introduction of a parameter Λ^{-1} with dimensions of distance. The delta functions become smeared over distances $r \lesssim \Lambda^{-1}$, which I denote by $\delta_\Lambda^{(3)}(r_{ij})$. This change must be compensated by a dependence of interaction strengths on Λ, in such a way that observables are not sensitive to the arbitrary regularization procedure. This second step—renormalization—is crucial to ensure that no detailed assumptions—other than symmetries—are being made about the short-range dynamics.

For $N = 2$, the action (5) gives rise to a potential

$$V_2(r_{ij}; \Lambda) = \frac{4\pi}{m} C_0(\Lambda) \, \delta_\Lambda^{(3)}(r_{ij}) \tag{6}$$

among the two particles labeled i and j, separated by the distance r_{ij}. It can be shown [14–17] that after renormalization the scattering amplitude is equivalent to the effective range expansion (ERE) [18], Fermi's pseudopotential [19], and boundary conditions at short distances [20,21]. (The equivalence to the ERE in dimensional regularization with a particular subtraction scheme was shown in Refs. [22,23].) In particular, we can write $C_0(\Lambda)$ as

$$C_0(\Lambda) = -\frac{1}{\theta_0 \Lambda} \left[1 + \frac{1}{\theta_0 \Lambda a_2} + \mathcal{O}\left(\frac{1}{(\Lambda a_2)^2}\right) \right], \tag{7}$$

where a_2 is the two-body scattering length (in the appropriate S-wave channel, if more than one exists), and $\theta_0 = \mathcal{O}(1)$ depends on the specific regularization employed. By dimensional analysis we expect the effective range and higher ERE parameters to be set by the potential range R, $|r_2| \sim R$, etc. In general the same is true for the scattering length, but $|a_2| \gg R$ when there exists a shallow real or virtual S-wave bound state. The scattering amplitude for on-shell momentum $k \ll 1/R$ is

$$T_2(k) = \frac{4\pi}{m} \left(a_2^{-1} + ik \right)^{-1} \left[1 + \mathcal{O}\left(kR, k\Lambda^{-1} \right) \right]. \tag{8}$$

There is a pole at a binding energy $B_2 = (ma_2^2)^{-1}$. The unitarity limit $a_2^{-1} \to 0$ (and $r_2 \to 0$, etc.) is an idealization of the region $|a_2^{-1}| \ll k \ll 1/R$, which I will refer to as the "unitarity window", where there is no important dimensionful parameter other than k itself. The unitarity limit corresponds to a non-trivial fixed point of the RG [24].

The vanishing of the binding energy in the unitarity limit is a reflection of scale invariance. Under a change of scales [25] with parameter $\alpha > 0$,

$$r \to \alpha r, \quad t/m \to \alpha^2 t/m, \quad \Lambda \to \alpha^{-1}\Lambda, \quad \psi \to \alpha^{-3/2}\psi, \tag{9}$$

the first two terms in Eq. (5) are invariant, but only when $a_2^{-1} \to 0$ in Eq. (7). Under a scale change, $mE \to \alpha^{-2}mE$, but in the unitarity limit there is no scale, so $B_2 = 0$. In this limit the $N = 2$ system is also conformally invariant [26].

Away from the unitarity limit, scale symmetry is explicitly broken by a dimensionful parameter, a_2. We can use the "spurion field" method [27], which is designed to exploit the consequences of an approximate symmetry, to determine the dependence of B_2 on a_2. The idea is that *if* under scale invariance a_2 changed to αa_2, then the first two terms in Eq. (5) would be invariant. In that case, the energy after the transformation should equal the transformed energy: $B_2(\alpha a_2) = \alpha^{-2} B_2(a_2)$. This implies $B_2(a_2) \propto (ma_2^2)^{-1}$. Now, since a_2 is fixed, $B_2(a_2)$ represents the specific way a_2 breaks scale invariance. In this particular case the spurion method is just dimensional analysis, since by allowing a_2 to vary we are changing all dimensionful quantities according to their (inverse mass) dimension. Of course this relation can be obtained directly from Eq. (8), but the spurion method illustrates how considerations of symmetries underlie dynamical results.

For $N \geq 2$ two-state fermions, the potential is approximately

$$V = \sum_{\{ij\}} V_2(r_{ij}; \Lambda), \tag{10}$$

where $V_2(r_{ij}; \Lambda)$ is given by Eq. (6) and the sum extends over pairs of particles in different states. The N-body T matrix is properly renormalized without further interactions [14, 16]. At unitarity, scale invariance then ensures that there cannot be any finite, negative energy B_N. In order to bind the system, some sort of "external" interaction must be imposed, such as gravity in a neutron star. More simply, we can place the system in a cubic box of volume V. The corresponding breaking of scale invariance can be treated like the breaking from a_2. The system now can have a ground state with energy [28]

$$E_N = \xi_N \left(\left(\frac{V}{3\pi^2 N a_2^3} \right)^{1/3} \right) \varepsilon_{FG} = \left[\xi_N(0) + \xi'_N(0) \left(\frac{V}{3\pi^2 N a_2^3} \right)^{1/3} + \cdots \right] \varepsilon_{FG}, \tag{11}$$

where

$$\varepsilon_{FG} = \frac{3}{10m} \left(\frac{3\pi^2 N}{V} \right)^{2/3} \tag{12}$$

is the energy per particle of a free Fermi gas, and ξ_N is a dimensionless, universal function of its dimensionless argument, which I expanded within the unitarity window $R^3 \ll V/(3\pi^2 N) \ll |a_2|^3$. Monte Carlo calculations [29–31] with simple potentials give a nearly constant even-odd oscillation in N (a pairing gap) superimposed on a linear growth,

$$\xi_N(0) \approx N\xi, \quad \xi'_N(0) \approx -N\zeta, \ldots, \tag{13}$$

with $\xi \simeq 0.4$, $\zeta \simeq 1$, *etc.* At unitarity, a large system described by the number density $\rho = N/V = k_F^3/(3\pi^2)$ does not collapse, and obeys Eq. (1). Away from unitarity, $\xi'_N(0) < 0$ means the energy increases as $-a_2^{-1}$ increases for $a_2 < 0$. Conversely, as $a_2^{-1} > 0$ increases the energy decreases and is consistent with that of a gas of spin-0 dimers interacting with a scattering length $a_{dd} \simeq 0.6a_2$ [32]. The $a_2 > 0$ Fermi gas is metastable due to dimer formation.

Obviously one can consider additional scales such as the effective range, temperature and polarization, leading to a dimensionless function of several dimensionless ratios and rich phase diagrams. For the effective range, for example, a generalization of the above argument leads to

$$\lim_{N \to \infty} \frac{E_N}{N} \equiv \frac{3k_F^2}{10m} \left(\xi - \frac{\zeta}{a_2 k_F} + \eta \, r_2 k_F + \cdots \right), \tag{14}$$

where $\eta \simeq 0.1$ [33].

How relevant are these results for nuclear physics? Evidence was presented in Ref. [34] that the momenta relevant for the ground states of light nuclei are within the unitarity window of the $^1S_0$ channel, $|a_{1S_0}^{-1}| \ll k \ll r_{1S_0}^{-1}$, where $a_{1S_0} \simeq -23.7$ fm and $r_{1S_0} \simeq 2.73$ fm. Starting from unitarity in this channel, binding energies can be obtained in an expansion in both $(k\, a_{1S_0})^{-1}$ and $k\, r_{1S_0}$—and for Z protons, also in $Z\alpha m_N/k$, where α is the fine-structure constant and $m_N \simeq 940$ MeV the nucleon mass, and other isospin-breaking corrections. Likewise, at finite density, Eq. (14) provides an approximation for neutron matter in the unitarity window (*cf.*, for example, Ref. [35]).

3 Three Bodies and Discrete Scale Invariance

Physics is quite different for multi-state fermions (such as protons and neutrons of spin up or down) or bosons (such as atomic ^{4}He), where the exclusion principle does not forbid three particles in the same state. The first surprise was probably the observation by Thomas [36] of a "collapse" in the $N = 3$ ground state: under a potential like (10) that gives rise to a shallow two-body (real or virtual) bound state, the ground-state binding energy $B_3 \propto \Lambda^2/m$ for $\Lambda \gg |a_2^{-1}|$. As Λ increases, excited bound states emerge and collapse as well. This is a consequence of a bizarre behavior of the half-off-shell amplitude for scattering of a particle on the two-particle bound state, which oscillates as a function of the off-shell momentum with a phase that depends on $\ln \Lambda$ [1–3]. Small cutoff variations result in large changes at low momentum. This regulator dependence indicates that the first two terms in Eq. (5) are not renormalizable beyond $N = 2$. Although they differ in detail, the situation is the same for multi-state fermions and bosons. For nucleons, for example, the coupled three-body integral equations for the half-off-shell amplitude reduce [3] at unitarity to an equation with a well-defined solution, plus an equation identical to that for bosons.

Since two-body interactions with more derivatives are smaller at low momentum, the appropriate counterterm must be a three-body force. Indeed, if we add in the potential the non-derivative three-body force with parameter D_0 (the third term in Eq. (5)),

$$V = \sum_{\{ij\}} V_2(r_{ij}; \Lambda) + \sum_{\{ijk\}} V_3(r_{ijk}; \Lambda), \tag{15}$$

where $\{ijk\}$ stands for a sum over all triplets and

$$V_3(r_{ijk}; \Lambda) = \frac{(4\pi)^2}{m} D_0(\Lambda)\, \delta_\Lambda^{(3)}(r_{ij}) \delta_\Lambda^{(3)}(r_{jk}), \tag{16}$$

then $D_0(\Lambda)$ can exactly counterbalance the cutoff variation for $N = 3$ if [1–3]

$$D_0(\Lambda) \propto \frac{1}{\Lambda^4} \frac{\sin\left(s_0 \ln(\Lambda/\Lambda_\star) - \arctan s_0^{-1}\right)}{\sin\left(s_0 \ln(\Lambda/\Lambda_\star) + \arctan s_0^{-1}\right)} \left[1 + \mathcal{O}\left((a\Lambda)^{-1}\right)\right], \tag{17}$$

where $s_0 \simeq 1.00624$ and $\Lambda_\star$ is a physical parameter, which is however only defined up to a factor $\exp(n_\star\pi/s_0)$, with $n_\star$ an integer. This is an RG limit cycle. Once one low-momentum datum is reproduced by a choice of $\Lambda_\star$, the phase of the half-off-shell scattering amplitude oscillation is fixed and all other low-momentum observables attain finite values as Λ increases. This is, in particular, true of bound states. Instead of the periodic emergence of bound states at zero energy before renormalization, after renormalization one observes the periodic emergence of *deeper* bound states, which achieve finite binding energies as Λ increases. At unitarity, in addition, the tower of states is geometric, extending down to threshold, with successive states having a ratio of binding energies $B_{3;n+1}/B_{3;n} = \exp(-2n\pi/s_0) \simeq 1/515$ [4]. At finite a_2 and R, only a few ($\sim \ln(|a_2|/R)/\pi$ [37]) of these states will be in the range of applicability of the theory. For atomic ^{4}He, for example, both the ground [7] and first-excited [8] states have been detected. For nucleons, only the ground state (triton/^{3}He) is observed, but one can show that there is a virtual state in neutron-deuteron (nd) scattering that becomes the triton excited state as the deuteron binding energy is decreased [38,39]. Here and below, I label bound states with an integer $n \geq 0$, starting at the ground state, that is, the lowest state within the EFT.

Like in the two-body sector, derivative three-body interactions provide relatively small effects [2,40–44]. Because the three-body force (17) contains a single parameter, one expects correlations among the observables that are sensitive to this force. Clearly that is not the case for every three-body observable; for example, the scattering length for nd scattering in the spin-3/2 channel, where the two neutrons are aligned, is determined to a very good accuracy by two-nucleon physics [14,16]. But observables in channels not affected by the exclusion principle are sensitive to this one parameter. The classic example is the Phillips line [45]: a line in the plane spanned by the triton binding energy and the spin-1/2 nd scattering length. This correlation was first discovered empirically, as a line formed by points representing various phenomenological potentials, which describe two-nucleon data up to relatively high momenta. As $\Lambda_\star$ is varied, the EFT also produces a line [1–3], which lies close not only to the experimental point but also to the empirical line. This means that the various phenomenological potentials, each with its many parameters, are equivalent to the same EFT with different values of $\Lambda_\star$. The conclusion that three-body systems at or near unitarity are determined by a single three-body force is confirmed.

The dimensionful parameter $\Lambda_\star$ arises from renormalization, even though at unitarity we start without any scale—an example of "dimensional transmutation". Scale (as well as conformal) invariance is "anomalously" broken, but a discrete scale invariance [46] remains [1–3]. Because of the characteristic dependence on Λ in Eq. (17), the first *three* terms in Eq. (5) are invariant under a transformation (9), but only for discrete values

$$\alpha_l = e^{l\pi/s_0} \simeq (22.7)^l, \qquad (18)$$

with l an integer. $\Lambda_\star$ offers a dimensionful scale upon which binding energies can be built in the unitarity limit. By dimensional analysis, we can write

$$B_{3;n} = \frac{\kappa_\star^2}{m} \beta_{3;n}\big((a_2\kappa_\star)^{-1}\big) = \frac{\kappa_\star^2}{m} \left[\beta_{3;n}(0) + \frac{\beta'_{3;n}(0)}{a_2\kappa_\star} + \cdots \right], \quad \beta_{3;0}(0) = 1, \qquad (19)$$

where $\beta_{3;n}((a_2\kappa_\star)^{-1})$ are universal, dimensionless functions and $\beta_{3;n}(0)$, $\beta'_{3;n}(0)$, *etc.* are pure numbers. Because $\Lambda_\star$ is only defined up to a factor, it was traded above by a fixed scale $\kappa_\star$ defined from the ground-state energy at unitarity:

$$s_0 \ln(\kappa_\star) = s_0 \ln(\beta \Lambda_\star) \mod \pi, \qquad (20)$$

with $\beta \simeq 0.383$ [47].

The energy of a bound state after a discrete scale transformation should equal the transformed energy but not necessarily of the same level, so that

$$\beta_{3;n+l}(0) = \alpha_l^{-2} \beta_{3;n}(0) \quad \Leftrightarrow \quad \beta_{3;n}(0) = e^{-2n\pi/s_0} \beta_{3;0}(0). \qquad (21)$$

Thus discrete scale invariance leads to Efimov's geometric tower. Using the spurion field method, the deviation from unitarity due to the two-body scattering length gives additionally

$$\beta_{3;n+l}\big((\alpha_l a_2\kappa_\star)^{-1}\big) = \alpha_l^{-2} \beta_{3;n}\big((a_2\kappa_\star)^{-1}\big). \qquad (22)$$

This relation gives information about how Efimov's tower evolves as $|a_2^{-1}|$ grows.[2] For example, taking a derivative and expanding in $(a_2\kappa_\star)^{-1}$, we see the leading effect of tower deformation:

$$\beta'_{3;n+l}(0) = \alpha_l^{-1} \beta'_{3;n}(0) \quad \Leftrightarrow \quad \beta'_{3;n}(0) = e^{-n\pi/s_0} \beta'_{3;0}(0), \qquad (23)$$

where $\beta'_{3;0}(0) \simeq 2.11$ [47]. Note that here the spurion method is not simply dimensional analysis because $\Lambda_\star$ is kept fixed. It instead tracks how the two-body scattering length explicitly breaks discrete scale invariance. Equation (22) is only valid to the extent that the three-body force retains discrete scale invariance except for $(a_2\Lambda)^{-1}$ terms—that is, as long as Eq. (17) contains no $a_2\Lambda_\star$ dependence, which would require in the spurion method that we scaled $\Lambda_\star$ as well.

Corrections in $r_2\kappa_\star$ can be introduced as in Eq. (14). As long as Eq. (17) contains no $r_2\Lambda_\star$ dependence, the coefficients of linear corrections should scale with α_l^{-3} [42], as can be easily verified with the spurion method. (However, an explicit calculation [42] says that these coefficients vanish.) Other corrections can be handled similarly.

4 More Bodies but No More Scales?

The crucial issue now is whether a new essential scale appears for $N \geq 4$. By "essential" I mean, not accountable for perturbatively: it is essential for the very formulation of the theory. A four-body force carrying a new dimensionful parameter, for example, is not forbidden by symmetries, thus it appears at some level of precision. The question is thus whether $N \geq 4$ systems are well defined with the potential (15), in the absence of higher-body forces. If so, higher-body forces and their scales can be accounted for in perturbation theory at higher orders. In this section I review the evidence which suggests that no new essential scales are needed

---

[2] I thank H.-W. Hammer for stressing to me the importance of this relation.

for systems with short-range forces near unitarity. If that is the case, the spectrum of bound states is entirely fixed by $\Lambda_\star$, except for small corrections.

A many-body force is generated by the elimination of short-distance physics, and if it is an essential ingredient it will show up as did the three-body force (17): many-body observables will not be renormalized properly without such a force; instead, they will fail to converge as the cutoff Λ increases. If there is one regulator for which lack of convergence is seen, renormalizability is disproved. By now, there exist a few calculations with different regulators, which display no lack of convergence.

Let me start with $N = 4$. The pioneering calculations of Refs. [48–50] have found apparent convergence in the binding energies of the ground states for bosons and nucleons (away from unitarity), as well as of the first excited state for bosons. A separable Gaussian regulator was used with the cutoff varied over values that correspond to a full cycle of the three-body force. Such a result was confirmed (i) with the same regulator in Ref. [51] (at unitarity), and (ii) with a local Gaussian regulator (under a similar cutoff variation) for bosons (very close to unitarity) in Ref. [11] and for nucleons (away from unitarity) in Refs. [52–54]. The apparent convergence in these calculations implies a correlation between four- and three-body observables. The classic example is the Tjon line [55] in the plane of the four- and three-body ground-state energies, $B_{4;0}$ and $B_{3;0}$. As with the Phillips line, this correlation was discovered empirically by plotting results of phenomenological nuclear potentials. It also exists for ^{4}He atoms [56]. It materializes in EFT as a variation of $\Lambda_\star$ [48–50].

In contrast, in Refs. [57–59] a contact model suggested the emergence of a four-body scale. However, a subtraction procedure was employed which has no transparent connection to field theory. It is unclear at this point whether the new scale is essential in the sense defined above, or a consequence of an unessential subtraction. If the former possibility is true, one has no reason to expect a Tjon line. Nor would one expect the EFT Tjon line, which would presumably be cutoff dependent, to be close to the empirical Tjon line. The fact that it is means that the diverse physics of all the phenomenological potentials that form the Tjon line is captured by the single scale $\Lambda_\star$. There is at least a very large class of potentials that do not seem to have an extra, essential parameter. This result is confirmed by other potential-model calculations [60–63] and an independent RG analysis [64]. However, one cannot, of course, exclude a priori the existence of an exotic system where a many-body force is very large.

This analysis has been extended to $N = 5, 6, 16$. A calculation [65] of the ground-state energy for $N = 6$ nucleons, with a cutoff provided by the number of shells in a harmonic-oscillator basis, converged without five- or six-body forces. Similarly, there was no sign of extraneous cutoff dependence for $N = 16$ nucleons with a local Gaussian regulator [54]. The four spin–isospin states of nucleons require five- and more-body forces to include derivatives, which should suppress them. This suppression is absent for bosons, but calculations [11] with a local Gaussian regulator showed no evidence of the need for those forces, either. The absence of higher-body forces at leading order implies the existence of correlations of five- and six-body observables with three-body observables. For example, variation of $\Lambda_\star$ creates "generalized Tjon lines" in the planes spanned by the ground-state energies $B_{5;0}$ or $B_{6;0}$ and $B_{3;0}$ [11]. Again, such correlations had already been discovered in the context of potential models [60,66,67].

A limitation in the renormalization argument for the absence of new essential scales has been the range of cutoff values used. This limitation[3] has been mitigated recently. As indicated in Eq. (8) for $N = 2$, one expects the fall off of observables with cutoff to be governed by the characteristic momentum of the process. One possible way to estimate the binding momentum per particle is

$$Q_N \equiv \sqrt{2mB_N/N}. \tag{24}$$

The rationale for this estimate is that Q_2 defined this way agrees with the location of the two-body bound state in the complex plane for the relative on-shell momentum; $Q_{N\gg 1}$ amounts to each particle contributing equally to the binding energy in the center of mass; and Q_∞ gives a finite momentum for a saturating system, such as nuclear matter. Thus we would expect for the nth state of the N-body system

$$B_{N;n}(\Lambda) = B_{N;n}\left[1 + \alpha_N \frac{Q_{N;n}}{\Lambda} + \beta_N \left(\frac{Q_{N;n}}{\Lambda}\right)^2 + \cdots\right], \tag{25}$$

where the dimensionless parameters α_N, β_N, etc. are expected to be of $\mathcal{O}(1)$ if there is convergence to the asymptotic value $B_{N;n}$ (from which $Q_{N;n}$ is obtained via Eq. (24)). Results for the ground states of $N \leq 6$ ^{4}He atoms display exactly such a behavior [11]. Thus, within the range of cutoffs studied, these observables

---

[3] Frequently emphasized by T. Frederico.

are entirely consistent with convergence at higher cutoff values. Although one cannot completely exclude the opposite, there is quite a bit of evidence against the emergence of new, essential scales in these systems.

5 How Many is Too Many for Universality?

This is a remarkably broad universality, which extends far beyond the three-body system. It means that all states within the validity of the EFT, i.e., those states that are insensitive to the details of physics at distances $r \lesssim R$, are described essentially by a single parameter $\Lambda_\star$. This should include not only cold atoms around Feshbach resonances, but also naturally occurring systems near unitarity, such as atomic ^{4}He systems where the ratio $B_2/B_{3;0} \simeq 10^{-3}$ is close enough to vanishing for the trimer to accommodate an excited state [8]. A comparison between results for $N \leq 6$ in EFT and potential models [11] suggests that these systems are within this universality class. Recently we have argued [51], for example showing that the Tjon line can be obtained in perturbation theory from the Tjon line at unitarity, that even nuclei are part of this class of systems—even though $B_2/B_{3;0} \simeq 0.26$ is not as small in this case.

The absence at leading order of more-body forces with additional dimensionful parameters means that $N \geq 4$ systems should also display discrete scale invariance. Reference [50] discovered that an $N = 3$ state spawns two $N = 4$ states, one very close to the $N = 3$ threshold and one about four times deeper. According to the accurate calculation of Ref. [63], for the two lowest $N = 4$ states at unitarity, $B_{4;0;0}/B_{3;0} \simeq 4.611$ and $B_{4;0;1}/B_{3;0} \simeq 1.002$. These states have been spotted in atomic systems [68]. Note that the alpha particle has two states, $B_{4;0;0}/B_{3;0} \simeq 3$ and $B_{4;0;1}/B_{3;0} \simeq 1$, which are qualitatively similar to unitarity states: another evidence that light nuclei are perturbatively close to unitarity [51].

Remarkably, potential-model calculations show that this doubling process continues with increasing number of bosons [9,69–71], so that for a given N there are $2^{(N-3)}$ "interlocking" towers of states. Generalizing Eq. (19),

$$B_{N;n;\{i\}} = \frac{\kappa_\star^2}{m} \, \beta_{N;n;\{i\}}\big((a_2\kappa_\star)^{-1}\big), \tag{26}$$

by labeling each state with the $N = 3$ ancestor state (n) and a set $\{i\} = \{i_1, i_2, \ldots, i_{N-3}\}$ of $i_j = 0, 1$ labels tracking the doubling, with $i_j = 0$ (1) denoting the lower (higher) state. Just as before, the dimensionless functions $\beta_{N;n;\{i\}}$ of $(a_2\kappa_\star)^{-1}$ reduce at unitarity to pure numbers $\beta_{N;n;\{i\}}(0)$, which obey

$$\beta_{N;n+l;\{i\}}(0) = \alpha_l^{-2} \, \beta_{N;n;\{i\}}(0) \quad \Leftrightarrow \quad \beta_{N;n;\{i\}}(0) = e^{-2n\pi/s_0} \, \beta_{N;0;\{i\}}(0). \tag{27}$$

Again the spurion method gives

$$\beta_{N;n+l;\{i\}}\big((\alpha_l a_2\kappa_\star)^{-1}\big) = \alpha_l^{-2} \, \beta_{N;n;\{i\}}\big((a_2\kappa_\star)^{-1}\big), \tag{28}$$

with similar implications as for $N = 3$.

I am not aware of an explanation for the doubling, which has a topological interpretation [72], but the replicating towers are a reflection of the surviving discrete scale invariance. For $N \geq 4$, all but the two lower states appear as resonances in the scattering of a particle on the $(N-1)$-particle ground state. Because of the tower structure, we can focus on these two lower states. The higher one is near the ground state of the system with one less particle,

$$\beta_{N;0;0,\ldots,0,1}(0) \simeq \beta_{N-1;0;0,\ldots,0}(0), \tag{29}$$

and thus can be thought of as a two-body system of a particle and an $(N-1)$ cluster. More generally, this state and its cousins up the tower are the analog of halo nuclei: they have a clusterized structure with a certain number of "valence" particles orbiting a tight cluster of the remaining particles. Their low-energy properties can be described by Halo EFT [73,74]. In contrast, the ground states close to unitarity get deeper and deeper as summarized by

$$B_N \equiv B_{N;0;\{0\}} \simeq \beta_{N;0;\{0\}}(0) \, B_{3;0} \equiv \kappa_N B_3, \tag{30}$$

which are the generalized Tjon lines calculated for atomic ^{4}He in Ref. [11].

For $N > 4$ multi-state fermions the pattern of states is not clear, but towers must also exist. We can retain the above notation, leaving the structure of the set $\{i\}$ unspecified. For the ground states, Eq. (30) still holds, but with a different set of numbers κ_N. The four-state fermion system, for example, reduces to a bosonic system [75], similarly to the three-nucleon system [3]. κ_4 is then the same in both cases, but $\kappa_{N\geq5}$ must differ on account of the exclusion principle. For example, in nuclei the $N = 5$ ground state is a shallow resonance, that is, the two poles that are bound states for bosons are instead in the lower half-plane of the complex nucleon-^{4}He relative-momentum plane. Halo EFT [73,74] provides a description of this type of state that is completely parallel to the Contact (or Pionless) EFT that describes identical particles near unitarity.

Equation (30) is the counterpart for bosons and multi-state fermions of a relation like (11) for two-state fermions. The set of numbers κ_N encapsulates the dynamical information about the ground states at unitarity, and similarly to Eq. (19) deviations from unitarity can be considered in a $(a_2 \kappa_\star)^{-1}$ expansion. There have been some attempts to determine the N dependence of B_N, mostly for bosons. Reference [10] extracted an N^2 growth from the statistical distribution of the two-particle propagator in Monte Carlo simulations. With a particular potential, Refs. [9,70] obtained values that for $N \leq 6$ are well described by the number of triplets, $N(N-1)(N-2)/6$, slowing down to an N^2 growth for $7 \leq N \leq 10$ and even slower for $N \geq 11$. Ground-state ratios calculated from other potentials [76,77] show different trends as N increases, but tend to agree with the form proposed in Ref. [78],

$$B_{N \geq 3} \approx \left[(N-3)\sqrt{\frac{B_4}{B_3}} + 4 - N \right]^2 B_3. \tag{31}$$

This relation is more general than (30), since it accommodates a separate four-body scale through B_4, but of course is consistent with Eq. (30) if $B_4/B_3 = \kappa_4$ is a pure number. In EFT, Ref. [11] found Eq. (3) for $N \leq 6$, which is consistent with Eq. (31) for $B_4/B_3 \approx 4$.

The question immediately arises,[4] what is the large-N behavior of κ_N? As κ_N quickly increases at small N, the Q_N in Eq. (24) increases, and the density increases. In EFT, finite energies for few-body systems arise from a balance between two- and three-body forces, and the latter cannot be separated from the high-momentum components of the former in a cutoff-independent way. But effectively the three-body force stops the Thomas collapse. We might expect the effectively repulsive effect of three-body forces to cause saturation, that is, that there is a finite limit

$$\lim_{N \to \infty} \frac{B_N}{N} = \kappa \frac{B_3}{3} \tag{32}$$

with a universal, pure number κ. At this point there is no strong support for this possibility, other than the tapering off in growth observed in Ref. [9]. It could be, for example, that at some N the binding momentum per particle becomes large enough that we can no longer neglect short-range interaction details, and new essential emerge. (For bosons in two spatial dimensions, for example, it has been argued [79] that B_{N+1}/B_N approaches a constant for large N, quickly leading to the loss of universality.) In fact, the above estimate for Q_N would indicate that one is already probing distances comparable to the Van der Waals length in ^{4}He atomic systems for $N \sim 6$. Still, in Ref. [11] it was found that EFT and potential-model results are in reasonably good agreement, suggesting that the above estimate for Q_N is an overestimate. The limit of validity of the description in terms of the single parameter $\Lambda_\star$ and its associated discrete scale invariance remains an open question.

The arguments in the previous paragraph need indeed to be taken with a truck of salt. For one thing, we have little experience with treating the many-body problem from an EFT perspective where renormalization ties two- and three-body forces. But such a "superfluid Efimov liquid" phase of a Bose fluid has been argued for in Ref. [80] on the basis of a Monte Carlo simulation for a short-range two-body interaction supplemented by a hard-core three-body potential. (In contrast, much of the literature on the Bose gas, for example Ref. [81], focuses on the atomic phase. While for $a_2 < 0$ such a phase is mechanically unstable, for $a_2 > 0$ it is metastable, and the dynamics before collapse is universal in the sense of the two-component Fermi gas.)

To see the potential differences stemming from an EFT approach, let me assume that not only the many-body ground state is within the EFT, but also that the energy per particle, $\lim_{N \to \infty} E_N/N$, of the infinite problem at zero temperature, when excited states do not contribute, can in a first approximation be described solely by the number density ρ. This is the case in density functional theory [82] and in a class of Skyrme forces where nuclear-matter saturation arises from (unrenormalized) two- and three-body forces [83]. Such is also the case for Eq. (1), but there it is a consequence of scale invariance. Here we can ask instead for the constraints from discrete scale invariance under the extra assumption of dependence solely on

$$\rho = \frac{g}{6\pi^2} k_\rho^3, \tag{33}$$

where g is the degeneracy factor ($g = 1$ for one boson species and $g = 4$ for a nucleon) and k_ρ^{-1} measures the average particle separation ($k_\rho \equiv (9\pi/2)^{1/3} r_0^{-1}$ for bosons and $k_\rho = k_F$ for fermions). I am assuming here

---

[4] A discussion on this topic with J. Carlson and M. J. Savage is gratefully acknowledged.

no polarization in the space of fermion states, e.g. no spin or isospin polarization for nucleons. Polarization and/or temperature can be accounted for along the lines of what is done for two-state fermions.

From dimensional analysis [84],

$$\lim_{N \to \infty} \frac{E_N[k_\rho]}{N} = \frac{\kappa_\star^2}{m} \, \varepsilon[k_\rho/\kappa_\star], \qquad (34)$$

with ε a dimensionless function of the dimensionless combination $k_\rho/\kappa_\star$. Since under a scale transformation $\rho \to \alpha^{-3}\rho$, discrete scale invariance implies additionally:

$$\varepsilon\left[\alpha_l^{-1} k_\rho/\kappa_\star\right] = \alpha_l^{-2} \, \varepsilon[k_\rho/\kappa_\star] \quad \Leftrightarrow \quad \varepsilon[k_\rho/\kappa_\star] = \frac{k_\rho^2}{\kappa_\star^2} \, \daleth\left[s_0 \ln(k_\rho/\kappa_\star)\right], \qquad (35)$$

or

$$\lim_{N \to \infty} \frac{E_N[k_\rho]}{N} = \frac{k_\rho^2}{m} \, \daleth\left[s_0 \ln(k_\rho/\kappa_\star)\right], \qquad (36)$$

where $\daleth$ is (real) periodic with period π,

$$\daleth[x + l\pi] = \daleth[x] = \frac{3\gamma_0}{10} - \sum_{n=1}^{\infty} \gamma_n \sin\left(2nx + \phi_n\right). \qquad (37)$$

In the last step I expanded $\daleth$ in a Fourier series in terms of real coefficients $\gamma_{n \geq 0}$ and phases $\phi_{n \geq 1}$. Corrections in $(a_2\kappa_\star)^{-1}$ and $r_2\kappa_\star/2$ can be incorporated by the spurion formalism as before.

I do not see a general argument to determine the form of $\daleth$, which could be different for bosons and fermions. γ_0 is the direct analog of the Bertsch parameter ξ in Eq. (1), and m/γ_0 can be thought of as an in-medium or effective mass. If there were no three-body effects, $\gamma_{n \geq 1} = 0$ and the pressure (related to the first derivative of Eq. (36)) would not vanish anywhere—there would be no saturation. Thus, for stability at least one of the $\gamma_{n \geq 1}$ is non-vanishing and the dependence on k_ρ is non-analytic—qualitatively very different from the power-law dependence on ρ of simple mean-field models with two- and three-body forces [83]. The energy (36) is a curve in the $\lim_{N \to \infty}(m E_N/N) \times k_\rho$ plane, which is a minimum at the saturation point $k_\rho = k_0$. Since $\kappa_\star$ provides the scale for both axis on this plane, as $\Lambda_\star$ changes the minima lie on a curve $\lim_{N \to \infty}(m E_N/N)_{k_0} \propto -k_0^2$. Again, a correlation between the minimum energy per nucleon and the saturation density—the so-called Coester line—has been found [85] for phase-shift-equivalent two-nucleon potentials. The line is, indeed, approximately quadratic in k_0 [84,85]. The EFT at unitarity, with its one essential parameter $\Lambda_\star$, provides a possible mechanism for this line, which would be displaced in nuclear systems by $(Qa_2)^{-1}$ and $r_2 Q/2$ effects—just like the Phillips and Tjon lines.

As a minimal model to explore the saturation effects of the three-body force, let me consider the first two terms in Eq. (37),

$$\lim_{N \to \infty} \frac{E_N[k_\rho]}{N} = \frac{k_\rho^2}{m} \left[\frac{3\gamma_0}{10} - \gamma_1 \sin\left(2s_0 \ln(k_\rho/\kappa_\star) + \phi_1\right)\right]. \qquad (38)$$

The position of the energy minimum and the energy at the saturation point are overdetermined by $\gamma_{0,1}$ and ϕ_1. These parameters can be used to fix the compressibility,

$$K_\infty = \left(k_\rho^2 \frac{d^2}{dk_\rho^2} \lim_{N \to \infty} \frac{E_N[k_\rho]}{N}\right)_{k_\rho = k_0} = 4\left[s_0^2 \gamma_0 \frac{3k_0^2}{10m} - (1 + s_0^2)\left(\lim_{N \to \infty} \frac{E_N[k_0]}{N}\right)\right], \qquad (39)$$

as well. If I additionally take $\phi_1 = 0$, saturation in (unpolarized, isospin-symmetric) nuclear matter, where $\kappa_\star \simeq 90$ MeV, $k_0 \simeq 260$ MeV, and $\lim_{N \to \infty}(E_N[k_0]/N) \simeq -16$ MeV, results from $\gamma_0 = 0.50$ and $\gamma_1 = 0.43$. Both values are natural and, interestingly, γ_0 turns out to be numerically close to the Bertsch parameter. For the compressibility one finds $K_\infty \simeq 170$ MeV, or about 30% off the empirical value $K_\infty \simeq 250$ MeV. This is reasonable agreement considering that corrections are $\mathcal{O}((\kappa_\star a_2)^{-1}, \kappa_\star r_2/2) \sim 0.5$. At this moment, however, there is nothing but optimism to justify Eq. (38).

Given the large period $\simeq 22.7\kappa_\star$ in momentum, the periodic behavior of $\daleth$ would be difficult to observe, as are excited Efimov states.[5] I do not believe anybody has seen a hint of oscillation in the nuclear-matter

---

[5] This paragraph benefited from an interesting exchange with H. W. Grießhammer.

equation of state, but it would have been surprising if they did: half of the period corresponds to a momentum above the maximum relative momentum used in the usual fits of data with phenomenological two-nucleon potentials. At such momenta it is unlikely that an expansion around unitarity converges. Minima that are deeper than the physical minimum must be outside the EFT, just as the states in the few-body towers that are below the physical ground states. To see the log-periodic behavior we would need more than one minimum within the EFT domain. In any case, we would be lucky if the expansion converges even at saturation. It would be interesting if for nuclear matter, for example, one could exclude Eq. (36), or else find a specific form that produces a realistic equation of state around the saturation density.

6 No Scales Left to Shed

There has been a lot of interest in systems that are universal in the sense of having dynamics depending in a first approximation on no parameter, and thus being ruled by scale invariance. Nowhere near as extensive is our understanding of the remarkable features of those systems that are characterized essentially by a single, three-body parameter, and ruled by discrete scale invariance.

Much of what I said above—with the possible exception of the speculation at the end—has been known to some people for a while. Still, my goal was to show that the latter class of systems is perhaps even more interesting than the former, because they present richer structures from a still pretty simple action. The discrete version of scale invariance allows for geometric towers of bound states in few-body systems and is conjectured to yield saturation in larger systems, which retains the log-periodic behavior observed in smaller systems. The simple model of Eq. (38), for example, should be contrasted with Eq. (1): while both contain a repulsive term that can be interpreted as a simple reduction of the kinetic term via a quadruplication of the effective mass, discrete scale invariance allows for an additional log-periodic attraction which stems from the three-body force (17). Beyond the intrinsic theoretical beauty of the structures that emerge from discrete scale invariance, there is some tantalizing evidence that these structures survive in bosonic atom clusters and in nuclei. Hopefully this *Comment* encourages further study of the many-body consequences of discrete scale invariance.

Acknowledgements I thank Harald Grießhammer and Hans Hammer for useful discussions and comments on the manuscript, and Betzalel Bazak, Joe Carlson, Tobias Frederico, Sebastian König, Denis Lacroix, Martin Savage, and Mike Wagman for additional discussions. This material is based upon work supported in part by the National Science Foundation under Grant No. NSF PHY-1125915, by the US Department of Energy, Office of Science, Office of Nuclear Physics, under award DE-FG02-04ER41338, and by the European Union Research and Innovation program Horizon 2020 under Grant No. 654002.

References

1. P.F. Bedaque, H.-W. Hammer, U. van Kolck, Phys. Rev. Lett. **82**, 463 (1999). [arXiv:nucl-th/9809025]
2. P.F. Bedaque, H.-W. Hammer, U. van Kolck, Nucl. Phys. A **646**, 444 (1999). [arXiv:nucl-th/9811046]
3. P.F. Bedaque, H.-W. Hammer, U. van Kolck, Nucl. Phys. A **676**, 357 (2000). [arXiv:nucl-th/9906032]
4. V. Efimov, Phys. Lett. B **33**, 563 (1970)
5. T. Kraemer et al., Nature **440**, 315 (2006)
6. B. Huang et al., Phys. Rev. Lett. **112**, 190401 (2014)
7. W. Schöllkopf, J.P. Toennies, J. Chem. Phys. **104**, 1155 (1996)
8. M. Kunitski et al., Science **348**, 551 (2015). [arXiv:1512.02036 [physics.atm-clus]]
9. J. von Stecher, J. Phys. B **43**, 101002 (2010)
10. A.N. Nicholson, Phys. Rev. Lett. **109**, 073003 (2012). [arXiv:1202.4402 [cond-mat.quant-gas]]
11. B. Bazak, M. Eliyahu, U. van Kolck, Phys. Rev. A **94**, 052502 (2016). [arXiv:1607.01509 [cond-mat.quant-gas]]
12. T. Mehen, I.W. Stewart, M.B. Wise, Phys. Rev. Lett. **83**, 931 (1999). [arXiv:hep-ph/9902370]
13. J. Vanasse, D.R. Phillips, Few Body Syst. **58**, 26 (2017). [arXiv:1607.08585 [nucl-th]]
14. P.F. Bedaque, U. van Kolck, Phys. Lett. B **428**, 221 (1998). [arXiv:nucl-th/9710073]
15. U. van Kolck, Lect. Notes Phys. **513**, 62 (1998). [arXiv:hep-ph/9711222]
16. P.F. Bedaque, H.-W. Hammer, U. van Kolck, Phys. Rev. C **58**, R641 (1998). [arXiv:nucl-th/9802057]
17. U. van Kolck, Nucl. Phys. A **645**, 273 (1999). [arXiv:nucl-th/9808007]
18. H.A. Bethe, Phys. Rev. **76**, 38 (1949)
19. E. Fermi, Ric. Sci. **7**, 13 (1936)
20. H.A. Bethe, R. Peierls, Proc. R. Soc. Lond. A **148**, 146 (1935)
21. H.A. Bethe, R. Peierls, Proc. R. Soc. Lond. A **149**, 176 (1935)
22. D.B. Kaplan, M.J. Savage, M.B. Wise, Phys. Lett. B **424**, 390 (1998). [arXiv:nucl-th/9801034]
23. D.B. Kaplan, M.J. Savage, M.B. Wise, Nucl. Phys. B **534**, 329 (1998). [arXiv:nucl-th/9802075]
24. S. Weinberg, Nucl. Phys. B **363**, 3 (1991)
25. C.R. Hagen, Phys. Rev. D **5**, 377 (1972)

26. T. Mehen, I.W. Stewart, M.B. Wise, Phys. Lett. B **474**, 145 (2000). [arXiv:hep-th/9910025]
27. B. d'Espagnat, J. Prentki, Nuovo Cim. **3**, 1045 (1956)
28. A. Bulgac, G.F. Bertsch, Phys. Rev. Lett. **94**, 070401 (2005). [arXiv:cond-mat/0404687]
29. J. Carlson, S.-Y. Chang, V.R. Pandharipande, K.E. Schmidt, Phys. Rev. Lett. **91**, 050401 (2003)
30. S.-Y. Chang, V.R. Pandharipande, J. Carlson, K.E. Schmidt, Phys. Rev. A **70**, 043602 (2004). [arXiv:physics/0404115 [physics.atom-ph]]
31. G.E. Astrakharchik, J. Boronat, J. Casulleras, S. Giorgini, Phys. Rev. Lett. **93**, 200404 (2004)
32. D.S. Petrov, C. Salomon, G.V. Shlyapnikov, Phys. Rev. Lett. **93**, 090404 (2004)
33. J. Carlson, S. Gandolfi, K.E. Schmidt, S. Zhang, Phys. Rev. A **84**, 061602 (2011). [arXiv:1107.5848 [cond-mat.quant-gas]]
34. S. König, H.W. Grießhammer, H.-W. Hammer, U. van Kolck, J. Phys. G **43**, 055106 (2016). [arXiv:1508.05085 [nucl-th]]
35. E. Epelbaum, H. Krebs, D. Lee, U.-G. Meißner, Eur. Phys. J. A **40**, 199 (2009). [arXiv:0812.3653 [nucl-th]]
36. L.H. Thomas, Phys. Rev. **47**, 903 (1935)
37. R.D. Amado, J.V. Noble, Phys. Rev. D **5**, 1992 (1972)
38. S.K. Adhikari, L. Tomio, Phys. Rev. C **26**, 83 (1982)
39. A. Vaghani, R. Higa, G. Rupak, U. van Kolck, In preparation
40. H.-W. Hammer, T. Mehen, Phys. Lett. B **516**, 353 (2001). [arXiv:nucl-th/0105072]
41. P.F. Bedaque, G. Rupak, H.W. Grießhammer, H.-W. Hammer, Nucl. Phys. A **714**, 589 (2003). [arXiv:nucl-th/0207034]
42. L. Platter, C. Ji, D.R. Phillips, Phys. Rev. A **79**, 022702 (2009). [arXiv:0808.1230 [cond-mat.other]]
43. C. Ji, D.R. Phillips, Few-Body Syst. **54**, 2317 (2013). [arXiv:1212.1845 [nucl-th]]
44. J. Vanasse, Phys. Rev. C **88**, 044001 (2013). [arXiv:1305.0283 [nucl-th]]
45. A.C. Phillips, Nucl. Phys. A **107**, 209 (1968)
46. D. Sornette, Phys. Rept. **297**, 239 (1998). [arXiv:cond-mat/9707012 [cond-mat.stat-mech]]
47. E. Braaten, H.-W. Hammer, Phys. Rept. **428**, 259 (2006). [arXiv:cond-mat/0410417]
48. L. Platter, H.-W. Hammer, U.-G. Meißner, Phys. Rev. A **70**, 052101 (2004). [arXiv:cond-mat/0404313]
49. L. Platter, H.-W. Hammer, U.-G. Meißner, Phys. Lett. B **607**, 254 (2005). [arXiv:nucl-th/0409040]
50. H.-W. Hammer, L. Platter, Eur. Phys. J. A **32**, 113 (2007). [arXiv:nucl-th/0610105]
51. S. König, H.W. Grießhammer, H.-W. Hammer, U. van Kolck, arXiv:1607.04623 [nucl-th]
52. J. Kirscher, H.W. Grießhammer, D. Shukla, H.M. Hofmann, Eur. Phys. J. A **44**, 239 (2010). [arXiv:0903.5538 [nucl-th]]
53. J. Kirscher, N. Barnea, D. Gazit, F. Pederiva, U. van Kolck, Phys. Rev. C **92**, 054002 (2015). [arXiv:1506.09048 [nucl-th]]
54. L. Contessi, A. Lovato, F. Pederiva, A. Roggero, J. Kirscher, U. van Kolck, arXiv:1701.06516 [nucl-th]
55. J.A. Tjon, Phys. Lett. B **56**, 217 (1975)
56. S. Nakaichi, Y. Akaishi, H. Tanaka, T.K. Lim, Phys. Lett. A **68**, 36 (1978)
57. M.T. Yamashita, L. Tomio, A. Delfino, T. Frederico, Europhys. Lett. **75**, 555 (2006)
58. M.R. Hadizadeh, M.T. Yamashita, L. Tomio, A. Delfino, T. Frederico, Phys. Rev. Lett. **107**, 135304 (2011). [arXiv:1101.0378 [physics.atm-clus]]
59. M.R. Hadizadeh, M.T. Yamashita, L. Tomio, A. Delfino, T. Frederico, Phys. Rev. A **85**, 023610 (2012). [arXiv:1110.5214 [cond-mat.soft]]
60. G.J. Hanna, D. Blume, Phys. Rev. A **74**, 063604 (2006)
61. J. von Stecher, J.P. D'Incao, C.H. Greene, Nat. Phys. **5**, 417 (2009)
62. J.P. D'Incao, J. von Stecher, C.H. Greene, Phys. Rev. Lett. **103**, 033004 (2009)
63. A. Deltuva, Phys. Rev. A **82**, 040701 (2010). [arXiv:1009.1295 [physics.atm-clus]]
64. R. Schmidt, S. Moroz, Phys. Rev. A **81**, 052709 (2010). [arXiv:0910.4586 [cond-mat.quant-gas]]
65. I. Stetcu, B.R. Barrett, U. van Kolck, Phys. Lett. B **653**, 358 (2007). [arXiv:nucl-th/0609023]
66. S. Nakaichi, T.K. Lim, Y. Akaishi, H. Tanaka, J. Chem. Phys. **71**, 4430 (1979)
67. T.K. Lim, S. Nakaichi, Y. Akaishi, H. Tanaka, Phys. Rev. A **22**, 28 (1980)
68. F. Ferlaino et al., Phys. Rev. Lett. **102**, 140401 (2009)
69. M. Gattobigio, A. Kievsky, M. Viviani, Phys. Rev. A **84**, 052503 (2011). [arXiv:1106.3853 [physics.atm-clus]]
70. J. von Stecher, Phys. Rev. Lett. **107**, 200402 (2011). [arXiv:1106.2319 [cond-mat.quant-gas]]
71. M. Gattobigio, A. Kievsky, M. Viviani, Phys. Rev. A **86**, 042513 (2012). [arXiv:1206.0854 [physics.atm-clus]]
72. Y. Horinouchi, M. Ueda, Phys. Rev. A **94**, 050702 (2016). [arXiv:1603.05328 [cond-mat.quant-gas]]
73. C.A. Bertulani, H.-W. Hammer, U. van Kolck, Nucl. Phys. A **712**, 37 (2002). [arXiv:nucl-th/0205063]
74. P.F. Bedaque, H.-W. Hammer, U. van Kolck, Phys. Lett. B **569**, 159 (2003). [arXiv:nucl-th/0304007]
75. L. Platter, Ph.D. dissertation, University of Bonn (2005)
76. A. Kievsky, N.K. Timofeyuk, M. Gattobigio, Phys. Rev. A **90**, 032504 (2014). [arXiv:1405.2371 [cond-mat.quant-gas]]
77. Y. Yan, D. Blume, Phys. Rev. A **92**, 033626 (2015)
78. M. Gattobigio, A. Kievsky, Phys. Rev. A **90**, 012502 (2014). [arXiv:1309.1927 [cond-mat.quant-gas]]
79. H.-W. Hammer, D.T. Son, Phys. Rev. Lett. **93**, 250408 (2004). [arXiv:cond-mat/0405206]
80. S. Piatecki, W. Krauth, Nat. Commun. **5**, 3503 (2014)
81. D. Borzov, M.S. Mashayekhi, S. Zhang, J.-L. Song, F. Zhou, Phys. Rev. A **85**, 023620 (2012)
82. J.E. Drut, R.J. Furnstahl, L. Platter, Prog. Part. Nucl. Phys. **64**, 120 (2010). [arXiv:0906.1463 [nucl-th]]
83. T.H.R. Skyrme, Nucl. Phys. **9**, 615 (1959)
84. A. Delfino, T. Frederico, V.S. Timóteo, L. Tomio, Phys. Lett. B **634**, 185 (2006). [arXiv:0704.0481 [nucl-th]]
85. F. Coester, S. Cohen, B. Day, C.M. Vincent, Phys. Rev. C **1**, 769 (1970)

Few-Body Syst (2017) 58:1–10
DOI 10.1007/s00601-016-1165-2

Ronen Weiss · Ehoud Pazy · Nir Barnea

Short Range Correlations: The Important Role of Few-Body Dynamics in Many-Body Systems

Received: 10 September 2016 / Accepted: 18 November 2016 / Published online: 16 December 2016
© Springer-Verlag Wien 2016

Abstract For many-body systems with short range interaction a series of relations were derived connecting many properties of the system to the dynamics of a closely packed few-body subsystems. Some of these relations were experimentally verified in ultra cold atomic gases. Here we shall review the implications of these developments on our understanding of nuclear one and two-body momentum distributions, and on the electron scattering Coulomb sum rule.

1 Introduction

Recent advances in the study of many-body systems suggest that the behavior of systems composed of particles interacting via short-range force are governed by the probability of finding a particle pair or triplet in a close proximity. Considering a system of two-component fermions interacting via "zero-range" s-wave forces, Tan [1–3] and later others (see e.g. [4,5] and references therein) have established a series of relations between the amplitude of the high-momentum tail of the momentum distribution $n(k)$ and many properties of the system. These relations, commonly known as the "Tan relations", are expressed through a new state variable, the "Contact" C, that for the aforementioned system dominates the tail of the momentum distribution $C = \lim_{k \to \infty} k^4 n(k)$. The contact C is a measure for the probability of finding a particle pair close to each other, or in other words, a measure for the short range correlations (SRCs) in the system. The Tan relations are universal, they hold for any few-body or many-body system where the interparticle distance d, and the magnitude of the scattering length a_s are both much larger than the potential range R_{pot}, i.e. $a_s, d \gg R_{pot}$, and the average particle momentum k is much smaller than $1/R_{pot}$.

Experiments done in two-component ultra-cold atomic Fermi gases, such as ^{40}K [6,7] and ^{6}Li [8–10], have tested the Tan relations. Using different experimental techniques and comparing several observables with the predictions of the universal theory, the Tan relations were verified.

Following Tan's seminal work on two-component fermions, similar relations were derived for one-component fermions, and for bosons. Due to the Pauli principle, in a system of one-component fermions, the low energy interaction is predominantly p-wave. In this case, the momentum distribution $n(k)$ falls as $1/k^2$ for large momentum and the corresponding contact is given by $C = \lim_{k \to \infty} k^2 n(k)$ [11].

The Tan relations for bosons interacting via zero-range force, have much in common with a system of two-component fermions. The main difference is the emergence of important 3-body correlations associated

This article belongs to the special issue "30th anniversary of Few-Body Systems".

R. Weiss · N. Barnea (✉)
Racah Institute of Physics, The Hebrew University, 91904 Givaat-Ram, Jerusalem, Israel
E-mail: nir@phys.huji.ac.il

E. Pazy
Department of Physics, NRCN, P.O.B. 9001, 84190 Beer-Sheva, Israel

with the 3-body force [12,13], that appear in low energy effective theory (For a review, see e.g. Ref. [14]) to prevent the Thomas collapse [15]. In this case the high momentum tail of the momentum distribution aquires a $1/k^5$ component modulated by a log-periodic function associated with the Efimov effect [16].

The atomic nuclei exhibit some key properties similar to these idealized systems. They are made of fermions, the nuclear force is short range, and the scattering length is much larger than the potential range. Therefore one expects SRCs to play an important role in nuclear physics (for literature see for example [17–25]), and the notion of the contact to be a useful concept in this context.

To apply Tan's ideas to nuclear systems we need to note also the differences between the atomic nuclei and the aforementioned systems. The nucleon can be regarded as a four-component fermion $N = (p\uparrow, p\downarrow, n\uparrow, n\downarrow)$, the protons are charged particles interacting via the long range Coulomb force, and most importantly the range of the nuclear force (dictated by the pion mass m_π) in not much shorter than the average interparticle distance, therefore the "zero range" approximation doesn't really hold.

The similarities between atomic nuclei and "zero-range" fermionic or bosonic systems suggest that the contact formalism can be a good starting point to study nuclear systems. The differences suggest that Tan's ideas must be generalized to accommodate nuclear systems, and that different or modified "relations" are expected to hold. For example, there is no obvious reason to assume that the asymptotic tail of the nuclear momentum distribution will take on Tan's form $\lim_{k \to \infty} n(k) = C/k^4$.

In the last couple of years the utility of the contact formalism in nuclear physics was started to be explored. The neutron-proton s-wave nuclear contacts have been defined and evaluated [26,27], relating their value to the Levinger photoabsorption constant [17,28,29], and to high energy inclusive electron scattering [27,30]. Considering not only s-wave but all partial waves, as well as finite-range interactions instead of zero-range, it was found that the nuclear SRCs are governed by a set of contact matrices [31]. Using this generalized contact formalism to relate the nuclear contacts to the one-nucleon and two-nucleon momentum distributions, an asymptotic relation between these two distributions was established [31], which is relevant to the study of SRCs in nuclei. This relation was verified using available numerical data [20]. The contact formalism was further applied to study the nuclear equation of state and symmetry energy (see e.g. [32] and references therein).

In this paper we first review Tan's relations for one-body and two-body momentum distributions, and their generalization to different ideal systems, Sect. 2. Then in Sect. 3 we present the generalization of these relations to nuclear systems [31]. Finally we utilize this formalism to derive the asymptotic behavior of the electron scattering Coulomb sum rule, Sect. 4.

Given our limited understanding of the role of 3-body effects in nuclear SRCs, throughout this paper we focus on the nuclear two-body contacts, and their applications.

2 Tan's Contact: The Single Channel Case

Consider an N–particle system that obeys Tan's assumptions (i.e. $a_s, d \gg R_{pot}$, and $k \ll 1/R_{pot}$), and is dominated by a single interaction channel such as an s-wave interaction in low energy Bose gas, or p-wave in a single-component Fermi gas. In such a gas, when interacting particle pair (ij) get close together, the many-body wave function Ψ can be factorized into a product of an asymptotic pair wave function $\varphi(\mathbf{r}_{ij})$, $\mathbf{r}_{ij} = \mathbf{r}_i - \mathbf{r}_j$, and a function A, also called the regular part of Ψ, describing the residual $N - 2$ particle system and the pair's center of mass $\mathbf{R}_{ij} = (\mathbf{r}_i + \mathbf{r}_j)/2$ motion [1–3,5],

$$\Psi \xrightarrow[r_{ij} \to 0]{} \varphi(\mathbf{r}_{ij}) A(\mathbf{R}_{ij}, \{\mathbf{r}_k\}_{k \neq i,j}) . \tag{1}$$

The asymptotic pair wave function φ depends on the interparticle potential and the dominant channel. In particular, in the zero-range model [33] the s-wave function is given by $\varphi = (1/r_{ij} - 1/a_s)$, where a_s is the scattering length. The contact C is then defined by [1–3,5]

$$C = 16\pi^2 N_{pairs} \langle A|A \rangle, \tag{2}$$

where

$$\langle A|A \rangle = \int \prod_{k \neq i,j} d\mathbf{r}_k \, d\mathbf{R}_{ij} \, A^\dagger \left(\mathbf{R}_{ij}, \{\mathbf{r}_k\}_{k \neq i,j} \right) \cdot A \left(\mathbf{R}_{ij}, \{\mathbf{r}_k\}_{k \neq i,j} \right) \tag{3}$$

and N_{pairs} is the number of interacting pairs. For a Bose or single-component Fermi gas $N_{pairs} = N(N-1)/2$, for two-component fermions $N_{pairs} = N_\uparrow N_\downarrow$.

Working in the momentum space,

$$\tilde{\Psi}(\boldsymbol{k}_1, \dots, \boldsymbol{k}_A) = \int \prod_{n=1}^{A} d^3 \boldsymbol{r}_n \, e^{i \sum_n \boldsymbol{k}_n \cdot \boldsymbol{r}_n} \, \Psi(\boldsymbol{r}_1, \dots, \boldsymbol{r}_A), \tag{4}$$

the short range factorization takes on the form of high momentum factorization. When a particle pair ij approach each other $r_{ij} \longrightarrow 0$ the relative momentum $\boldsymbol{k}_{ij} = (\boldsymbol{k}_i - \boldsymbol{k}_j)/2$ diverges $k_{ij} \longrightarrow \infty$. In this limit

$$\tilde{\Psi} \xrightarrow[k_{ij} \to \infty]{} \tilde{\varphi}(\boldsymbol{k}_{ij}) \tilde{A}(\boldsymbol{K}_{ij}, \{\boldsymbol{k}_n\}_{n \neq i,j}) , \tag{5}$$

where $\boldsymbol{K}_{ij} = \boldsymbol{k}_i + \boldsymbol{k}_j$, $\tilde{\varphi}$ is the Fourier transform of φ, and $\tilde{A}$ is the Fourier transform of A.

Using these definitions it is easy to see that the contact can be equivalently written as

$$C = 16\pi^2 N_{pairs} \langle \tilde{A} | \tilde{A} \rangle, \tag{6}$$

since

$$\langle \tilde{A} | \tilde{A} \rangle = \int \frac{d\boldsymbol{K}_{ij}}{(2\pi)^3} \prod_{n \neq i,j} \frac{d\boldsymbol{k}_n}{(2\pi)^3} \tilde{A}^\dagger \left(\boldsymbol{K}_{ij}, \{\boldsymbol{k}_n\}_{n \neq i,j} \right) \cdot \tilde{A} \left(\boldsymbol{K}_{ij}, \{\boldsymbol{k}_n\}_{n \neq i,j} \right)$$

$$= \langle A | A \rangle. \tag{7}$$

For finite systems such as nuclei and for non homogeneous infinite matter it can be important to consider two-body center of mass (CM) effects. Therefore it is convenient to introduce the *contact function* [1–3]

$$C(\boldsymbol{K}_{ij}) = 16\pi^2 N_{pairs} \int \prod_{n \neq i,j} \frac{d\boldsymbol{k}_n}{(2\pi)^3} \tilde{A}^\dagger \left(\boldsymbol{K}_{ij}, \{\boldsymbol{k}_n\}_{n \neq i,j} \right) \cdot \tilde{A} \left(\boldsymbol{K}_{ij}, \{\boldsymbol{k}_n\}_{n \neq i,j} \right). \tag{8}$$

An equivalent expression can also be derived for $C(\boldsymbol{R}_{ij})$. Comparing Eqs. (7) and (8) we see that the relation between the contact and the cotact function is given through the integral

$$C = \int \frac{d\boldsymbol{K}}{(2\pi)^3} C(\boldsymbol{K}). \tag{9}$$

In the following sections we will present the application of these definitions to derive Tan's relations for the one and two-body momentum distributions. For simplicity we first start with the two–body case.

2.1 The Two-Nucleon Momentum Distribution

Consider now a pair of particles ij with relative momentum $\boldsymbol{k}$ and CM momentum $\boldsymbol{K}$. We denote by $f(\boldsymbol{k} + \boldsymbol{K}/2, -\boldsymbol{k} + \boldsymbol{K}/2)$ the density probability to find particle i with momentum $\boldsymbol{k}_i = \boldsymbol{k} + \boldsymbol{K}/2$ and particle j with momentum $\boldsymbol{k}_j = -\boldsymbol{k} + \boldsymbol{K}/2$. We immediately see that

$$f(\boldsymbol{k}_i, \boldsymbol{k}_j) = N_{pairs} \int \prod_{m \neq i,j} \frac{d^3 \boldsymbol{k}_m}{(2\pi)^3} \left| \tilde{\Psi}(\boldsymbol{k}_1, \dots \boldsymbol{k}_A) \right|^2. \tag{10}$$

Here f is normalized in such away that $\int \frac{d^3 \boldsymbol{k}_i}{(2\pi)^3} \frac{d^3 \boldsymbol{k}_j}{(2\pi)^3} f(\boldsymbol{k}_i, \boldsymbol{k}_j) = N_{pairs}$.

For very large relative momentum the main contribution to $f(\boldsymbol{k}_i, \boldsymbol{k}_j)$ comes from the asymptotic $k \to \infty$ part of the wave function, given in Eq. (5). All other terms will vanish due to the fast oscillating $\exp(i\boldsymbol{k} \cdot \boldsymbol{r}_{ij})$ factor. Using Eq. (5) and substituting $\tilde{\Psi}$ into Eq. (10), we get

$$f(\boldsymbol{k} + \boldsymbol{K}/2, -\boldsymbol{k} + \boldsymbol{K}/2) = \frac{C(\boldsymbol{K})}{16\pi^2} |\tilde{\varphi}(\boldsymbol{k})|^2 . \tag{11}$$

Thus we see that in the limit $k \longrightarrow \infty$ the two-body momentum distribution is given by a product of the contact function and the universal momentum distribution $|\tilde{\varphi}(\boldsymbol{k})|^2$. For particles interacting via zero-range s-wave potential $\varphi \approx (1/r - 1/a)$ and $\tilde{\varphi} \approx 4\pi/k^2$. Similarly for the p-wave case $\tilde{\varphi} \approx 4\pi/k$.

The probability density to find a pair with relative momentum k is obtained by integrating over the CM momentum

$$F(k) = \int \frac{dK}{(2\pi)^3} f(k + K/2, -k + K/2).$$
(12)

Utilizing now the relation between the contact function and the contact we can now substitute the asymptotic form of f, Eq. (11), to get

$$F(k) = \frac{C}{16\pi^2} |\tilde{\varphi}(k)|^2$$
(13)

2.2 The One-Body Momentum Distribution

We would like to relate now the contact also to the single particle momentum distribution. To this end we will follow Tan's derivation for the two-body case [1–3]. To simplify the notation for the moment we will only consider the case of bosons and one-component fermions.

Normalized to the number of particles in the system, $\int \frac{dk}{(2\pi)^3} n(k) = N$, the single particle momentum distribution $n(k)$ is given by

$$n(k) = N \int \prod_{l \neq p} \frac{dk_l}{(2\pi)^3} \left| \tilde{\Psi}(k_1, \ldots, k_p = k, \ldots, k_A) \right|^2,$$
(14)

where p is any particle.

In the $k \longrightarrow \infty$ limit the main contribution to $n(k)$ emerges from the asymptotic parts of the wave function, i.e. from $r_{ps} = |r_p - r_s| \to 0$, for any particle $s \neq p$. In this limit

$$\tilde{\Psi}(k_1, \ldots, k_p = k, \ldots, k_A) = \sum_{s \neq p} \tilde{\varphi}\left(k - K_{ps}/2\right) \tilde{A}\left(K_{ps}, \{k_j\}_{j \neq p,s}\right),$$
(15)

where $K_{ps} = k_p + k_s$ is the center of mass momentum of the ps pair. We note that k is fixed in (14) while integrating over all other momenta. Therefore we can replace the integration $\int dk_s$ with integration over the pair's center of mass momentum $\int dK_{ps}$. Substituting this result into Eq. (14), we get summations over particles s and s' different from p. The contribution of non diagonal $s \neq s'$ terms, will be significant only for $k_s \approx k_{s'} \approx -k$, due to the regularity of A. In this case $k, k_s, k_{s'} \to \infty$ together, which is clearly a three body effect, and we expect it to be less important than the leading two-body contribution [34]. Consequently, we only consider the diagonal elements and obtain

$$n(k) = N \sum_{s \neq p} \int \frac{dK_{ps}}{(2\pi)^3} \prod_{l \neq p,s} \frac{dk_l}{(2\pi)^3} |\tilde{\varphi}(k - K_{ps}/2)|^2 |\tilde{A}(K_{ps}, \{k_j\}_{j \neq p,s})|^2$$

$$= 2 \int \frac{dK}{(2\pi)^3} \frac{C(K)}{16\pi^2} |\tilde{\varphi}(k - K/2)|^2$$
(16)

Deriving this result we have utilized the definition of the contact function, Eq. (8). The prefactor 2, results from the number of interacting pairs that for bosons or one-component fermions is given by $N_{pairs} = N(N-1)/2$. Consequently, for two-component fermions one obtains the same result up to this factor of 2.

Since A is regular, we expect $C(K)$ to be significant only for K_{ps} of the order of the average interparticle distance $1/d$. Therefore K can be considered to be much smaller than k. Expanding $|\tilde{\varphi}|^2$ around k,

$$|\tilde{\varphi}(k - K/2)|^2 \cong |\tilde{\varphi}(k)|^2 - \frac{K}{2} \cdot \left(\tilde{\varphi}^\dagger(k) \nabla_k \tilde{\varphi}(k) + \nabla_k \tilde{\varphi}^\dagger(k) \tilde{\varphi}(k)\right) + \ldots$$
(17)

and keeping only the leading order, which is a good approximation for any power-law function, we obtain

$$n(k) = 2 \frac{C}{16\pi^2} |\tilde{\varphi}(k)|^2 .$$
(18)

Substituting now the universal s-wave function $\tilde{\varphi}(k) \approx 4\pi/k^2$ we obtain Tan's result for two-component Fermi gas $n(k) = C/k^4$, recalling of course that for a two-component Fermi gas the factor of 2 disappears.

Comparing Eqs. (18) and Eq. (13), we can see that for $k \longrightarrow \infty$ there is a simple relation between the one-body and the two-body momentum distributions. For bosons or one-component fermions:

$$n(\mathbf{k}) = 2F(\mathbf{k}), \tag{19}$$

and for two-component fermions

$$n(\mathbf{k}) = F(\mathbf{k}). \tag{20}$$

3 The Nuclear Contact Matrices: The Multi Channel Case

Turning now to consider nuclear physics, we regard the nucleons as four-component fermions, which are the protons and neutrons with their spin being either up or down ($p\uparrow, p\downarrow, n\uparrow, n\downarrow$). As a result in the most simplistic model of the nuclear interaction one needs to consider at least two contacts [26], and in reality one needs to consider strong coupling between channels, such as s-wave and d-wave mixture in the deuteron. As a result, when extending the contact formalism to nuclear physics one needs to consider the different interaction channels and the possible couplings between them. Here we shall follow reference [31].

When a nucleon i gets close to nucleon j, we must abandon the factorization ansatz and write the wave function as a sum of products of two-body terms φ_{ij} and $A-2$-body terms A_{ij}, taking into account all possible channels. The asymptotic form of the wave function is then given by

$$\Psi \xrightarrow[r_{ij} \to 0]{} \sum_\alpha \varphi_{ij}^\alpha (\mathbf{r}_{ij}) A_{ij}^\alpha (\mathbf{R}_{ij}, \{\mathbf{r}_k\}_{k \neq i,j}). \tag{21}$$

We note that due to symmetry the asymptotic functions are invariant under same particle permutations. Therefore the index ij is an indicator to the particle pair type, i.e. proton-proton (pp), neutron-neutron (nn) or neutron-proton (np). The pair wave functions depend on the total spin of the pair s_2, and its angular momentum quantum number ℓ_2 (with respect to the relative coordinate $\mathbf{r}_{ij}$) which are coupled to create the total pair angular momentum j_2 and projection m_2. The quantum numbers (s_2, ℓ_2, j_2, m_2) define the pair's channel. In general, the expansion (21) may contain more then one term per channel, however in the limit $r_{ij} \to 0$ only the leading term survives. In short, the sum over α denotes a sum over the four channel quantum numbers (s_2, ℓ_2, j_2, m_2).

To ensure an A-body wave function with total angular momentum J and projection M the regular functions A_{ij}^α are given by

$$A_{ij}^\alpha = \sum_{J_{A-2}, M_{A-2}} \langle j_2 m_2 J_{A-2} M_{A-2} | J M \rangle A_{ij}^{\{s_2, \ell_2, j_2\} J_{A-2}, M_{A-2}}, \tag{22}$$

where J_{A-2} and M_{A-2} are the angular momentum quantum numbers with respect to the sum $\mathbf{J}_{A-2} + \mathbf{L}_{2,CM}$ of the total angular momentum of the residual ($A-2$) particles $\mathbf{J}_{A-2}$, and the orbital angular momentum $\mathbf{L}_{2,CM}$ corresponding to $\mathbf{R}_{ij}$. $A_{ij}^{\{s_2, \ell_2, j_2\} J_{A-2}, M_{A-2}}$ is a set of functions with angular momentum quantum numbers J_{A-2} and M_{A-2}, which depends also on the numbers s_2, ℓ_2, j_2.

$$\varphi_{ij}^\alpha \equiv \varphi_{ij}^{(\ell_2 s_2) j_2 m_2} = \phi_{ij}^{\ell_2, s_2, j_2}(r_{ij}) [Y_{\ell_2}(\hat{r}_{ij}) \otimes \chi_{s_2}]^{j_2 m_2}, \tag{23}$$

where $\chi_{s_2 \mu_s}$ is the two-body spin function, and $Y_{\ell m}$ are the spherical harmonics.

An important property of the set of asymptotic functions $\{\varphi_{ij}^\alpha\}$ is that they are "universal", in the limited sense that they do not depend on a specific nucleus or on a specific nuclear state. However, they can depend on the details of the nuclear potential and therefore cannot be related in a simple manner to the low energy scattering parameters.

Since the A_{ij}^α functions are not generally orthogonal for different α, we are led to define matrices of nuclear contacts in the following way [31]:

$$C_{ij}^{\alpha\beta} = \frac{16\pi^2 N_{ij}}{2J+1} \sum_M \langle A_{ij}^\alpha | A_{ij}^\beta \rangle = \frac{16\pi^2 N_{ij}}{2J+1} \sum_M \langle \tilde{A}_{ij}^\alpha | \tilde{A}_{ij}^\beta \rangle. \tag{24}$$

As before, $\tilde{A}_{ij}^\alpha$ is the Fourier transform of A_{ij}^α, ij stands for one of the pairs: pp, nn or np, N_{ij} is the number of ij pairs, and α and β are the matrix indices. Since the magnetic quantum number M is usually unknown in experiments, it is useful to define the averaged nuclear contacts.

In a similar fashion we can generalize the *contact function* (Eq. 8) and define the *contact matrix function* taking into account CM effects

$$C_{ij}^{\alpha\beta}(\boldsymbol{K}_{ij}) = \frac{16\pi^2 N_{ij}}{2J+1} \sum_M \int \prod_{n\neq i,j} \frac{d\boldsymbol{k}_n}{(2\pi)^3} \tilde{A}_{ij}^{\alpha\dagger}\left(\boldsymbol{K}_{ij}, \{\boldsymbol{k}_n\}_{n\neq i,j}\right) \cdot \tilde{A}_{ij}^{\beta}\left(\boldsymbol{K}_{ij}, \{\boldsymbol{k}_n\}_{n\neq i,j}\right). \tag{25}$$

In general the matrices $C_{ij}^{\alpha\beta}$ are built from 2×2 blocks, except for the two 1×1 blocks associated with the $j_2 = 0$ case. Each block has well defined j_2, m_2 values. A detailed discussion of the structure of the matrices $C_{ij}^{\alpha\beta}$ is given in [31].

It should be mentioned that the factorization of the wave function given in Eq. (21) was used before in the study of nuclear SRCs, see e.g. [25] and references therein. In these works the relation between the asymptotic many-body wave function and the deuteron wave function was utilized, and the corresponding contact was defined, see e.g. [23] Eq. (29). However, it was assumed that the nuclear contact is a single number and the general structure of nuclear contact matrix was not defined or analyzed. In the following we will demonstrate the utility of the general contact formalism, deriving analytic relations between the nuclear one-body and two-body momentum distributions.

3.1 The Nuclear Two-Nucleon Momentum Distribution

Now we can utilize the generalized contact formalism to find a relation between the two-nucleon momentum distribution and the nuclear contacts. To this end we define $f_{ij}^{JM}(\boldsymbol{k}_i, \boldsymbol{k}_j)$ to be the two-body momentum distribution of the ij pair associated with a nuclear state Ψ with magnetic projection M, and $f_{ij}(\boldsymbol{k}_i, \boldsymbol{k}_j) = 1/(2J+1) \sum_M f_{ij}^{JM}(\boldsymbol{k}_i, \boldsymbol{k}_j)$ to be the averaged two-body momentum distribution.

Following the footsteps and the arguments presented in Sect. 2.1 we can immediately get for $k \longrightarrow \infty$

$$f_{ij}(\boldsymbol{k} + \boldsymbol{K}/2, \boldsymbol{k} - \boldsymbol{K}/2) = \sum_{\alpha,\beta} \frac{C_{ij}^{\alpha\beta}(\boldsymbol{K})}{16\pi^2} \tilde{\varphi}_{ij}^{\alpha\dagger}(\boldsymbol{k}) \tilde{\varphi}_{ij}^{\beta}(\boldsymbol{k}). \tag{26}$$

Accordingly, we see that if

$$F_{ij}(\boldsymbol{k}) = \int \frac{d^3K}{(2\pi)^3} f_{ij}(\boldsymbol{k} + \boldsymbol{K}/2, -\boldsymbol{k} + \boldsymbol{K}/2), \tag{27}$$

then asymptotically [31]

$$F_{ij}(\boldsymbol{k}) = \sum_{\alpha,\beta} \frac{C_{ij}^{\alpha\beta}}{16\pi^2} \tilde{\varphi}_{ij}^{\alpha\dagger}(\boldsymbol{k}) \tilde{\varphi}_{ij}^{\beta}(\boldsymbol{k}). \tag{28}$$

3.2 The Nuclear One-Nucleon Momentum Distribution

In a similar way we can adress the asymptotic one-nucleon momentum distribution. We consider first the proton's momentum distribution $n_p(\boldsymbol{k})$. For convenience we will define $n_p(\boldsymbol{k}) = \sum_M n_p^{JM}(\boldsymbol{k})/(2J+1)$ to be the magnetic projection averaged proton momentum distribution. Normalized to the number of protons in the system Z, $\int \frac{d^3k}{(2\pi)^3} n_p^{JM}(\boldsymbol{k}) = Z$, n_p^{JM} is given by

$$n_p^{JM}(\boldsymbol{k}) = Z \int \prod_{l\neq p} \frac{d\boldsymbol{k}_l}{(2\pi)^3} \left| \tilde{\Psi}(\boldsymbol{k}_1, \dots, \boldsymbol{k}_p = \boldsymbol{k}, \dots, \boldsymbol{k}_A) \right|^2, \tag{29}$$

where p is any proton.

Following the arguments leading to Eq. (16) we see that

$$n_p^{JM}(\boldsymbol{k}) = Z \sum_{s\neq p} \sum_{\alpha,\beta} \int \prod_{l\neq p,s} \frac{d\boldsymbol{k}_l}{(2\pi)^3} \frac{d\boldsymbol{K}_{ps}}{(2\pi)^3} \tilde{\varphi}_{ps}^{\alpha\dagger}(\boldsymbol{k} - \boldsymbol{K}_{ps}) \tilde{\varphi}_{ps}^{\beta}(\boldsymbol{k} - \boldsymbol{K}_{ps})$$

$$\times \tilde{A}_{ps}^{\alpha\dagger}(\boldsymbol{K}_{ps}, \{\boldsymbol{k}_j\}_{j\neq p,s}) \tilde{A}_{ps}^{\beta}(\boldsymbol{K}_{ps}, \{\boldsymbol{k}_j\}_{j\neq p,s}). \tag{30}$$

110

We will now divide the sum $\sum_{s \neq p}$ into a sum over protons and a sum over neutrons $\sum_{p' \neq p} + \sum_n$, and average over M. Since the asymptotic functions $A^\alpha_{pp'}$ and $\varphi^\alpha_{pp'}$ are the same for all pp' pairs we can take them out of the sum. The same holds for the np pairs. As a result we get

$$n_p(\mathbf{k}) = 2 \sum_{\alpha, \beta} \int \frac{d\mathbf{K}}{(2\pi)^3} \frac{C^{\alpha\beta}_{pp}(\mathbf{K})}{16\pi^2} \tilde{\varphi}^{\alpha\dagger}_{pp}(\mathbf{k} - \mathbf{K}/2) \tilde{\varphi}^\beta_{pp}(\mathbf{k} - \mathbf{K}/2)$$

$$+ \sum_{\alpha, \beta} \int \frac{d\mathbf{K}}{(2\pi)^3} \frac{C^{\alpha\beta}_{pn}(\mathbf{K})}{16\pi^2} \tilde{\varphi}^{\alpha\dagger}_{pn}(\mathbf{k} - \mathbf{K}/2) \tilde{\varphi}^\beta_{pn}(\mathbf{k} - \mathbf{K}/2) \tag{31}$$

Expanding $\tilde{\varphi}^\alpha_{pn}(\mathbf{k} - \mathbf{K}/2)$ around $\mathbf{k}$ and keeping the leading term we obtain

$$n_p(\mathbf{k}) = 2 \sum_{\alpha, \beta} \frac{C^{\alpha\beta}_{pp}}{16\pi^2} \tilde{\varphi}^{\alpha\dagger}_{pp}(\mathbf{k}) \tilde{\varphi}^\beta_{pp}(\mathbf{k})$$

$$+ \sum_{\alpha, \beta} \frac{C^{\alpha\beta}_{pn}}{16\pi^2} \tilde{\varphi}^{\alpha\dagger}_{pn}(\mathbf{k}) \tilde{\varphi}^\beta_{pn}(\mathbf{k}). \tag{32}$$

Similarly, for the neutrons:

$$n_n(\mathbf{k}) = 2 \sum_{\alpha, \beta} \frac{C^{\alpha\beta}_{nn}}{16\pi^2} \tilde{\varphi}^{\alpha\dagger}_{nn}(\mathbf{k}) \tilde{\varphi}^\beta_{nn}(\mathbf{k})$$

$$+ \sum_{\alpha, \beta} \frac{C^{\alpha\beta}_{pn}}{16\pi^2} \tilde{\varphi}^{\alpha\dagger}_{pn}(\mathbf{k}) \tilde{\varphi}^\beta_{pn}(\mathbf{k}). \tag{33}$$

Comparing Eqs. (32) and (33) to Eq. (28), we can see that for $k \longrightarrow \infty$ there is a simple relation between the one-nucleon and the two-nucleon momentum distributions [31]:

$$n_p(\mathbf{k}) = 2F_{pp}(\mathbf{k}) + F_{pn}(\mathbf{k}) \tag{34}$$

$$n_n(\mathbf{k}) = 2F_{nn}(\mathbf{k}) + F_{pn}(\mathbf{k}). \tag{35}$$

These connections are the nuclear analog of Eqs. (19) and (20). Their validity was verified in [31] utilizing the numerical data of Ref. [20].

4 The Coulomb Sum Rule

Sum rules are useful tools in many fields of physics. In nuclear physics, they typically involve an integration of the response function, associated with transitions between the ground state and excited states due to an external probe, over the spectrum with a weight function composed of integer powers of the energy. The Coulomb sum rule (CSR) is the integral over the inelastic part of the longitudinal electron scattering nuclear response function. The CSR dates back to 1931 when Heisenberg studied the photoabsorption cross-section of x-rays with momentum q by atoms. Its formulation for electron scattering is due to Drell and Schwartz [35], and McVoy and Van Hove [36].

Assuming point-like particles, the CSR can be expressed as [37]

$$CSR \equiv \langle \Psi | \hat{\rho}^\dagger_c(\mathbf{q}) \hat{\rho}_c(\mathbf{q}) | \Psi \rangle - \left| \langle \Psi | \hat{\rho}_c(\mathbf{q}) | \Psi \rangle \right|^2, \tag{36}$$

where

$$\hat{\rho}_c(\mathbf{q}) = \sum_{j=1}^{Z} e^{i\mathbf{q} \cdot \mathbf{r}_j} \tag{37}$$

is the Fourier transform of the charge density operator $\hat{\rho}_c(\mathbf{r}) = \sum_{j=1}^{Z} \delta(\mathbf{r} - \mathbf{r}_j)$. The sum in Eq. (37) is understood to include protons only as essentially the CSR is a measure of charge fluctuations in the nucleus. Substituting the charge density operator (37) into (36), one gets

$$\langle \Psi | \hat{\rho}_c^\dagger(\mathbf{q}) \hat{\rho}_c(\mathbf{q}) | \Psi \rangle = Z + \langle \Psi | \sum_{i \neq j} e^{i\mathbf{q}\cdot(\mathbf{r}_i - \mathbf{r}_j)} | \Psi \rangle, \tag{38}$$

where the first contribution on the right-hand side (rhs) comes from the $i = j$ term in the sum. Due to the rapidly oscillating exponent, in the $q \longrightarrow \infty$ limit the matrix element on the rhs will be dominated by the behavior of the wave-function when two protons approach each other. In this limit we can replace the wave-function by its asymptotic form (21) and obtain

$$\langle \Psi | \hat{\rho}_c^\dagger(\mathbf{q}) \hat{\rho}_c(\mathbf{q}) | \Psi \rangle = Z + \sum_{i \neq j} \sum_{\alpha\beta} \int d\mathbf{r}_{ij} d\mathbf{R}_{ij} \prod_{k \neq i,j} d\mathbf{r}_k \varphi_{pp}^{\alpha\dagger}(\mathbf{r}_{ij}) e^{i\mathbf{q}\cdot\mathbf{r}_{ij}} \varphi_{pp}^{\beta}(\mathbf{r}_{ij})$$

$$\times A_{pp}^{\alpha\dagger}(\mathbf{R}_{ij}, \{\mathbf{r}_k\}_{k \neq i,j}) A_{pp}^{\beta}(\mathbf{R}_{ij}, \{\mathbf{r}_k\}_{k \neq i,j}),$$

$$= Z + Z(Z-1) \sum_{\alpha\beta} \langle A_{pp}^{\alpha\dagger} | A_{pp}^{\beta} \rangle h_{pp}^{\alpha\beta}(\mathbf{q}) \tag{39}$$

where

$$h_{pp}^{\alpha\beta}(\mathbf{q}) = \int d\mathbf{r} \varphi_{pp}^{\alpha\dagger}(\mathbf{r}) e^{i\mathbf{q}\cdot\mathbf{r}} \varphi_{pp}^{\beta}(\mathbf{r}) \tag{40}$$

is a universal proton-proton function independent of the particular nucleus or its quantum state. Averaging this result over the magnetic projection M and utilizing Eq. (24) we finally get for $q \to \infty$

$$CSR = Z + \sum_{\alpha\beta} \frac{2C_{pp}^{\alpha\beta}}{16\pi^2} h_{pp}^{\alpha\beta}(\mathbf{q}) - \rho_c^2(\mathbf{q}), \tag{41}$$

where $\rho_c(\mathbf{q}) = \langle \Psi | \hat{\rho}_c(\mathbf{q}) | \Psi \rangle$ is the nuclear charge distribution. Equation (41) relates the CSR to the proton-proton contact, as ultimately it should, since the CSR is a measure of charge fluctuations in the nucleus and on short length scales these are completely dominated by two protons coming close together. The CSR can be effectively probed experimentally in deep inelastic electron scattering experiments in which the virtual photon explores medium and short internucleon distances. In this regime the longitudinal nuclear structure function is sensitive to the pp SRCs which can modify the way the CSR approaches its model independent limit. The connection between the CSR and the contact provides a principle way to extract the proton-proton contacts from the available experimental electron scattering data. We note that the CSR for point-like particles is closely related to a quantity called the static structure factor, which was previously shown to be related to the single-channel s-wave contact for two-component fermions in the zero-range case [10,38].

5 Summary

Summing up, we have reviewed the derivation of Tan's relation for the one and two-body momentum distributions and their generalization to nuclear physics. Though Tan's relations were originally experimentally established for cold atomic systems [6–10], due to their universality they are also of significance in nuclear physics in which the energy scales are orders of magnitude larger and length scales are many orders of magnitude smaller. We have seen that the asymptotic $k \longrightarrow \infty$ tail of these distributions can be written as a product of the contact matrix and a universal two-body function, highlighting the role of few-body dynamics within a larger many-body system.

In general, the contact is a measure for the probability of finding a particle pair close to each other, it is thus a measure of the nuclear SRCs. While mean-field calculations provide important information on the nuclear shell structure, they do not constitute a full description of the structure of nuclei. Especially, deviations are large for small, dense nuclear structures. It is long been known that the strong, short-range component of the nucleon-nucleon potential generates a high-momentum tail in the nucleon momentum distribution. We have demonstrated how this phenomena can be described in terms of the contact. Employing high-energy electromagnetic probes, short scale fluctuations in the nucleus can be studied and information can thus be

⚛ Springer

obtained regarding the different nuclear contacts. Whereas in [26,27,30,31,39] a connection was found between the different nuclear *np* contacts and electromagnetic experimental rates, in this paper we have shown how short range charge fluctuations are connected to the pp contact. Thus different experiments can help extract information on the different nuclear contacts which can in turn help us better understand the details of the short range nuclear structure.

Acknowledgements The authors would like to thank Giuseppina Orlandini for useful discussions regarding the Coulomb sum-rule. This work was supported by the Pazy foundation.

References

1. S. Tan, Energetics of a strongly correlated Fermi gas. Ann. Phys. (N.Y.) **323**, 2952 (2008)
2. S. Tan, Large momentum part of a strongly correlated Fermi gas. Ann. Phys. (N.Y.) **323**, 2971 (2008)
3. S. Tan, Generalized virial theorem and pressure relation for a strongly correlated Fermi gas. Ann. Phys. (N.Y.) **323**, 2987 (2008)
4. E. Braaten, in *BCS-BEC Crossover and the Unitary Fermi Gas*, Chapter 6, ed. by W. Zwerger. Universal Relations for Fermions with Large Scattering Length (Springer, New York, 2012)
5. F. Werner, Y. Castin, General relations for quantum gases in two and three dimensions: two-component fermions. Phys. Rev. A **86**, 013626 (2012)
6. J.T. Stewart, J.P. Gaebler, T.E. Drake, D.S. Jin, Verification of universal relations in a strongly interacting Fermi gas. Phys. Rev. Lett. **104**, 235301 (2010)
7. Y. Sagi, T.E. Drake, R. Paudel, D.S. Jin, Measurement of the homogeneous contact of a unitary Fermi gas. Phys. Rev. Lett. **109**, 220402 (2012)
8. G.B. Partridge, K.E. Strecker, R.I. Kamar, M.W. Jack, R.G. Hulet, Molecular probe of pairing in the BEC-BCS crossover. Phys. Rev. Lett. **95**, 020404 (2005)
9. F. Werner, L. Tarruel, Y. Castin, Number of closed-channel molecules in the BEC-BCS crossover. Eur. Phys. J. B **68**, 401 (2009)
10. E.D. Kuhnle, H. Hu, X.-J. Liu, P. Dyke, M. Mark, P.D. Drummond, P. Hannaford, C.J. Vale, Universal behavior of pair correlations in a strongly interacting Fermi gas. Phys. Rev. Lett. **105**, 070402 (2010)
11. C. Luciuk, S. Trotzky, S. Smale, Z. Yu, S. Zhang, J.H. Thywissen, Evidence for universal relations describing a gas with p-wave interactions. Nat. Phys. **12**, 599 (2016)
12. P.F. Bedaque, H.-W. Hammer, U. van Kolck, Renormalization of the three-body system with short-range interactions. Phys. Rev. Lett. **82**, 463 (1999)
13. P.F. Bedaque, H.-W. Hammer, U. van Kolck, Effective theory of the triton. Nucl. Phys. A **676**, 357 (2000)
14. P.F. Bedaque, U. van Kolck, Effective field theory for few-nucleon systems. Annu. Rev. Nucl. Part. Sci. **52**, 339 (2002)
15. L. Thomas, The interaction between a neutron and a proton and the structure of H3. Phys. Rev. **47**, 903 (1935)
16. V. Efimov, Energy levels arising from resonant two-body forces in a three-body system. Phys. Lett. B **33**, 563 (1970)
17. J.S. Levinger, The high energy nuclear photoeffect. Phys. Rev. **84**, 43 (1951)
18. O. Hen et al., (CLAS Collaboration), Momentum sharing in imbalanced Fermi systems. Sci. **346**, 614 (2014)
19. J. Arrington, D. Higinbotham, G. Rosner, M. Sargsian, Hard probes of short-range nucleon-nucleon correlations. Prog. Part. Nucl. Phys. **67**, 898 (2012)
20. R.B. Wiringa, R. Schiavilla, S.C. Pieper, J. Carlson, Nucleon and nucleon-pair momentum distributions in $A \le 12$ nuclei. Phys. Rev. C **89**, 024305 (2014)
21. N. Fomin et al., New measurements of high-momentum nucleons and short-range structures in nuclei. Phys. Rev. Lett. **108**, 092502 (2012)
22. M. Alvioli, C.C. Degli Atti, H. Morita, Proton-neutron and proton-proton correlations in medium-weight nuclei and the role of the tensor force. Phys. Rev. Lett. **100**, 162503 (2008)
23. C.C. Degli Atti, S. Simula, Realistic model of the nucleon spectral function in few- and many-nucleon systems. Phys. Rev. C **53**, 1689 (1996)
24. M. Alvioli, C.C. Degli Atti, L.P. Kaptari, C.B. Mezzetti, H. Morita, Nucleon momentum distributions, their spin-isospin dependence, and short-range correlations. Phys. Rev. C **87**, 034603 (2013)
25. C.C. Degli Atti, In-medium short-range dynamics of nucleons: recent theoretical and experimental advances. Phys. Rep. **590**, 1 (2015)
26. R. Weiss, B. Bazak, N. Barnea, Nuclear neutron-proton contact and the photoabsorption cross section. Phys. Rev. Lett. **114**, 012501 (2015)
27. O. Hen, L.B. Weinstein, E. Piasetzky, G.A. Miller, M.M. Sargsian, Y. Sagi, Correlated fermions in nuclei and ultracold atomic gases. Phys. Rev. C **92**, 045205 (2015)
28. M.L. Terranova, D.A. De Lima, J.D. Pinheiro Filho, Evaluation of total nuclear photoabsorption cross-sections by the modified quasi-deuteron model. Europhys. Lett. **9**, 523 (1989)
29. O.A.P. Tavares, M.L. Terranova, Nuclear photoabsorption by quasi-deuterons and an updated evaluation of Levinger's constant. J. Phys. G **18**, 521 (1992)
30. M. Alvioli, C. Ciofi degli Atti, H. Universality of nucleon-nucleon short-range correlations: the factorization property of the nuclear wave function, the relative and center-of-mass momentum distributions, and the nuclear contacts. Morita, Phys. Rev. C **94**, 044309 (2016)
31. R. Weiss, B. Bazak, N. Barnea, Generalized nuclear contacts and momentum distributions. Phys. Rev. C **92**, 054311 (2015)

32. C. Bao-Jun, Li Bao-An, Symmetry energy of cold nucleonic matter within a relativistic mean field model encapsulating effects of high-momentum nucleons induced by short-range correlations. Phys. Rev. C **93**, 014619 (2016)
33. H.A. Bethe, R. Peierls, Quantum theory of the diplon. Proc. R. Soc. **148**, 146 (1935)
34. E. Braaten, D. Kang, L. Platter, Universal relations for identical bosons from three-body physics. Phys. Rev. Lett. **106**, 153005 (2011)
35. S.D. Drell, C.L. Schwartz, Sum rules for inelastic electron scattering. Phys. Rev. **112**, 568 (1958)
36. K.W. McVoy, L. Van Hove, Inelastic electron-nucleus scattering and nucleon-nucleon correlations. Phys. Rev. **125**, 1034 (1962)
37. G. Orlandini, M. Traiui, Sum rules for electron-nucleus scattering. Rep. Prog. Phys. **54**, 257 (1991)
38. H. Hu, X.-J. Liu, P.D. Drummond, Static structure factor of a strongly correlated Fermi gas at large momenta. Europhys. Lett. **91**, 20005 (2010)
39. R. Weiss, B. Bazak, N. Barnea, The generalized nuclear contact and its application to the photoabsorption cross-section. Eur. Phys. J. A **52**, 92 (2016)

Few-Body Syst (2017) 58:1–12
DOI 10.1007/s00601-016-1179-9

P. Giannakeas · Chris H. Greene

Van der Waals Universality in Homonuclear Atom-Dimer Elastic Collisions

Received: 29 August 2016 / Accepted: 7 December 2016 / Published online: 30 December 2016
© Springer-Verlag Wien 2016

Abstract The universal aspects of atom-dimer elastic collisions are investigated within the framework of Faddeev equations. The two-body interactions between the neutral atoms are approximated by the separable potential approach. Our analysis considers a pure van der Waals potential tail as well as soft-core van der Waals interactions permitting us in this manner to address the universally general features of atom-dimer resonant spectra. In particular, we show that the atom-dimer resonances are solely associated with the *excited* Efimov states. Furthermore, the positions of the corresponding resonances for a soft-core potentials with more than 5 bound states are in good agreement with the corresponding results from an infinitely deep pure van der Waals tail potential.

1 Introduction

Vitaly Efimov in the 1970s conveyed the concept of three-body bosonic bound states where the two-body subsystems are unbound [1]. Furthermore, within a zero-range model Efimov showed that at unitarity there is an infinity of such trimer states whereas the corresponding binding energies scale exponentially. Despite the appealing simplicity of the concept of Efimov states from a theoretical viewpoint, their experimental observation only occurred about four decades later through advances in the realm of ultracold atomic physics [2]. This experimental realization sparked extensive theoretical [3–18] and experimental [19–39] interest in order to develop a comprehensive understanding of the underlying physics of Efimov states. One intriguing aspect of the Efimov physics is that within the zero-range approximation the corresponding spectrum is not bounded from below, which is sometimes referred to as the "Thomas collapse" which is associated with the zero-range model Hamiltonian. By introducing an additional length scale, i.e. the three body parameter, the Efimov spectrum possesses a ground state with finite energy. Various theoretical studies [1,9,10] pointed out that the Efimov spectrum is inherently system dependent due to the three body parameter. However, experimental evidences [36] on three-body collisions between ultracold atoms manifest that the resulted three-body spectrum possesses universal characteristics associated with the van der Waals interactions which the neutral atoms experience. The microscopic mechanism associated with the universality of the three-body parameter in ultracold collisions was shown theoretically to be related with the sudden drop of the two-body van der Waals interaction potential as the interparticle distance decreases, which in return produces an effective repulsive three-body potential

---

This article belongs to the special issue "30th anniversary of Few-Body Systems".

P. Giannakeas (✉) · C. H. Greene
Department of Physics and Astronomy, Purdue University, West Lafayette, IN 47907, USA
E-mail: pgiannak@purdue.edu

C. H. Greene
E-mail: chgreene@purdue.edu

that prevents the three neutral atoms from approaching to hyperradial distances $R < 2\ell_{\text{vdW}}$ (ℓ_{vdW} is the van der Waals length scale) [14,16,17].

In view of the theoretical efforts to understand the underlying universal mechanisms of three-body bound systems it is of interest to extent the concept of van der Waals universality to atom-dimer collisional systems where a boson elastically collides on a pair of bound particles. In particular, one main question that immediately arises is whether an atom-dimer system possesses resonant spectra that exhibit universal characteristics. In order to address this question in this work the elastic atom-dimer collisions are investigated beyond the zero-range approach [9,10,40–42]. In the spirit of a simple model, the theoretical framework of Faddeev equations is employed in our study and the two body interactions are treated within the separable potential approximation. In particular two types of interactions are considered, namely a van der Waals tail potential as well as soft-core van der Waals two body interactions. The case of the pure van der Waals tail potential is modeled by the separable potential approximation introduced by Naidon *et al.* [16] whereas the soft-core van der Waals interactions are factorized by employing properties of the Hilbert–Schmidt approach [43–45]. Note that for the case of the universal three body bound states the Naidon *et al.* separable potential approximation was shown to encompass the key aspects of the van der Waals physics that constrain the 3-body parameter. Therefore, these types of potentials allow us to systematically explore the universal aspects of the corresponding atom-dimer resonant spectra. More specifically, our calculations show that in the regime of low energies the atom-dimer resonances occur only for the excited Efimov states and not for the ground Efimov state. In addition, we observe that for deep soft-core potential containing more than 5 bound states the corresponding atom-dimer resonances are in good agreement with the resonances for the Naidon *et al.* separable potential approximation which corresponds to an infinitely deep van der Waals potential. Namely, in this case the first atom-dimer resonance occurs at scattering lengths $a_s \sim 5.618 \pm 0.013\ell_{vdw}$ whereas the second one occurs at $a_s \sim 192.5 \pm 0.3\ell_{vdw}$.

2 Low Energy Atom-Dimer Collisions

Consider a system of three identical neutral bosonic atoms which collide at low energies. More specifically, it is assumed that a bosonic pair forms a bound state with dimer energy E_d, whereas the third atom has just barely enough kinetic energy to collide with the dimer. The two-body interactions between the neutral atoms are described by a van der Waals potential $V_{\text{vdW}}(r)$. Due to low collisional energies only the s-wave interactions are assumed to be important between each pair of atoms. In addition, the total colliding energy of the atom-dimer system is considered to be below the three-body breakup threshold and of course just above the dimer threshold, namely $E_d \leq E < 0$.

In order to explore the universal aspects of the atom-dimer collisions the Faddeev equations are employed. In particular, the framework of the Faddeev equations permits us to derive the full off-shell scattering amplitudes for the atom-dimer system [44–51]. Moreover, in the spirit of a simple model our analysis is restricted to the separable potential approximation [52], in which the two-body interaction for s-wave collisions is expressed as follows:

$$\hat{V}_{\text{vdW}} = -\frac{\hbar^2 \lambda}{m} |\chi\rangle \langle\chi| \tag{1}$$

where λ denotes the strength of the two body potential, m indicates the mass of the atom and $|\chi\rangle$ represents the form factor. Note that the explicit expression of the form factor $|\chi\rangle$ is constrained by the nature of the low energy two-body physics. In particular, for $\langle q|\chi\rangle \to 1$, where q refers to the relative momentum between two particles, the corresponding potential describes the collisional aspects of two particles interacting with zero-range interactions. Note that in this approximation the corresponding three-body Faddeev equations reduce to the Skorniakov-Ter-Martirosian equation [53]. In order to include the two-body effective range corrections a different expression for the form factors must be adopted [54]. In our study, the focus is on neutral atoms which interact via a van der Waals potential, hence the form factor $|\chi\rangle$ is constructed such that the long range behavior of the $V_{\text{vdW}}(r)$ is included. A more detailed discussion about this aspect is presented in subsection 2.2. An important attribute of the separable potential approximation is that the corresponding Faddeev equations are significantly simplified yielding one dimensional integral equations that can be solved efficiently and accurately.

For completeness, the following subsection briefly reviews the three body off-shell scattering amplitude integral equations for elastic atom-dimer collisions, adopting the notation based mainly on Ref. [55].

  Springer

2.1 Faddeev Equations for Elastic Atom-Dimer Collisions

In order to tackle the elastic atom-dimer collisions, the three-body scattering amplitudes are expressed in momentum space. In particular, the relative motion between dimer bosons is described by the q momentum whereas the p is the momentum of the spectator boson with respect to the pair. Then the integral equation for the three-body scattering amplitude reads:

$$X(\boldsymbol{p}, \boldsymbol{p}'; E^+) = 2Z(\boldsymbol{p}, \boldsymbol{p}'; E^+) + 2 \int \frac{d\boldsymbol{p}''}{(2\pi)^3} Z(\boldsymbol{p}, \boldsymbol{p}''; E^+) \tau(E^+ - \frac{p''^2}{2M}) X(\boldsymbol{p}'', \boldsymbol{p}'; E^+), \quad (2)$$

where $M = 2m/3$ is the reduced mass of the spectator boson with respect to the dimer and m is the mass of the atom. $\tau(\cdot)$ indicates the two-body transition matrix embedded in the three-body momentum space. $E^+ = E + i\epsilon$ denotes the total colliding energy including an positive infinitesimal quantity ϵ. Note that the infinitesimal ϵ enforces outgoing wave boundary conditions and permits closed-form contour integration of the poles which are contained in the amplitude $Z(\cdot, \cdot; \cdot)$ and the dressed atom-dimer propagator $\tau(\cdot)$. Note that the amplitude $X(\cdot, \cdot; \cdot)$ describes the elastic collision of an atom with a dimer.

The atom-dimer scattering amplitude in Eq. (2) can be expanded in partial-waves for total angular momentum $L = 0$ [44–46]. Since all the three atoms interact with only s-wave interactions the corresponding form of the atom-dimer scattering amplitude is given by the following expression:

$$X_0(p, p'; E^+) = 2Z_0(p, p'; E^+) + 2 \int_0^\infty \frac{dp''}{2\pi^2} p''^2 Z_0(p, p''; E^+) \tau_0(E^+ - \frac{p''^2}{2M}) X_0(p'', p'; E^+). \quad (3)$$

After implementing the separable potential given in Eq. (1), the first Born amplitude $Z_0(\cdot, \cdot; \cdot)$ in Eq. (3) is given by the following expression:

$$Z_0(p, p'; E^+) = \frac{1}{2} \int_{-1}^{1} d\xi \, \frac{\chi(|\boldsymbol{p} + \frac{\boldsymbol{p}'}{2}|) \chi(|\boldsymbol{p}' + \frac{\boldsymbol{p}}{2}|)}{E^+ - \frac{p^2}{m} - \frac{p'^2}{m} - \frac{\boldsymbol{p} \cdot \boldsymbol{p}'}{m}}, \quad (4)$$

where $\xi = \hat{\boldsymbol{p}} \cdot \hat{\boldsymbol{p}}'$ is the angle between the momenta $\boldsymbol{p}$ and $\boldsymbol{p}'$.

As was mentioned at the beginning of the section, the total colliding energy is always negative in this study, and below the breakup threshold. This implies that the Z_0 Born amplitudes are free of poles; however, the energy is above the dimer threshold meaning that the dressed two-body propagator τ_0 (see Eq. (3)) has poles which should be integrated out. In order to show this, the two-body transition matrix for a separable potential is demonstrated whereas its pole structure is explicitly isolated. The full off-shell two-body transition matrix with separable interactions reads:

$$\langle \boldsymbol{q} | \hat{t} | \boldsymbol{q}' \rangle = \chi(\boldsymbol{q}) \tau_0(E) \chi(\boldsymbol{q}'), \quad (5)$$

where the dressed propagator $\tau_0(E)$ is given by the following expression:

$$\tau_0(E) = \frac{S(E)}{E - E_d} \quad \text{with} \quad S^{-1}(E) = \int_0^\infty \frac{dq}{2\pi^2} q^2 \frac{\chi(q)^2}{(E - \frac{q^2}{m})(E_d - \frac{q^2}{m})}, \quad (6)$$

where at $E = E_d$ the dressed propagator diverges, while the quantity $S(E)$ for $E \rightarrow E_d$ is free of poles since $E_d \leq 0$.

Since the pole structure of the dressed propagator has been isolated, Eq. (6) is substituted in Eq. (3), which produces the following expression for the atom-dimer scattering amplitude:

$$X_0(p, p'; E^+) = 2Z_0(p, p'; E^+) + 2 \int_0^\infty \frac{dp''}{2\pi^2} p''^2 Z_0(p, p''; E^+) \frac{S(E^+ - \frac{3p''^2}{4m})}{E^+ - \frac{3p''^2}{4m} - E_d} X_0(p'', p'; E^+). \quad (7)$$

By inspection of Eq. (7) it is evident that the second term always diverges. Specifically, when the spectator particle is far from the dimer either before or after the collision the total energy is distributed into the binding energy of the dimer and the kinetic energy of the spectator atom, namely $E^+ = E_d + \frac{3k^2}{4m} + i0$, where $k > 0$.

Because of the latter the denominator is proportional to $\sim (k^2 - q^2 + i0)$ yielding a singular integrand in Eq. (7). One way to treat the pole of equation is to convert the integral equation in Eq. (7) into a principal value integral equation through the identity $\frac{1}{E + - \frac{3p''^2}{4m} - E_d} = \mathcal{P} \frac{1}{E - \frac{3p''^2}{4m} - E_d} - i\pi\delta(E - \frac{3p''^2}{4m} - E_d)$. Then the atom-dimer scattering amplitude in Eq. (7) is converted into an off-shell K-matrix for elastic atom-dimer collisions. Note that the above mentioned K-matrix is not the *conventional* reaction or K-matrix of formal scattering theory. Indeed, as we show below the K-matrix of Eq. (8) is proportional to the tangent of the atom-dimer phase shift, i.e. $\tan \delta_{AD}$. The principal value integral equation for this atom-dimer K-matrix is given by the following expression:

$$K(p, p'; E) = 2Z_0(p, p'; E) + 2\mathcal{P} \int_0^\infty \frac{dp''}{2\pi^2} p''^2 Z_0(p, p''; E) \frac{S(E - \frac{3p''^2}{4m})}{E - \frac{3p''^2}{4m} - E_d} K(p'', p'; E), \qquad (8)$$

where the off-shell K-matrix is real, symmetric and free of poles. These attributes make its numerical implementation and solution rather straightforward.

As stated above the atom-dimer K-matrix is the off-shell one, however only its on-shell part is of physical significance and can be associated with the corresponding atom-dimer scattering length. The notion of the on-shell K-matrix basically means that the Jacobi momenta p and p' are equal to the relative momentum k between the spectator atom and the interacting pair, namely $p = p' = k$. Recall, that the total colliding energy of the atom-dimer system is $E = E_d + \frac{3k^2}{4m}$ and that we are interested in the elastic scattering of an atom from a pair of particles. Assuming that the s-wave atom-dimer phase shift is δ_{AD}, then the on-shell K-matrix obeys the following relation:

$$K(k, k; E) = -\frac{3}{2m\pi} \frac{\tan \delta_{AD}}{k} S^{-1}(E_d). \qquad (9)$$

Here the quantity $-k^{-1} \tan \delta_{AD}$ defines a *generalized energy-dependent* atom-dimer scattering length, $a_{AD}(k)$, for total collision energy $E = E_d + \frac{3k^2}{4m}$. Using Eq. (9) the atom-dimer s-wave scattering length $a_{AD}(k)$ can be expressed in terms of the on-shell K-matrix.

$$a_{AD}(k) = \frac{2m\pi}{3} S(E_d) K(k, k; E). \qquad (10)$$

Having briefly reviewed the off-shell K-matrix principal value integral equations for elastic atom-dimer collisions, and having introduced the energy dependent s-wave atom-dimer scattering length in terms of the on-shell K-matrix, the following subsection focuses on the construction of the separable potential which fully encapsulates the two-body s-wave van der Waals physics of the neutral atoms.

2.2 Separable Potentials for Two-Body van der Waals Interactions

The separable potential approximation plays a key role in the simplification of the Faddeev equations. However, it should be noted that a separable potential is inherently non-local in contrast to the *true* two-body interactions which are local. Difficulties associated with the nonlocality can largely be circumvented by designing the $|\chi\rangle$ form factors of a separable potential such that they describe as much of the two-body physics of two interacting neutral atoms as is manageable. Indeed, Naidon *et al* in Refs. [16,17] highlight the role of the separable potential approximation designed specifically to include the two-body van der Waals physics. These two studies provide further physical insight on the universal aspects of the spectrum of Efimov states [36] for three neutral atoms with equal masses.

Therefore, in order to study the universal characteristics of elastic atom-dimer collisions, the following subsection reviews the separable potential of Naidon *et al.* and adopts an alternative method to construct van der Waals separable potentials based on the Hilbert-Schmidt expansion [43–45]. The latter technique is applied in the case of a soft-core van der Waals potential, which enables a straightforward comparison with the Naidon *et al.* approach. Using this implementation, we study in detail how the atom-dimer scattering length depends on the *total* number of bound states which the soft-core two-body interactions support. Note that both types of separable potentials contain information only for the least bound state which is supported from the *real* van der Waals interactions. This constraint arises from the fact that the corresponding atom-dimer system experiences only elastic collisions, since inelastic processes associated with deeper two-body bound states are not included in the model. Consequently, both separable potentials have the same functional form as is given in Eq. (1)

2.2.1 The Naidon Et Al. separable potential for van der Waals interactions

Naidon et al. in Refs. [16,17] consider only a van der Waals potential tail, namely $V_{vdW}(r) = -C_6/r^6$, where C_6 is the dispersion coefficient and r refers to the relative distance between two neutral atoms. The corresponding separable potential is constructed by employing the *zero-energy* two-body wavefunction which is known analytically [56] and obeys the following relation:

$$\phi(r) = \Gamma\left(\frac{5}{4}\right)\sqrt{\frac{r}{\ell_{vdW}}} J_{\frac{1}{4}}\left(2\frac{\ell_{vdW}^2}{r^2}\right) - \frac{\ell_{vdW}}{a_s}\Gamma\left(\frac{3}{4}\right)\sqrt{\frac{r}{\ell_{vdW}}} J_{-\frac{1}{4}}\left(2\frac{\ell_{vdW}^2}{r^2}\right), \tag{11}$$

where $\ell_{vdW} = \frac{1}{2}(mC_6/\hbar^2)^{1/4}$ indicates the van der Waals length scale and a_s is the s-wave scattering length. The quantities $\Gamma(\cdot)$ and $J_{\pm\frac{1}{4}}(\cdot)$ represent the Gamma and Bessel functions, respectively. Note that this wavefunction at large distances behaves as $\phi(r) \to 1 - r/a_s$, whereas at short length scales it possesses fast oscillations and vanishes at the origin.

In momentum space a separable potential of the form given in Eq. (1) obeys the relation

$$V_{NP}(\boldsymbol{q}, \boldsymbol{q}') = -\frac{\hbar^2\lambda}{m}\chi(\boldsymbol{q})\chi(\boldsymbol{q}'), \tag{12}$$

where the abbreviation "NP" stands for the Naidon et al. potential, λ indicates the strength of the potential, m denotes the mass of the atom, and the form factor $\chi(\boldsymbol{q})$ is expressed in terms of the zero-energy two-body wavefunction (see Eq. (11))

$$\chi(\boldsymbol{q}) \equiv \chi(q) = 1 - q\int_0^\infty dr\left[1 - \frac{r}{a_s} - \phi(r)\right]\sin(qr). \tag{13}$$

Due to the s-wave character of the two-body interactions the form factor depends only on the magnitude of the momentum. The strength of V_{NP} is also self consistently expressed in terms of the zero-energy two-body wavefunction.

$$\lambda = \left[-\frac{1}{4\pi a_s} + \frac{1}{2\pi^2}\int_0^\infty dq|\chi(q)|^2\right]^{-1}. \tag{14}$$

2.2.2 Separable Potentials for Soft-Core van der Waals Interactions

Consider that two neutral atoms interact with the following soft-core potential:

$$V_{SC}(r) = -\frac{C_6}{r^6 + \sigma^6}, \tag{15}$$

where the abbreviation "SC" stands for "soft-core" σ indicates a quantity that controls the depth of the potential and *regularizes* it for $r \to 0$.

Moreover, assuming only s partial wave interactions the soft-core potential in momentum space obtains the form

$$V_{SC}^s(q, q') = 4\pi\int_0^\infty dr\, r^2 j_0(qr)V(r)j_0(q'r), \tag{16}$$

where $j_0(\cdot)$ denotes the s-wave spherical Bessel function, and superscript s indicates that the s-wave orbital angular momentum of the two interacting atoms, namely $\ell = 0$.

The local potential of Eq. (16) is desirable to be expressed in terms of separable potentials, i.e. factorized in momentum space. As was mentioned previously the main aim is to construct a separable potential for the soft-core van der Waals interactions which contains *all* the relevant information of the *least* bound state supported by the potential of Eq. (15). This is achieved here by exploiting the Hilbert-Schmidt approach and its properties. More specifically, our goal is to obtain the eigenfunctions and eigenvalues of the non-symmetric two-body Lippmann-Schwinger integral equation which reads:

$$\hat{V}_{SC}^s \hat{G}_0(E) \, |g_\nu; E\rangle = \eta_\nu(E) \, |g_\nu; E\rangle, \tag{17}$$

where $\hat{G}_0(E) = [E - H_0]^{-1}$ is the two-body Green's function and H_0 indicates the kinetic energy operator. The quantity E is assumed to be negative, because we are interested in the case where the two particles can form a bound state. Recall that the main purpose is to derive a separable van der Waals potential which then is used to describe elastic atom-dimer collisions. Finally, $\eta_\nu(E)$ refers to the ν-th eigenvalue of the non-symmetric Lippmann-Schwinger kernel, where ν is an integer labeling the different eigenvalues in a descending order.

The orthonormalization condition for the eigenvectors $|g_\nu; E\rangle$ is given by the following relation:

$$\langle g_{\nu'}; E|G_0(E)|g_\nu; E\rangle = -\delta_{\nu'\nu}. \tag{18}$$

One main property of the eigenvalues of Eq. (17) is that they should fulfill the condition $\eta_\nu(E) = 1$ when $E = E_n$, where E_n corresponds to the energy of the *least* bound state of a soft-core van der Waals potential with n bound states in total. Therefore, by fixing the parameter σ of the potential in Eq. (16) the total number of bound states is defined. Then, in the momentum representation the integral equation in Eq. (17) is diagonalized by varying the energy E in order to obtain the eigenvalue η_ν which obey the condition $\eta_\nu(E = E_n) = 1$. The corresponding eigenvector $|g_\nu; E_n\rangle$ is the form factor associated with the n-th *least* bound state of V_{SC}. The two-body wavefunction of the shallowest bound state and the eigenvectors $|g_\nu; E_n\rangle$ are connected in the momentum representation according to the following relation:

$$\phi_{n,\nu}(q) = N_n \frac{g_\nu(q; E_n)}{\frac{\hbar^2 q^2}{m} - E_n}, \quad \text{with } E_n < 0 \tag{19}$$

where N_n is the normalization constant of the n-th bound state.

The separable potential associated with the wavefunction in Eq. (19) then has the following form:

$$\bar{V}_{SC}(q, q') = -g_\nu(q; E_n)g_\nu(q; E_n), \tag{20}$$

where it should be noted that the form factors in the preceding equation are proportional to two-body wavefunction at energy E_n whereas in the Naidon *et al.* separable potential the corresponding form factor is proportional to the zero-energy two-body wavefunction only.

Figure 1 compares the form factors obtained for the separable potentials $V_{NP}(q, q')$ and $\bar{V}_{SC}(q, q')$ with the s-wave scattering length set to $a_s = 32.3873\ell_{vdW}$. In particular, the orange dashed line denotes the scaled form factor $\sqrt{\frac{\lambda}{\ell_{vdW}}}\chi(q\ell_{vdW})$ [see Eq. (13)] for the Naidon *et al.* separable potential. The blue solid line indicates

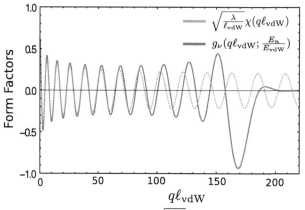

Fig. 1 (color online) An illustration of the scaled form factor $\sqrt{\frac{\lambda}{\ell_{vdW}}}\chi(q\ell_{vdW})$ given in Eqs. (13) and (14) (*orange dashed line*) in the Naidon *et al.* separable approximation. $g_\nu(q\ell_{vdW}; \frac{E_n}{E_{vdW}})$ (*blue solid line*) indicates the form factor for the soft-core vdW potential which supports 22 bound states. In both cases the scattering length is $a_s = 32.3873\ell_{vdW}$. Note that $\ell_{vdW} = \frac{1}{2}(mC_6/\hbar^2)^{1/4}$ is the van der Waals length scale and $E_{vdW} = \frac{\hbar^2}{m\ell_{vdW}^2}$ refers to the van der Waals energy scale

the form factor $g_\nu(q\ell_{\text{vdW}}; \frac{E_n}{E_{\text{vdW}}})$ which corresponds to the least bound state of a soft-core potential with 22 bound states in total. At large momenta q the $g_\nu(q\ell_{\text{vdW}}; \frac{E_n}{E_{\text{vdW}}})$ vanishes whereas $\sqrt{\frac{\lambda}{\ell_{\text{vdW}}}}\chi(q\ell_{\text{vdW}})$ possesses an oscillatory behavior with vanishing amplitude. This behavior of $\sqrt{\frac{\lambda}{\ell_{\text{vdW}}}}\chi(q\ell_{\text{vdW}})$ mainly arises from the fact that the form factors in Naidon *et al.* separable potential correspond to the least bound state of an infinitely deep vdW potential tail. In addition, at small momenta q we observe that both form factors, i.e. $\sqrt{\frac{\lambda}{\ell_{\text{vdW}}}}\chi(q\ell_{\text{vdW}})$ and $g_\nu(q\ell_{\text{vdW}}; \frac{E_n}{E_{\text{vdW}}})$, possess the same nodal structure. This is indicative of the van der Waals universality, since the small momenta behavior in spatial space refers to large separation distances which is strongly characterized by the attractive van der Waals potential tail.

Summarizing, in this subsection two types of separable potentials for vdW interactions are constructed, one is based on a hard wall, i.e. Naidon's separable potential [(see Eq. 12)] and the other one is based on a soft-core vdW potential [See Eq. (20)]. In the following, these two types of separable potentials will be employed to calculate atom-dimer elastic collisions in order to extract their universal aspects.

3 Results and Discussion

Figure 2 illustrates the dimer energies which are compared for different separable potentials, in order to highlight their regimes of validity. In particular, the dimer energies which are computed within the separable potential based on the Naidon *et al.* approach are indicated by the blue solid line in Fig. 2. Moreover, the red dotted and orange solid lines denote the dimer energies for a soft-core vdW potential which contains a total number number of 22 and 10 bound states, respectively. Note, that the depicted dimer energies (red dotted and orange solid line) correspond to the least bound state which can be supported by the soft-core vdW potential. The black dashed line denotes the universal dimer binding energies, i.e. $E = -\hbar^2/(ma_s^2)$. For $a_s > 25\ell_{\text{vdW}}$ the two-body binding energies calculated within the separable approximation are in excellent agreement with the binding energies for the universal dimer indicating in this manner the region where the two-body collisions are considered to be universal, namely independent of effective range corrections. For scattering lengths $a_s < 25\ell_{\text{vdW}}$ the effective range corrections become important and thus the binding energies of the separable potentials deviate from those of the universal dimer. In addition, as the s wave scattering length becomes $a_s < 2.77\ell_{\text{vdW}}$ the binding energies for the Naidon *et al.* separable potential differ from those of a soft-core vdW interaction. This designates the region of validity of the Naidon *et al.* separable potential. More specifically, for $a_s < \ell_{\text{vdW}}$ the Naidon *et al.* approach yields dimer energies which become shallower as the s-wave scattering length

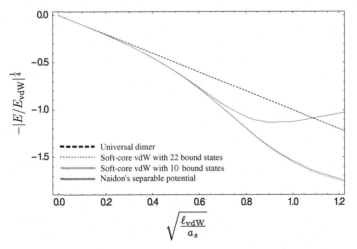

Fig. 2 (color online) The dimer binding energies as a function of the two-body s-wave scattering length. The *black dashed line* indicates the universal dimer energies. The *blue solid line* corresponds to the dimer energies based on the Naidon *et al.* separable potential. The *red dotted* and *orange solid line* refer to the calculations for soft-core vdW potential which contain in total 22 and 10 bound states, respectively. Note that $\ell_{\text{vdW}} = \frac{1}{2}(mC_6/\hbar^2)^{1/4}$ is the van der Waals length scale and $E_{\text{vdW}} = \frac{\hbar^2}{m\ell_{\text{vdW}}^2}$ indicates the van der Waals energy scale

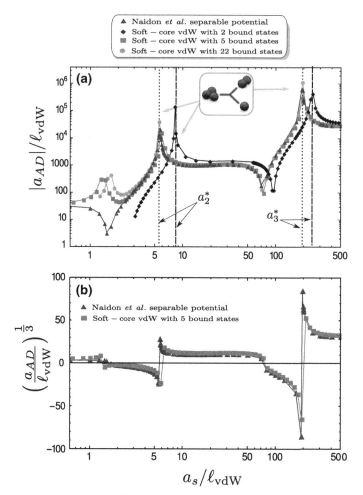

Fig. 3 (color online) **a** The absolute value atom-dimer scattering length, $|a_{AD}|/\ell_{vdW}$, as a function of the two-body s-wave scattering length for different types of two-body potentials. The *blue triangles* correspond to Naidon *et al.* separable potential. The *black diamonds*, *red squares* and *orange circles* refer to the calculations for a soft-core vdW potential which contains in total 2, 5, and 22 bound states, respectively. The *vertical dotted*, *dashed* and *dashed-dotted lines* indicate the positions of the atom-dimer resonances. **b** An illustration of the sign dependence of the atom-dimer scattering length, $|a_{AD}|/\ell_{vdW}$ for the Naidon *et al.* separable approach (*blue triangles*) and for the soft-core vdW potential which contains 5 bound states (*red squares*). Note that $\ell_{vdW} = \frac{1}{2}(mC_6/\hbar^2)^{1/4}$ is the van der Waals length scale

vanishes. The main reason is that the separable potential in the Naidon *et al.* approach is constructed by the zero-energy two-body wavefunction. Thus, it is expected that at energies away from the zero-energy bound state the potential cannot be regarded as non-local.

On the other hand, we observe that the soft-core vdW potential does not suffer from this restriction since the corresponding binding energies are obtained by diagonalizing the Lippmann-Schwinger kernel according to the prescription given in the previous subsection.

Figure 3(a) depicts the absolute value of the elastic atom-dimer scattering length, $|a_{AD}|$ as function of the s-wave scattering length a_s in van der Waals units. In particular the blue triangles indicate the atom-dimer scattering length where the the Naidon *et al.* separable potential is used. The black diamonds denote the a_{AD} scattering length for the separable potential given in Eq. (20) which is constructed by a soft-core vdW interaction which contains 2 bound states in total. Similarly, the red squares and the orange circles refer to the atom-dimer scattering lengths which correspond to soft-core potentials which contain 5 and 22 bound states, respectively.

Figure 3(a) in a compact manner provides insights on the universal behavior of the atom-dimer scattering length. Namely, the dependence of the atom-dimer scattering length for soft-core vdW interactions and van der Waals potential tail within the separable potential approximation. Furthermore, for values of s-wave scattering length larger than $2\ell_{\text{vdW}}$, $a_s > 2\ell_{\text{vdW}}$, the atom-dimer scattering length possesses resonant features for both types of separable potentials. For $a_s < 2\ell_{\text{vdW}}$ the atom-dimer scattering length exhibits pronounced features in the case of soft-core potentials which contain 5 (red squares) and 22 (orange circles) bound states. These particular features are not associated with a resonant process and arise due to numerical instabilities.

In Fig. 3(a) we observe that the resonant features of a_{AD} for soft-core vdW potentials with only 2 bound states (black diamonds) are shifted to higher values of a_s and as the number of total bound states is increased, i.e. red squares and orange circles, the atom-dimer resonances approach the results based on the Naidon *et al.* separable approximation which corresponds to a van der Waals tail containing an infinity of bound states. This behavior mainly arises by the van der Waals character of the two-body potential. Namely, from a WKB viewpoint, the classical allowed region of soft-core potentials, see Eq. (15) can be divided into two regimes: (i) $r < \sigma$ the region close to origin where the potential is constant and (ii) the region of $r > \sigma$ where the attractive vdW potential tail prevails, i.e. the vdW region. Therefore, as the soft-core potential becomes deeper the vdW region extends towards the origin mimicking in return a potential that possesses a pure van der Waals tail. Thus, for deep soft-core potentials the atom-dimer collisions are strongly governed by the vdW physics exhibiting universal characteristics. More specifically, as Fig. 3(a) shows the atom-dimer scattering length for a soft-core potential with more than 5 bound states (red squares and orange circles) are in good agreement with the a_{AD} calculations of a pure van der Waals potential tail (blue triangles).

The underlying physical mechanism of the resonant features depicted in Fig. 3a at $a_s > 2\ell_{\text{vdW}}$ are associated with Efimov states. Namely, as a_s increases an Efimov state intersects with the dimer threshold; hence, the atom-dimer scattering length diverges at a_2^* or a_3^*. Panel(b) of Fig. 3 demonstrates the sign change of the atom-dimer scattering length for the Naidon *et al.* separable potential (blue triangles) whereas the red squares indicate the corresponding calculation for the soft-core vdW potential with 5 bound states. In particular, Fig. 3b shows that for an increasing a_s a trimer state is accessed, binding the atom and the dimer. For $a_s < 2 \ell_{\text{vdW}}$ in the case of soft-core vdW potential with 5 bound states we observe that the corresponding enhancement possesses a positive a_{AD} at small a_s and as the s-wave scattering length increases a_{AD} becomes negative. Note that this behavior is not associated with a pole of the a_{AD} but due to numerical instabilities.

In order to identify the Efimov states which are associated with the atom-dimer resonances of Fig. 3a in Fig. 4 the first three Efimov states are illustrated. More specifically, Fig. 4 shows the Efimov trimer energies as a function of $\text{Sign}(a_s)/\sqrt{a_s}$ in van der Waals units. The bound spectrum is calculated by numerically solving the corresponding Faddeev equation employing the Naidon *et al.* separable potential [16,17]. The ground Efimov state, in Fig. 4a, is indicated by the blue solid line, and the 1st (2nd) excited Efimov state is denoted by the orange (green) solid line whereas the black solid line refers to the dimer threshold. Evidently, in Fig. 4a we observe that, in contrast with the universal zero-range theory, the ground Efimov state does not cross the dimer threshold whereas the excited Efimov states [for the 2nd excited state see Fig. 4b] merge with the dimer threshold at a_2^* or a_3^*. This suggests that the atom-dimer resonances in Fig. 3a are associated with excited Efimov states only within our model.

In addition, table 1 contains information about the position of the atom-dimer resonances for both the soft-core and pure van der Waals potential tail which are depicted in Fig. 3a manifesting the universal behavior. In particular for the position of the atom-dimer resonances which are associated with the 1st excited Efimov state we observe that the results for soft-core vdW potentials with more than 5 bound states possess a difference less than 5.4 % from the corresponding calculations for a pure van der Waals tail, i.e. the Naidon *et al.* separable potential. Similarly, for the atom-dimer resonances which are associated with the 2nd excited Efimov state the corresponding difference in the position of the resonances becomes less than ~ 2 %. In addition, the third column of Table 1 indicates the ratio $\frac{a_3^*}{a_2^*}$ which differs from the corresponding ratio within the zero-range theory, namely $\frac{a_3^*}{a_2^*} = 22.69$. This difference between the zero-range theory and vdW separable potential approach is indicative of the effective range corrections which are absent in the zero-range potential approximation. Furthermore, for the a_2^* atom-dimer resonance the theoretical values are in reasonable agreement with the experimental ones. Recall that in the ^{133}Cs experiment of Ref. [37] two resonances are observed for two different Feshbach resonances (FR). Namely for a narrow FR the position of the atom-dimer resonance is $a_2^* = (6.48 \pm 0.24)\ell_{\text{vdW}}$ whereas the broad one the atom-dimer process is enhanced at $a_2^* = (4.15 \pm 0.09)\ell_{\text{vdW}}$.

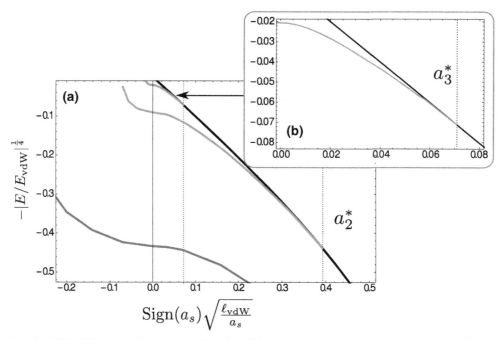

Fig. 4 (color online) The biding energies, E, for the first three Efimov states as a function of the s-wave scattering length, a_s. **a** The ground Efimov state is indicated by the *blue line*. The 1st and 2nd excited states are denoted by the *orange* and *green lines*, respectively. The *black solid line* indicates the dimer threshold. **b** Zoom-in plot for the 2nd excited Efimov state (*green line*) which crosses the dimer threshold (*black line*). The *black dotted lines* denote the position where the 1st (2nd) excited Efimov state crosses the dimer threshold, i.e. $a_s \equiv a_2^*$ ($a_s \equiv a_3^*$). Note that $\ell_{vdW} = \frac{1}{2}(mC_6/\hbar^2)^{1/4}$ is the van der Waals length scale and $E_{vdW} = \frac{\hbar^2}{m\ell_{vdW}^2}$ indicates the van der Waals energy scale

Table 1 The position of the atom-dimer resonances in van der Waals for the potentials which are used in Fig. 3a. The quantity a_2^* (a_3^*) indicates the position of the atom-dimer resonance due to the 1st (2nd) excited Efimov state

	a_2^*	a_3^*	a_3^*/a_2^*
Soft-core with 2 bound states	8.44	245.7	29.11
Soft-core with 5 bound states	5.930	197.6	33.32
Soft-core with 22 bound states	5.609	192.7	34.35
The Naidon *et al.* separable potential	5.628	192.3	34.16
^{7}Li in Ref. [39]	6.03 ± 0.12		
^{133}Cs in Ref. [37]	6.48 ± 0.24 and 4.15 ± 0.09		

4 Summary

In summary, the elastic atom-dimer collisions under the influence of two-body soft-core or pure van der Waals tail potential are studied. The theoretical framework of our analysis is based on the integral Faddeev equations whereas the two-body interactions are modeled via suitable separable potentials. The pure van der Waals potential tail is described using the Naidon *et al.* separable interaction where the zero-energy two-body wavefunction is utilized. On the other hand, the factorization of the soft-core vdW potentials is based on the properties of the Hilbert-Schmidt expansion. In particular, the form factor of the soft-core vdW potentials is obtained by calculating the two-body energy of the least bound state of the soft-core vdW interaction and its corresponding wavefunction.

Using these two types of separable potentials the universal aspects of the elastic atom-dimer collisions is then studied. It is shown that the atom-dimer scattering lengths for soft-core vdW potentials with more than 2 bound states are in good agreement with the corresponding calculations for an infinitely deep vdW potential exhibiting the same resonant structure. Furthermore, in order to identify the origin of the resonant features in the atom-dimer scattering length the Faddeev equation for a three-body bound system is solved. Our analysis shows that the ground Efimov state does not cross the dimer threshold thus the resonant features in atom-dimer

Springer

collisions are solely associated with the excited Efimov states. In addition, highlighting the universal behavior of the atom-dimer elastic collisions we observe that: in the case of the 1st excited Efimov states, the position of the resonances for soft-core vdW potential with more than 2 bound states differs by less than 5.4 % from the corresponding calculations with the Naidon et al. separable potential. This difference decreases in the case of the 2nd excited Efimov states where it becomes less than 2 %. Complementing our study similar conclusions were shown in Ref. [57] for a pure vdW potential which contains only a finite number of bound states. In addition, Ref. [57] shows that the Naidon et al. separable potential yields a three-body recombination rate for the ground Efimov state which does not follow the a^4 scaling law. Beyond the simple model calculations presented here and in Ref. [57], Mestrom et al. in Ref. [58] illustrates that the first excited Efimov states are strongly influenced by the higher partial wave interactions nearby the atom-dimer threshold. In particular, Mestrom et al. demonstrates that an atom-dimer system colliding in the presence of s- and d-wave interactions force the ground and first excited Efimov states to not merge with the atom-dimer continuum.

Acknowledgements We thank Jose D'Incao and Paul Mestrom for stimulating discussions related to the present investigation. This work was supported in part by NSF grant PHY-1607180.

References

1. V.N. Efimov, Weakly-bound states of 3 resonantly-interacting particles. Sov. J. Nucl. Phys. **12**(5), 589 (1971)
2. T. Kraemer, M. Mark, P. Waldburger, J.G. Danzl, C. Chin, B. Engeser, A.D. Lange, K. Pilch, A. Jaakkola, H.-C. Nägerl et al., Evidence for Efimov quantum states in an ultracold gas of caesium atoms. Nature **440**(7082), 315–318 (2006)
3. D.V. Fedorov, A.S. Jensen, Efimov effect in coordinate space Faddeev equations. Phys. Rev. Lett. **71**, 4103–4106 (1993)
4. B.D. Esry, C.H. Greene, J.P. Burke Jr., Recombination of three atoms in the ultracold limit. Phys. Rev. Lett. **83**(9), 1751 (1999)
5. P.F. Bedaque, E. Braaten, H.-W. Hammer, Three-body recombination in Bose gases with large scattering length. Phys. Rev. Lett. **85**(5), 908 (2000)
6. H. Suno, B.D. Esry, C.H. Greene, J.P. Burke Jr., Three-body recombination of cold helium atoms. Phys. Rev. A **65**(4), 042725 (2002)
7. H. Suno, B.D. Esry, C.H. Greene, Recombination of three ultracold fermionic atoms. Phys. Rev. Lett. **90**(5), 053202 (2003)
8. D.S. Petrov, Three-boson problem near a narrow Feshbach resonance. Phys. Rev. Lett. **93**, 143201 (2004). doi:10.1103/PhysRevLett.93.143201
9. E. Braaten, H.-W. Hammer, Efimov physics in cold atoms. Ann. Phys. **322**(1), 120–163 (2007)
10. E. Braaten, H.-W. Hammer, Universality in few-body systems with large scattering length. Phys. Rep. **428**(5), 259–390 (2006)
11. H.W. Hammer, L. Platter, Universal properties of the four-body system with large scattering length. Eur. Phys. J. A **32**(1), 113–120 (2007)
12. J. von Stecher, J.P. D'Incao, C.H. Greene, Signatures of universal four-body phenomena and their relation to the Efimov effect. Nat. Phys. **5**(6), 417–421 (2009)
13. J.P. D'Incao, C.H. Greene, The short-range three-body phase and other issues impacting the observation of Efimov physics in ultracold quantum gases. J. Phys. B: At. Mol. Opt. Phys. **42**(4), 044016 (2009)
14. J. Wang, J.P. D'Incao, B.D. Esry, C.H. Greene, Origin of the three-body parameter universality in Efimov physics. Phys.l Rev. Lett. **108**, 263001 (2012)
15. Y. Wang, J.P. D'Incao, C.H. Greene, Efimov effect for three interacting bosonic dipoles. Phys. Rev. Lett. **106**(23), 233201 (2011)
16. P. Naidon, S. Endo, M. Ueda, Physical origin of the universal three-body parameter in atomic Efimov physics. Phys. Rev. A **90**, 022106 (2014a)
17. P. Naidon, S. Endo, M. Ueda, Microscopic origin and universality classes of the Efimov three-body parameter. Phys. Rev. Lett. **112**, 105301 (2014b)
18. D. Blume, Efimov physics and the three-body parameter for shallow van der Waals potentials. Few-Body Syst. **56**(11—-12), 859–867 (2015)
19. M. Zaccanti, D. Deissler, C. D'Errico, M. Fattori, M. Jona-Lasinio, S. Müller, G. Roati, M. Inguscio, G. Modugno, Observation of an Efimov spectrum in an atomic system. Nat. Phys. **5**(8), 586–591 (2009)
20. J.R. Williams, E.L. Hazlett, J.H. Huckans, R.W. Stites, Y. Zhang, K.M. O'Hara, Evidence for an excited-state Efimov trimer in a three-component Fermi gas. Phys. Rev. Lett. **103**, 130404 (2009)
21. S. Knoop, J.S. Borbely, W. Vassen, S.J.J.M.F. Kokkelmans, Universal three-body parameter in ultracold ^{4}he*. Phys. Rev. A **86**, 062705 (2012)
22. F. Ferlaino, S. Knoop, M. Berninger, W. Harm, J.P. D'Incao, H.-C. Nägerl, R. Grimm, Evidence for universal four-body states tied to an Efimov trimer. Phys. Rev. Lett. **102**, 140401 (2009)
23. M. Berninger, A. Zenesini, B. Huang, W. Harm, H.-C. Nägerl, F. Ferlaino, R. Grimm, P.S. Julienne, J.M. Hutson, Universality of the three-body parameter for Efimov states in ultracold cesium. Phys. Rev. Lett. **107**, 120401 (2011). doi:10.1103/PhysRevLett.107.120401
24. T. Lompe, T.B. Ottenstein, F. Serwane, A.N. Wenz, G. Zürn, S. Jochim, Radio-frequency association of Efimov trimers. Science **330**(6006), 940–944 (2010a)

25. S. Nakajima, M. Horikoshi, T. Mukaiyama, P. Naidon, M. Ueda, Measurement of an Efimov trimer binding energy in a three-component mixture of ^{6} Li. Phys. Rev. Lett. **106**, 143201 (2011)

26. T. Lompe, T.B. Ottenstein, F. Serwane, K. Viering, A.N. Wenz, G. Zürn, S. Jochim, Atom-dimer scattering in a three-component Fermi gas. Phys. Rev. Lett. **105**, 103201 (2010b). doi:10.1103/PhysRevLett.105.103201

27. S. Nakajima, M. Horikoshi, T. Mukaiyama, P. Naidon, M. Ueda, Nonuniversal Efimov atom-dimer resonances in a three-component mixture of ^{6} Li. Phys. Rev. Lett. **105**, 023201 (2010)

28. S. Knoop, F. Ferlaino, M. Mark, M. Berninger, H. Schöbel, H.-C. Nägerl, R. Grimm, Observation of an Efimov-like trimer resonance in ultracold atom-dimer scattering. Nat. Phys. **5**(3), 227–230 (2009)

29. S.E. Pollack, D. Dries, R.G. Hulet, Universality in three- and four-body bound states of ultracold atoms. Science **326**(5960), 1683–1685 (2009)

30. N. Gross, Z. Shotan, S. Kokkelmans, L. Khaykovich, Observation of universality in ultracold ^{7} Li three-body recombination. Phys. Rev. Lett. **103**, 163202 (2009)

31. A.N. Wenz, T. Lompe, T.B. Ottenstein, F. Serwane, G. Zürn, S. Jochim, Universal trimer in a three-component Fermi gas. Phys. Rev. A **80**, 040702 (2009)

32. G. Barontini, C. Weber, F. Rabatti, J. Catani, G. Thalhammer, M. Inguscio, F. Minardi, Observation of heteronuclear atomic Efimov resonances. Phys. Rev. Lett. **103**, 043201 (2009)

33. J.H. Huckans, J.R. Williams, E.L. Hazlett, R.W. Stites, K.M. O'Hara, Three-body recombination in a three-state Fermi gas with widely tunable interactions. Phys. Rev. Lett, **102**, 165302 (2009)

34. A. Zenesini, B. Huang, M. Berninger, S. Besler, H.C. Ngerl, F. Ferlaino, R. Grimm, C.H. Greene, J. von Stecher, Resonant five-body recombination in an ultracold gas of bosonic atoms. New J. Phys. **15**(4), 043040 (2013)

35. N. Gross, Z. Shotan, S. Kokkelmans, L. Khaykovich, Nuclear-spin-independent short-range three-body physics in ultracold atoms. Phys. Rev. Lett. **105**, 103203 (2010)

36. F. Ferlaino, A. Zenesini, M. Berninger, B. Huang, H.C. Nägerl, R. Grimm, Efimov resonances in ultracold quantum gases. Few-Body Syst. **51**(2), 113–133 (2011)

37. A. Zenesini, B. Huang, M. Berninger, H.-C. Nägerl, F. Ferlaino, R. Grimm, Resonant atom-dimer collisions in cesium: testing universality at positive scattering lengths. Phys. Rev. A **90**, 022704 (2014)

38. M. Kunitski, S. Zeller, J. Voigtsberger, A. Kalinin, L.Ph.H. Schmidt, M. Schöffler, A. Czasch, W. Schöllkopf, R.E. Grisenti, T. Jahnke, T. Blume, Observation of the Efimov state of the helium trimer. Science **348**(6234), 551–555 (2015)

39. O. Machtey, Z. Shotan, N. Gross, L. Khaykovich, Association of Efimov trimers from a three-atom continuum. Phys. Rev. Lett. **108**, 210406 (2012)

40. K. Helfrich, H.-W. Hammer, Resonant atom-dimer relaxation in ultracold atoms. EPL (Europhys. Lett.) **86**(5), 53003 (2009)

41. A. Kievsky, M. Gattobigio, E. Garrido, Universality in few-body systems: from few-atoms to few-nucleons. J. Phys.: Conf. Ser. **527**, 012001 (2014). (IOP Publishing)

42. H.-W. Hammer, D. Kang, L. Platter, Efimov physics in atom-dimer scattering of Li 6 atoms. Phys. Rev. A **82**(2), 022715 (2010)

43. V.F. Kharchenko, S.A. Storozhenko, The three-nucleon problem with the square-well potential. Nucl. Phys. A **137**(2), 437–444 (1969)

44. A.G. Sitenko, *Lectures in Scattering Theory: International Series of Monographs in Natural Philosophy*, vol. 39 (Elsevier, Amsterdam, 2013)

45. E.W. Schmid, H. Ziegelmann, *The Quantum Mechanical Three-Body Problem: Vieweg Tracts in Pure and Applied Physics* (Elsevier, Amsterdam, 2013)

46. M.K. Watson, J. Nuttall, J.S.R. Chisholm, *Topics in several particle dynamics* (Holden-Day, San Francisco, 1967)

47. E. Braaten, H.-W. Hammer, Three-body recombination into deep bound states in a bose gas with large scattering length. Phys. Rev. Lett. **87**(16), 160407 (2001)

48. F.P. Bedaque, G. Rupak, H.W. Griesshammer, H.-W. Hammer, Low energy expansion in the three body system to all orders and the triton channel. Nucl. Phys. A **714**(3), 589–610 (2003)

49. F.P. Bedaque, H.-W. Hammer, U. Van Kolck, Renormalization of the three-body system with short-range interactions. Phys. Rev. Lett. **82**(3), 463 (1999a)

50. P.F. Bedaque, U. Van Kolck, Effective field theory for few-nucleon systems. Annu. Rev. Nucl. Part. Sci. **52**, 339–396 (2002)

51. P.F. Bedaque, H.-W. Hammer, U. Van Kolck, The three-boson system with short-range interactions. Nucl. Phys. A **646**(4), 444–466 (1999b)

52. Y. Yamaguchi, Two-nucleon problem when the potential is nonlocal but separable. I. Phys. Rev. **95**, 1628–1634 (1954)

53. G.V. Skorniakov, K.A. Ter-Martirosian, Three body problem for short range forces. I. Scattering of low energy neutrons by deuterons. Sov. Phys. JETP **4**, 648 (1957)

54. J.R. Shepard, Calculations of recombination rates for cold He 4 atoms from atom-dimer phase shifts and determination of universal scaling functions. Phys. Rev. A **75**(6), 062713 (2007)

55. I.R. Afnan, D.R. Phillips, Three-body problem with short-range forces: renormalized equations and regulator-independent results. Phys. Rev. C **69**(3), 034010 (2004)

56. V.V. Flambaum, G.F. Gribakin, C. Harabati, Analytical calculation of cold-atom scattering. Phys. Rev. A **59**, 1998–2005 (1999)

57. J.-L. Li, X.-J. Hu, Y.-C. Han, Shu-Lin Cong, Simple model for analyzing Efimov energy and three-body recombination of three identical bosons with van der Waals interactions. Phys. Rev. A **94**, 032705 (2016)

58. P. Mestrom, J. Wang, C.H. Greene, J.P. D'Incao. Efimov universality for ultracold atoms with positive scattering lengths. ArXiv:1609.02857 (2016)

Few-Body Syst (2017) 58:35
DOI 10.1007/s00601-016-1181-2

E. A. Kolganova · A. K. Motovilov · W. Sandhas

The ^{4}He Trimer as an Efimov System: Latest Developments

Dedicated to the 30th anniversary of the "Few-Body Systems" journal

Received: 4 November 2016 / Accepted: 8 December 2016 / Published online: 12 January 2017
© Springer-Verlag Wien 2017

Abstract Kolganova et al. (Few-Body Syst 51:249, 2011) reviewed the results that demonstrate the Efimov nature of the ^{4}He three-atom system. The present note represents an extension of that survey to the time period which passed since its publication.

This short note may be viewed as a complement to our review paper on the same subject [1]. We decided to write it because there was a significant progress in studying Efimov systems since the time of publication of [1]. The progress is based mainly on the recent experimental studies of dilute ultracold gases of alkali atoms in magnetic traps [2–5] but there are interesting news also on the ^{4}He trimers [6].

We recall that the genuine Efimov effect is a remarkable phenomenon which may occur in a system of three particles with short-range pairwise interactions provided that none of the two-body subsystems has bound states. If at least two of the two-body subsystems are formed of distinguishable particles or identical bosons and have infinite s-wave scattering lengths then the three-particle system has infinitely many binding energies that exponentially converge to the three-particle threshold. This is the essence of the Efimov effect in its "full-scale" form [7,8]. The corresponding bound states are called Efimov states. It should be emphasized that the asymptotic value of the ratio of the consecutive binding energies of the Efimov states is universal in the sense that it only depends on the ratios of particle masses (but not on the form of the pairwise interactions).

All known two-body systems (both nuclear and atomic/molecular) have finite scattering lengths. Thus, for an isolated three-body system, i.e., in the absence of external fields, it is rather impossible to observe the full-scale Efimov effect with an infinite set of binding energies. Nevertheless, three-body systems featuring at least some peculiarities of the Efimov effect are already of great interest. A qualitative analysis presented by Efimov himself [7] shows that the total number of bound states in the three-boson system is proportional to the logarithm of the ratio of the boson-boson scattering length and effective radius of the two-body forces, provided that this ratio is very large. In the case of the ^{4}He three-atom system the ratio of the atom-atom scattering length and the effective radius is quite large (about 25). However it is not "very large", so that the Efimov estimate implies the existence rather of a single excited state for the ^{4}He$_3$ molecule (see discussion and references in [1]). That such a situation should take place in reality, is confirmed by the results of numerous

---

This article belongs to the Topical Collection "30th anniversary of Few-Body Systems".

E. A. Kolganova (✉)· A. K. Motovilov
Bogoliubov Laboratory of Theoretical Physics, JINR, Dubna, Russia
E-mail: kea@theor.jinr.ru

E. A. Kolganova · A. K. Motovilov
Dubna State University, Dubna, Russia

W. Sandhas
Physikalisches Institut, Universität Bonn, Endenicher Allee 11-13, 53115 Bonn, Germany

Table 1 ^{4}He dimer binding energy ϵ_d, ^{4}He dimer bond length $\langle R \rangle$, ^{4}He–^{4}He scattering length $\ell_{sc}^{(1+1)}$, excited state energy E^* of the ^{4}He trimer, and the difference $|E^* - \epsilon_d|$ for various He–He potentials, as compared to the experimental values from [6,29,44]

Potential	ε_d (mK)	$\ell_{sc}^{(1+1)}$ (Å)	$\langle R \rangle$ (Å)	E^* (mK)	$|E^* - \varepsilon_d|$ (mK)
HFDHE2 [18]	−0.830	124.65	64.21[a]	1.67	0.84
LM2M2 [19]	−1.303	100.23	51.84[a]	2.27	0.97
TTY [20]	−1.309[d]	100.01	51.65[a]	2.28	0.97
CCSAPT07 [21]	−1.564[b]	91.82	47.78[a]	2.59[b]	1.02
PCKLJS [22]	−1.615[b]	90.42	47.09[a]	2.65[b]	1.03
HFD-B [23]	−1.685	88.50	46.46[a]	2.74	1.05
Jeziorska [24]	−1.728[c]	87.53		2.78[c]	1.06
SAPT96 [25]	−1.744[b]		45.45[a]	2.80[b]	1.06
Exp.	$1.1^{+0.3}_{-0.2}$ [29]	104^{+8}_{-18} [29]	52^{+4}_{-4} [29]		
	$1.76^{+0.15}_{-0.15}$ [44]				0.98 ± 0.2 [6]

[a] Results from [34]

[b] Results from [9,10]

[c] Results from [33]

[d] This result slightly differs from the one presented in [32] due to a different number of terms used in the dispersion series for the TTY potential [20]. Unmarked values in the second, third, and fifth columns were obtained by the authors (see, e.g., [1,35])

calculations of the ^{4}He trimer binding energies: For various realistic atom-atom potentials suggested in the last three decades, the ^{4}He trimer has exactly one excited state (see [1] and references therein; see also the recent papers [9–12]). The Efimov nature of this state was conjectured for the first time in [13]. Later on, this conjecture was strongly supported by several numerical calculations involving very small variation of the atom-atom potential strength but producing arbitrarily large change of the atom-atom scattering length [14–17]. For details and more references on this approach we again refer to [1].

By now it is rather well established, both theoretically and experimentally, that the system of two ^{4}He atoms possesses a single bound state. The energy of this state is very small in molecular scale. In particular, the mostly used realistic potential models [18–25] predict the ^{4}He dimer energy between 0.8 and 1.8 mK, which results in a very large scattering length around 100 Å (see Table 1). In an experiment, ^{4}He dimers have been observed for the first time in 1993 by the Minnesota group [26], and in 1994 by Schöllkopf and Toennies [27]. Along with the dimers, the experimental work [27] has also proved the existence of ^{4}He trimers. A first experimental estimate for the size of the ^{4}He$_2$ molecule has been presented in [28]. According to [28], the root mean square distance between ^{4}He nuclei in the ^{4}He dimer is equal to 62 ± 10 Å. Several years later, the bond length for ^{4}He$_2$ was measured again by Grisenti et al. [29] who found for this length the value of 52 ± 4 Å. The estimates of [28,29] imply that the ^{4}He dimer is the most extended known diatomic molecular ground state. The measurements [29] also allowed to evaluate a ^{4}He–^{4}He scattering length $\ell_{sc}^{(1+1)}$ of 104^{+8}_{-18} Å and a ^{4}He dimer energy ϵ_d of $1.1^{+0.3}_{-0.2}$ mK. The size of the ^{4}He$_3$ ground state has been estimated for the first time in the experiment [30]. According to [30] the He–He bond length in the ^{4}He$_3$ ground state is 11^{+4}_{-5} Å, in agreement with theoretical predictions.

Until 2015, there was no reliable experimental evidence for the existence of an excited state in the ^{4}He trimer. A good news [6] on the experimental observation of this long-predicted Efimov-type state came just in that year. The experiment [6] was based on a combination of the Coulomb explosion imaging technique [31] with cluster mass selection by matter wave diffraction [27]. The helium clusters were prepared in a molecular beam by expanding helium gas at a temperature of 8 K through a 5-μm nozzle. Helium trimers were extracted from the molecular beam by means of matter wave diffraction. Every ^{4}He atom of a cluster was then singly ionized by a strong ultrashort laser field, which led to the subsequent Coulomb explosion of the cluster. Momenta of the ions acquired during the explosion were measured by cold target recoil-ion momentum spectroscopy. These momenta were then used to reconstruct the initial pair-distance distribution and extract the cluster energy. As a result, the difference $|E^* - \epsilon_d|$ between the binding energy E^* of the exited state of the trimer and the ground state energy ϵ_d of the dimer was found to be equal to 0.98 ± 0.2 mK [6]. This result is close to the theoretical predictions for $|E^* - \epsilon_d|$ corresponding to various potentials (see column 6 of Table 1), although these potentials give quite different binding energies for the dimer and trimer relative to the breakup threshold (see columns 2 and 5 of Table 1, respectively). Moreover, the theoretical values for $|E^* - \epsilon_d|$ lie inside the experimental error bar. It is also worth recalling that, since the ^{4}He$_3$ system is almost Efimov, the dependence of the excited state energy on the dimer energy lies on a universal curve (see, e.g. [36–38]) and the difference between these energies varies much slower than the energies themselves. Furthermore, the energy values given in Table 1 represent only a very small piece of this curve.

The experimental technique used in [6] and earlier in [39] also gave information about geometrical structure of helium trimers. From the results of [39] it follows that there is no exceptional mutual position of helium atoms (like, say, equilateral triangle or a linear chain) in the ^{4}He$_3$ ground state. This state is described rather as a structureless random cloud. In [6] it was shown that the most probable geometry of the ^{4}He$_3$ exited state is completely different: two atoms in this state are close to each other and the third atom is far away. The conclusions of [6,39] agree with theoretical predictions of the trimer shapes made in [40,41]. The experimental images of pair-distance distribution initiated additional theoretical calculations of geometrical properties of trimer bound states [11,42,43]. The results of these calculations are rather in good agreement with [6,39].

Another recent news, already of 2016, concerns binding energy of the ^{4}He dimer. This energy was evaluated in the experiment [44], just for the second time and in about 15 years after the previous experimental evaluation in [29]. The investigation [44] was based on a technique very similar to the one used in [6]. The dimer energy of $1.76^{+0.15}_{-0.15}$ mK obtained in [44] differs significantly from the experimental value of $1.1^{+0.3}_{-0.2}$ mK established in [29]. Such an uncertainty in the experimental results does not allow one to make a choice in favor of a particular potential model. Clearly, further experiments dedicated to determining binding energies of helium dimer and trimer are very necessary in order to verify He–He potential models and to choose the most appropriate one.

Acknowledgements This work was supported in part by the Heisenberg-Landau Program and the Russian Foundation for Basic Research.

References

1. E.A. Kolganova, A.K. Motovilov, W. Sandhas, The ^{4}He trimer as an Efimov system. Few-Body Syst. **51**, 249 (2011)
2. J.R. Williams, E.L. Hazlett, J.H. Huckans, R.W. Stites, Y. Zhang, K.M. O'Hara, Evidence for an excited-state Efimov trimer in a three-component Fermi gas. Phys. Rev. Lett. **103**, 130404 (2009)
3. B.S. Rem, A.T. Grier, I. Ferrier-Barbut et al., Lifetime of the Bose gas with resonant interactions. Phys. Rev. Lett. **110**, 163202 (2013)
4. R.S. Bloom, M.-G. Hu, T.D. Cumby, D.S. Jin, Tests of universal three-body physics in an ultracold Bose–Fermi mixture. Phys. Rev. Lett. **111**, 105301 (2013)
5. J. Ulmanis, S. Häfner, R. Pires, F. Werner, D.S. Petrov, E.D. Kuhnle, M. Weidemüller, Universal three-body recombination and Efimov resonances in an ultracold Li–Cs mixture. Phys. Rev. A **93**, 022707 (2016)
6. M. Kunitski, S. Zeller, J. Voigtsberger et al., Observation of the Efimov state of the helium trimer. Science **348**, 551 (2015)
7. V.N. Efimov, Weakly-bound states of 3 resonantly-interacting particles. Sov. J. Nucl. Phys. **12**, 589 (1970) **[Yad. Fiz. 12, 1080 (1970)]**
8. V. Efimov, Energy levels of three resonantly interacting particles. Nucl. Phys. A. **210**, 157 (1973)
9. E. Hiyama, M. Kamimura, Variational calculation of ^{4}He tetramer ground and excited states using a realistic pair potential. Phys. Rev. A **85**, 022502 (2012)
10. E. Hiyama, M. Kamimura, Linear correlations between ^{4}He trimer and tetramer energies calculated with various realistic ^{4}He potentials. Phys. Rev. A **85**, 062505 (2012)
11. H. Suno, Geometrical structure of helium triatomic systems: comparison with the neon trimer. J. Phys. B **49**, 014003 (2016)
12. A. Deltuva, Momentum-space calculation of ^{4}He triatomic system with realistic potential. Few-Body Syst. **56**, 897 (2015)
13. T.K. Lim, S.K. Duffy, W.C. Damert, Efimov state in the ^{4}He trimer. Phys. Rev. Lett. **38**, 341 (1977)
14. T. Cornelius, W. Glöckle, Efimov states for three ^{4}He atoms? J. Chem. Phys. **85**, 3906 (1986)
15. B.D. Esry, C.D. Lin, C.H. Greene, Adiabatic hyperspherical study of the helium trimer. Phys. Rev. A **54**, 394 (1996)
16. E.A. Kolganova, A.K. Motovilov, S.A. Sofianos, Three-body configuration space calculations with hard-core potentials. J. Phys. B **31**, 1279 (1998)
17. R. Lazauskas, J. Carbonell, Description of ^{4}He tetramer bound and scattering states. Phys. Rev. A **73**, 062717 (2006)
18. R.A. Aziz, V.P.S. Nain, J.S. Carley, W.L. Taylor, G.T. McConville, An accurate intermolecular potential for helium. J. Chem. Phys. **79**, 4330 (1979)
19. R.A. Aziz, M.J. Slaman, An examination of ab initio results for the helium potential energy curve. J. Chem. Phys. **94**, 8047 (1991)
20. K.T. Tang, J.P. Toennies, Yiu, Accurate analytical He-He can der Waals potential based on perturbation theory. Phys. Rev. Lett. **74**, 1546 (1995)
21. T. Korona, H.L. Williams, R. Bukowski, B. Jeziorski, K. Szalewicz, Helium dimer potential from symmetry-adapted perturbation theory calculations using large Gaussian geminal and orbital basis sets. J. Chem. Phys. **106**, 5109 (1997)
22. M. Przybytek, W. Cencek, J. Komasa, G. Łach, B. Jeziorski, K. Szalewicz, Relativistic and quantum electrodynamics effects in the helium pair potential. Phys. Rev. Lett. **104**, 183003 (2010)
23. R.A. Aziz, F.R.W. McCourt, C.C.K. Wong, A new determination of the ground state interatomic potential for He$_2$. Mol. Phys. **61**, 1487 (1987)
24. M. Jeziorska, W. Cencek, K. Patkowski, B. Jeziorski, K. Szalewicz, Pair potential for helium from symmetry-adapted perturbation theory calculations and from supermolecular data. J. Chem. Phys. **127**, 124303 (2007)

25. A.R. Janzen, R.A. Aziz, An accurate potential energy curve for helium based on ab initio calculations. J. Chem. Phys. **107**, 914 (1997)
26. F. Luo, G.C. McBane, G. Kim, C.F. Giese, W.R. Gentry, The weakest bond: experimental observation of helium dimer. J. Chem. Phys. **98**, 3564 (1993)
27. W. Schöllkopf, J.P. Toennies, Nondestructive mass selection of small van der Waals clusters. Science **266**, 1345 (1994)
28. F. Luo, C.F. Giese, W.R. Gentry, Direct measurement of the size of the helium dimer. J. Chem. Phys. **104**, 1151 (1996)
29. R. Grisenti, W. Schöllkopf, J.P. Toennies, G.C. Hegerfeld, T. Köhler, M. Stoll, Determination of the bond length and binding energy of the helium dimer by diffraction from a transmission grating. Phys. Rev. Lett. **85**, 2284 (2000)
30. R. Brühl, A. Kalinin, O. Kornilov, J.P. Toennies, G.C. Hegerfeld, M. Stoll, Matter wave diffraction from an inclined transmission grating: searching for the elusive ^{4}He trimer Efimov state. Phys. Rev. Lett. **95**, 063002 (2005)
31. Z. Vager, R. Naaman, E.P. Kanter, Coulomb explosion imaging of small molecules. Science **244**, 426 (1989)
32. V. Roudnev, Ultra-low energy elastic scattering in a system of three He atoms. Chem. Phys. Lett. **367**, 95 (2003)
33. H. Suno, B.D. Esry, Adiabatic hyperspherical study of triatomic helium systems. Phys. Rev. A **78**, 062701 (2008)
34. A. Kievsky, E. Garrido, C. Romero-Redondo, P. Barletta, The helium trimer with soft-core potentials. Few-Body Syst. **51**, 259 (2011)
35. E.A. Kolganova, A.K. Motovilov, W. Sandhas, Ultracold collisions in the system of three helium atoms. Phys. Part. Nucl. **40**, 206 (2009)
36. E. Braaten, H.-W. Hammer, Universality in few-body systems with large scattering length. Phys. Rep. **428**, 259 (2006)
37. P. Naidon, E. Hiyama, M. Ueda, Universality and the three-body parameter of ^{4}He trimers. Phys. Rev. A **86**, 012502 (2012)
38. A. Kievsky, M. Gattobigio, Universal nature and finite-range corrections in elastic atom-dimer scattering below the dimer breakup threshold. Phys. Rev. A **87**, 052719 (2013)
39. J. Voigtsberger, S. Zeller, J. Becht et al., Imaging the structure of the trimer systems ^{4}He$_3$ and ^{3}He^{4}He$_2$. Nat. Commun. **5**, 5765 (2014)
40. D. Blume, C.H. Greene, B.D. Esry, Comparative study of He$_3$, Ne$_3$, and Ar$_3$ using hyperspherical coordinates. J. Chem. Phys. **113**, 2145 (2000)
41. D. Bressanini, G. Morosi, What is the shape of the helium trimer? A comparison with the neon and argon trimers. J. Phys. Chem. A **115**, 10880 (2011)
42. D. Bressanini, The structure of the asymmetric helium trimer ^{3}He^{4}He$_2$. J. Phys. Chem. A **118**, 6521 (2014)
43. P. Stipanović, L.V. Markić, J. Boronat, Elusive structure of helium trimers. J. Phys. B **49**, 185101 (2016)
44. S. Zeller, M. Kunitski, J. Voigtsberger et al., Imaging the He$_2$ quantum halo state using a free electron laser. arXiv:1601.03247

Few-Body Syst (2017) 58:26
DOI 10.1007/s00601-016-1173-2

Jared Vanasse · Daniel R. Phillips

Three-Nucleon Bound States and the Wigner-SU(4) Limit

Received: 1 August 2016 / Accepted: 13 October 2016 / Published online: 7 January 2017
© Springer-Verlag Wien 2017

Abstract We examine the extent to which the properties of three-nucleon bound states are well-reproduced in the limit that nuclear forces satisfy Wigner's SU(4) (spin–isospin) symmetry. To do this we compute the charge radii up to next-to-leading order (NLO) in an effective field theory that is an expansion in powers of R/a, with R the range of the nuclear force and a the nucleon–nucleon (NN) scattering lengths. In the Wigner-SU(4) limit, the triton and helium-3 point charge radii are equal. At NLO in the range expansion both are 1.66 fm. Adding the first-order corrections due to the breaking of Wigner symmetry in the NN scattering lengths gives a ^{3}H point charge radius of 1.58 fm, which is remarkably close to the experimental number, 1.5978 ± 0.040 fm (Angeli and Marinova in At Data Nucl Data Tables 99:69–95, 2013). For the ^{3}He point charge radius we find 1.70 fm, about 4% away from the experimental value of 1.77527 ± 0.0054 fm (Angeli and Marinova 2013). We also examine the Faddeev components that enter the tri-nucleon wave function and find that an expansion of them in powers of the symmetry-breaking parameter converges rapidly. Wigner's SU(4) symmetry is thus a useful starting point for understanding tri-nucleon bound-state properties.

1 Introduction

Quantum-mechanical systems in which the two-particle potential is short-ranged, and the two-body scattering length is large compared to that range, share "universal" features [2]. The most striking of these is the Efimov effect; the existence of an infinite series of three-body bound states . In the "unitary limit" the scattering length $|a| \rightarrow \infty$, and the three-body problem exhibits discrete scale invariance, with states in the Efimov tower related to one another through a rescaling of co-ordinates by a factor that is $e^{\pi/s_0} = 22.7$ [3,4] for the equal-mass case. The existence of two states related by this Efimov ratio has recently been demonstrated for Cesium atoms near a Feshbach resonance—i.e., essentially in the unitary limit [5]—and for clusters of Helium atoms that have a large, but finite, two-body scattering length [6].

Both of these systems consist of bosons, whereas the particles that make up nuclei are spin one-half fermions of two different isospins. This means that—even in the approximation that S-wave interactions dominate the formation of the three-nucleon bound state—nucleon–nucleon (NN) interactions in two different channels, the $^1S_0$ and the $^3S_1$, contribute to the binding of the three-nucleon system. Nevertheless, the Efimov [7] effect also occurs for three nucleons: the virtual state in doublet S-wave neutron-deuteron scattering becomes an excited Efimov state of the triton in appropriate limits [8–10]. Most recently, Kievsky and Gattobigio [10] studied the

This article belongs to the special issue "30th anniversary of Few-Body Systems".

J. Vanasse · D. R. Phillips (✉)
Department of Physics and Astronomy, Institute of Nuclear and Particle Physics, Ohio University, Athens, OH 45701, USA
E-mail: phillips@phy.ohiou.edu

J. Vanasse
E-mail: vanasse@ohio.edu

physics of the three-nucleon bound state with model Gaussian potentials, showing that Efimov states appear in the three-nucleon spectrum as the $^1S_0$ and $^3S_1$ scattering lengths tend towards the unitary limit. They argued that this means the triton is inside the "Efimov window" in that its structure is governed by 'a few control parameters, [such] as the two-body energies and scattering lengths', i.e., it can be described within the context of few-body universality.

An effective field theory (EFT) with only short-range interactions provides a systematic way to organize the treatment of three-body states in this universal/Efimov-window regime. It exploits the hierarchy of scales $R \ll |a|$, and in nuclear physics it is known as the pionless EFT (EFT($\pi\!\!\!/$)) [11–14]. At leading order (LO) in EFT($\pi\!\!\!/$) the particles interact via zero-range forces, whose strengths are tuned to reproduce, e.g. the $^1S_0$ and $^3S_1$ scattering lengths. At higher orders corrections to two-body observables due to the finite effective ranges, r, can be computed in perturbation theory [15], with a nominal expansion parameter of $r/a \approx 30\%$ in the $^3S_1$ channel.

The leading-order equations for the triton in this EFT were worked out in Ref. [16], and it was quickly apparent that those equations are equivalent to the (single) equation for bosons [17,18] in the limit that the $^3S_1$ and $^1S_0$ scattering lengths are equal, i.e., if the NN interaction displays a Wigner-SU(4) spin–isospin symmetry [16,19]. That equation, known as the Skornyakov–Ter–Martirosian (STM [20]) equation, must be regulated. Using a momentum-space cutoff Λ its solution is sensitive to the value of Λ, i.e., to short-distance physics in the three-body system; the STM equation does not posses a unique solution in the limit $\Lambda \to \infty$ [21]. These problems can be removed by adding a three-body force to the EFT at leading order [18]. The three-body force prevents Thomas collapse [22].

The leading-order EFT calculation recovers the prediction of the Efimovian spectrum in the unitary limit and also permits straightforward extension of that result to finite scattering lengths—and to finite, and different, $S = 0$ and $S = 1$ scattering lengths in the nuclear-physics case. This reproduces findings of Efimov [3,4,7] and others [8,9] for zero-range forces. Crucially, the LO three-body force in the three-nucleon problem is Wigner-SU(4) symmetric [19]—even for the situation where the $S = 0$ and $S = 1$ channels exhibit a different scattering length; Wigner-SU(4)-anti-symmetric three-body forces do not enter the EFT until much higher orders in the expansion [23–25]. Higher orders in the R/a expansion are calculated by considering perturbative corrections to three-body observables due to the finite range of the nuclear force. EFT calculations at next-to-leading (NLO) and next-to-next-to-leading-order (NNLO) in the range appeared in Refs. [26–29] (for three bosons) and [30–32] (for the three-nucleon system). Most recently, Vanasse has shown that the triton point charge radius is well described within EFT($\pi\!\!\!/$), obtaining $\langle r_{^3H}^2 \rangle_{\mathrm{pt}} = 1.14 + 0.45 + 0.03 = 1.62\,\mathrm{fm}$ at leading order and for NLO and NNLO corrections [33]. The NLO and NNLO results agree with the experimental value of $1.5978 \pm 0.040\,\mathrm{fm}$ [1]. While the NLO correction is sizable, the excellent agreement and reasonable convergence pattern support the contention of Ref. [10] that the triton is within the purview of few-body universality.

In this paper we use EFT($\pi\!\!\!/$) to answer the question of how relevant Wigner-SU(4) symmetry is to the physics of both the triton and ^{3}He. Naively the NN system seems far from the Wigner-SU(4) limit: the deuteron binding momentum is 45 MeV, while the corresponding scale in the $^1S_0$ channel, the inverse of the $^1S_0$ neutron-proton (np) scattering length, is $1/a_{np}^{S=0} = -8.3\,\mathrm{MeV}$. Thus the parameter that governs Wigner-SU(4) breaking:

$$\delta \equiv \frac{1}{2}\left(1/a_{np}^{S=1} - 1/a_{np}^{S=0}\right) \tag{1}$$

is not small compared to the average of $1/a_{np}^{S=1}$ and $1/a_{np}^{S=0}$. However, we shall see that an expansion around the Wigner-SU(4) limit, where $\delta = 0$, converges well. The triton binding energy changes by only 0.8 MeV due to Wigner-SU(4) breaking, and the triton charge radius in the Wigner-SU(4) limit is 1.66 fm at NLO in EFT($\pi\!\!\!/$), quite close to the average of the experimental ^{3}H and ^{3}He point charge radii. Perhaps most tellingly, the Wigner-SU(4)-odd component of the triton wave function is $<10\%$ the size of the SU(4)-even part, which implies that an expansion around the Wigner-SU(4) limit will be successful for all trinucleon bound-state observables—or at least for all observables that do not vanish in that limit.

Wigner-SU(4) (spin–isospin) symmetry has had considerable phenomenological success in nuclear physics, ever since, in 1937, Wigner classified nuclear states according to their SU(4) representation in order to explain the pattern of nuclear masses up to $A \approx 40$ [34]. Subsequently he worked out the consequences of such a symmetry for nuclear beta decays [35]. The "Wigner super-multiplet theory" was later applied to inelastic electron scattering from, and muon capture on, ^{12}C and ^{16}O [36–38]; the particle-hole states were usefully classified according to Wigner-SU(4), thereby explaining the existence of a family of giant resonances in these nuclei.

We note that the presence of Wigner-SU(4) symmetry in the three-nucleon problem is a weaker condition than that the three-nucleon problem exhibit the unitary ($|a| \to \infty$) limit. The unitary limit may be relevant for few-nucleon systems in large magnetic fields [39] or in a version of QCD with slightly larger and unequal up- and down-quark masses [40]. Recently, König et al. have argued that the binding energies of the $A = 3$ and $A = 4$ systems can be understood both qualitatively and quantitatively via an expansion around the unitary limit. We will comment specifically on this idea in Sect. 7. In the Wigner-SU(4) limit the four NN scattering lengths a_{nn}, a_{pp}, $a_{np}^{S=0}$, and $a_{np}^{S=1}$ are all equal, but could be finite. Efimovian towers can still occur for finite scattering lengths (e.g. the helium trimers), but they are related by a scaling factor which is smaller than the 22.7 that applies for equal masses when $|a| \to \infty$. In this situation the equations for the triton are those for a two-neutron halo with a neutron-core scattering length equal to the neutron–neutron scattering length [41]. Therefore the Wigner-SU(4) limit not only connects the trinucleons to the three-boson systems being investigated experimentally in Innsbruck [5], Frankfurt [6], and elsewhere, it also permits us to understand the triton as the lightest two-neutron halo.

Our discussion of this connection proceeds as follows. In Sect. 2 we introduce the basic formalism for Wigner-SU(4) symmetry and its breaking in the two-body sector, while Sect. 3 introduces this formalism in the three-body sector. Sections 4–6 discuss the effects of Wigner-SU(4) symmetry and its breaking on binding energy, charge and matter radii, and triton vertex functions. In Sect. 7 we examine the values obtained for three-nucleon charge radii in the unitary limit and in Sect. 8 we conclude.

2 Wigner-SU(4) Symmetry in the Two-Body Sector

The LO NN interaction in EFT($\not{\pi}$) can be written as [42]

$$\mathcal{L}_2 = -\frac{1}{2} C_0^T \hat{N}^\dagger \sigma_i \hat{N} \hat{N}^\dagger \sigma_i \hat{N} - \frac{1}{2} C_0^S \hat{N}^\dagger \hat{N} \hat{N}^\dagger \hat{N}. \tag{2}$$

A Wigner transformation $\hat{N} \to \hat{U} \hat{N}$ is a simultaneous transformation under spin and isospin given by the operator $\hat{U} = e^{i\alpha_{\mu\nu}\sigma_\mu\tau_\nu}$, where $\sigma_\mu = \{1, \sigma_i\}$ and $\tau_\nu = \{1, \tau_a\}$ are four vectors with $\mu, \nu = 0, 1, 2, 3$ and $i, a = 1, 2, 3$. The determinant of $\hat{U}$ is equal to one and $\alpha_{\mu\nu}$ is a 4×4 matrix of real numbers [19,34], with $\alpha_{00} = 0$. It is immediately obvious that the C_0^S term is invariant under a Wigner transformation while the C_0^T term is not. Thus EFT($\not{\pi}$) is Wigner symmetric at LO if and only if $C_0^T = 0$. The LO NN interaction can also be written in the partial-wave basis yielding

$$\mathcal{L}_2^{PW} = -C_0^{(^3S_1)} \left(\hat{N}^T P_i \hat{N} \right)^\dagger \left(\hat{N}^T P_i \hat{N} \right) - C_0^{(^1S_0)} \left(\hat{N}^T \bar{P}_a \hat{N} \right)^\dagger \left(\hat{N}^T \bar{P}_a \hat{N} \right), \tag{3}$$

where $P_i = \frac{1}{\sqrt{8}}\sigma_2\sigma_i\tau_2$ $\left(\bar{P}_a = \frac{1}{\sqrt{8}}\sigma_2\tau_2\tau_a \right)$ projects out the spin-triplet iso-singlet (spin-singlet iso-triplet) combination of nucleons. Parameters in Eq. (2) can be related to parameters in the partial wave basis via [19]

$$C_0^{(^1S_0)} = C_0^S - 3C_0^T, \quad C_0^{(^3S_1)} = C_0^S - C_0^T, \tag{4}$$

so the condition $C_0^T = 0$ for Wigner-SU(4) symmetry is equivalent to $C_0^{(^1S_0)} = C_0^{(^3S_1)}$ in the partial-wave basis. At LO in the EFT($\not{\pi}$) power counting the NN scattering amplitude is given by an infinite sum of bubble diagrams [12,13]. Fitting to the $^3S_1$ ($^1S_0$) bound (virtual bound) state pole gives

$$C_0^{(^3S_1)} = \frac{4\pi}{M_N} \frac{1}{\gamma_t - \mu}, \quad C_0^{(^1S_0)} = \frac{4\pi}{M_N} \frac{1}{\gamma_s - \mu}, \tag{5}$$

for the low-energy constants (LECs) in the partial-wave basis. (The scale μ comes from using the power divergence subtraction scheme with dimensional regularization [12,13].) If $\mu \gg \gamma_t$, γ_s then Wigner-SU(4) symmetry is approximate in the NN system. However, if $\gamma_t = \gamma_s$ then Wigner-SU(4) symmetry is exact for the NN system at LO. $\gamma_t = 45.7025$ MeV and $\gamma_s = -7.890$ MeV [43] correspond to the momenta at which poles of the NN scattering amplitude occur in the $^3S_1$ and $^1S_0$ channels, respectively. At LO in the EFT($\not{\pi}$) expansion they are equal to $1/a_{np}^{S=1}$ and $1/a_{np}^{S=0}$ [15,43,44]. Since $\gamma_s \neq \gamma_t$ Wigner-SU(4) symmetry is not exact. We will explore the extent to which an expansion in powers of $\gamma_s - \gamma_t$ gives access to the properties of three-nucleon bound states.

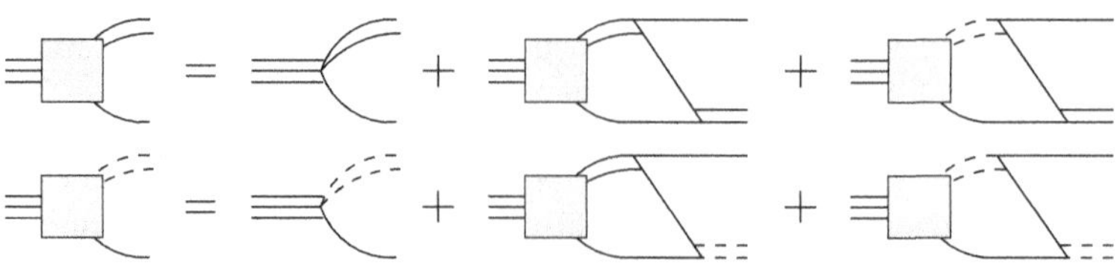

Fig. 1 Set of coupled integral equations for the LO tri-nucleon vertex function. *Single lines* are nucleons, *double lines* $^3S_1$ dibaryons, *double dashed lines* $^1S_0$ dibaryons, and *triple lines* tri-nucleons

Up to NLO in the EFT expansion the Wigner-SU(4) symmetric limit is attained if all effective-range expansion parameters occurring up to that order are equal in the $^3S_1$ and $^1S_0$ channels. This results in equal Lagrangian parameters in the $^1S_0$ and $^3S_1$ channels, thus guaranteeing symmetry of the Lagrangian under Wigner-SU(4) transformations. Tensor interactions complicate the definition at higher orders. But at NLO this means that Wigner-SU(4) symmetry is satisfied if and only if the $^1S_0$ and $^3S_1$ channels have equal scattering lengths and effective ranges.

3 Wigner-SU(4) Symmetry in the Three-Body Sector

The LO triton vertex function is the solution to a set of coupled integral equations shown in Fig. 1 [33]. The coupled set of integral equations can be written as

$$
\mathcal{G}_t^{(\mathrm{LO})}(p) = 1 + \frac{1}{\pi} \int_0^{\Lambda} dq\, q^2 R^{(\mathrm{LO})}(q, p, E) \left\{ D_t(q, E)\mathcal{G}_t^{(\mathrm{LO})}(q) + 3 D_s(q, E)\mathcal{G}_s^{(\mathrm{LO})}(q) \right\}
$$

$$
\mathcal{G}_s^{(\mathrm{LO})}(p) = 1 + \frac{1}{\pi} \int_0^{\Lambda} dq\, q^2 R^{(\mathrm{LO})}(q, p, E) \left\{ 3 D_t(q, E)\mathcal{G}_t^{(\mathrm{LO})}(q) + D_s(q, E)\mathcal{G}_s^{(\mathrm{LO})}(q) \right\}, \tag{6}
$$

where

$$
R^{(\mathrm{LO})}(q, p, E) = \frac{1}{qp} Q_0\left(\frac{q^2 + p^2 - M_N E}{qp} \right), \tag{7}
$$

and the dibaryon propagators are defined by

$$
D_{\{t,s\}}(q, E) = \frac{1}{\sqrt{\frac{3}{4}q^2 - M_N E} - \gamma_{\{t,s\}}}. \tag{8}
$$

Vertex functions are equivalent to Faddeev components. From them, the triton wave function can be reconstructed. For further details see Refs. [33,46,47]. The superscripts designate that we refer here to quantities that are LO in the EFT($\not\pi$) power counting. $Q_0(a)$ is a Legendre function of the second kind given by

$$
Q_0(a) = \frac{1}{2} \ln\left(\frac{1+a}{a-1} \right). \tag{9}
$$

The binding energy, $E = -B$, where the numerical value chosen for B is discussed in the next section. In order to investigate the consequences of the Wigner-SU(4) limit in the three-body system it is convenient to rewrite the LO triton vertex function in a Wigner-SU(4) basis, which is defined by

$$
\mathcal{G}_+^{(\mathrm{LO})}(p) = \mathcal{G}_t^{(\mathrm{LO})}(p) + \mathcal{G}_s^{(\mathrm{LO})}(p), \quad \mathcal{G}_-^{(\mathrm{LO})}(p) = \mathcal{G}_t^{(\mathrm{LO})}(p) - \mathcal{G}_s^{(\mathrm{LO})}(p). \tag{10}
$$

In this basis it is necessary to take the sum and difference of the ${}^3S_1$ and ${}^1S_0$ dibaryon propagators of Eq. (8). Defining $\gamma_{\text{AVE}} = (\gamma_t + \gamma_s)/2$ and $\delta = (\gamma_t - \gamma_s)/2$ the sum of dibyaron propagators is

$$
\frac{1}{\sqrt{\frac{3}{4}q^2 - M_N E} - \gamma_t} + \frac{1}{\sqrt{\frac{3}{4}q^2 - M_N E} - \gamma_s}
$$
$$
= \frac{2}{\sqrt{\frac{3}{4}q^2 - M_N E} - \gamma_{\text{AVE}}} \sum_{n=0}^{\infty} \frac{\delta^{2n}}{\left(\sqrt{\frac{3}{4}q^2 - M_N E} - \gamma_{\text{AVE}}\right)^{2n}} \tag{11}
$$

and the difference

$$
\frac{1}{\sqrt{\frac{3}{4}q^2 - M_N E} - \gamma_t} - \frac{1}{\sqrt{\frac{3}{4}q^2 - M_N E} - \gamma_s} = \frac{2}{\sqrt{\frac{3}{4}q^2 - M_N E} - \gamma_{\text{AVE}}} \sum_{n=0}^{\infty} \frac{\delta^{2n+1}}{\left(\sqrt{\frac{3}{4}q^2 - M_N E} - \gamma_{\text{AVE}}\right)^{2n+1}},
$$
$$
\tag{12}
$$

where we have expanded in powers of δ which parametrizes the distance from the Wigner-SU(4) limit. In addition to expanding the dibaryon propagators in powers of δ, the triton vertex functions are also expanded in powers of δ via

$$
\mathcal{G}_+^{(\text{LO})}(p) = \sum_{n=0}^{\infty} \mathcal{G}_+^{(2n)}(p)\delta^{2n}, \quad \mathcal{G}_-^{(\text{LO})}(p) = \sum_{n=0}^{\infty} \mathcal{G}_-^{(2n+1)}(p)\delta^{2n+1}. \tag{13}
$$

Equations (10)–(13) can then be used in Eq. (6), and equating terms order-by-order in δ yields the set of coupled integral equations

$$
\widetilde{\mathcal{G}}_+^{(2n)}(p) = 2\delta_{0n} + D(p, E)\widetilde{\mathcal{G}}_-^{(2n-1)}(p) + \frac{4}{\pi}\int_0^{\Lambda} dq\, q^2 D(q, E) R^{(\text{LO})}(q, p, E)\widetilde{\mathcal{G}}_+^{(2n)}(q)
$$

$$
\widetilde{\mathcal{G}}_-^{(2n+1)}(p) = D(p, E)\widetilde{\mathcal{G}}_+^{(2n)}(p) - \frac{2}{\pi}\int_0^{\Lambda} dq\, q^2 D(q, E) R^{(\text{LO})}(q, p, E)\widetilde{\mathcal{G}}_-^{(2n+1)}(q), \tag{14}
$$

where

$$
D(q, E) = \frac{1}{\sqrt{\frac{3}{4}q^2 - M_N E} - \gamma_{\text{AVE}}}. \tag{15}
$$

The functions $\widetilde{\mathcal{G}}_\pm^{(n)}(p)$ are defined by

$$
\widetilde{\mathcal{G}}_+^{(2n)}(p) = \mathcal{G}_+^{(2n)}(p) + D(p, E)\widetilde{\mathcal{G}}_-^{(2n-1)}(p)
$$
$$
\widetilde{\mathcal{G}}_-^{(2n+1)}(p) = \mathcal{G}_-^{(2n+1)}(p) + D(p, E)\widetilde{\mathcal{G}}_+^{(2n)}(p). \tag{16}
$$

Writing things in terms of $\widetilde{\mathcal{G}}$'s, rather than $\mathcal{G}$'s, means that the equations simplify considerably and the correction at a given order only depends on the order immediately preceeding it, and not all orders preceeding it. For $n = 0$ we note that $\widetilde{\mathcal{G}}_-^{(2n-1)}(p) = 0$ and therefore $\widetilde{\mathcal{G}}_+^{(0)}(p) = \mathcal{G}_+^{(0)}(p)$. Also in the limit $\delta = 0$ only the $\mathcal{G}_+^{(0)}(p)$ term gives a non-zero contribution and its integral equation is equivalent to that for three bosons [16].

In order to properly normalize the triton vertex function it must be multiplied by the triton wavefunction renormalization which is given by

$$
Z_\psi = \frac{1}{\Sigma'(E)}, \tag{17}
$$

where $\Sigma(E)$ is the triton self energy

$$\Sigma(E) = \frac{1}{2\pi} \int_0^{\Lambda} dq q^2 \left\{ \frac{1}{\sqrt{\frac{3}{4}q^2 - M_N E} - \gamma_t} \mathcal{G}_t(q) + \frac{1}{\sqrt{\frac{3}{4}q^2 - M_N E} - \gamma_s} \mathcal{G}_s(q) \right\}. \tag{18}$$

Again expanding the dibaryon propagators and the triton vertex functions in powers of δ we find that only even powers of δ enter in the expansion of Σ:

$$\Sigma(E) = \sum_{n=0}^{\infty} \Sigma^{(2n)}(E) \delta^{2n}, \tag{19}$$

where

$$\Sigma^{(2n)}(E) = \frac{1}{2\pi} \int_0^{\Lambda} dq q^2 D(q) \widetilde{\mathcal{G}}_+^{(2n)}(q). \tag{20}$$

Thus the triton wavefunction renormalization in the δ expansion is given by

$$Z_\psi = \frac{1}{\Sigma'(E)} = \frac{1}{\Sigma^{(0)'}(E)} - \frac{\Sigma^{(2)'}(E)}{\left(\Sigma^{(0)'}(E)\right)^2} + \cdots \tag{21}$$

3.1 Range Corrections

The $O(r)$ (NLO in the nuclear-force's range) correction to the triton vertex function in the Z-parametrization is given by [33]

$$\mathcal{G}_t^{(\text{NLO})}(p) = \mathcal{G}_t^{(\text{LO})}(p) R_t^{(\text{NLO})}(p, E) + \frac{1}{\pi} \int_0^{\Lambda} dq q^2 D_t(q, E) R^{(\text{LO})}(q, p, E) \mathcal{G}_t^{(\text{NLO})}(q)$$

$$+ \frac{3}{\pi} \int_0^{\Lambda} dq q^2 D_s(q, E) R^{(\text{LO})}(q, p, E) \mathcal{G}_s^{(\text{NLO})}(q)$$

$$\mathcal{G}_s^{(\text{NLO})}(p) = \mathcal{G}_s^{(\text{LO})}(p) R_s^{(\text{NLO})}(p, E) + \frac{3}{\pi} \int_0^{\Lambda} dq q^2 D_t(q, E) R^{(\text{LO})}(q, p, E) \mathcal{G}_t^{(\text{NLO})}(q)$$

$$+ \frac{1}{\pi} \int_0^{\Lambda} dq q^2 D_s(q, E) R^{(\text{LO})}(q, p, E) \mathcal{G}_s^{(\text{NLO})}(q), \tag{22}$$

where

$$R_{\{t,s\}}^{(\text{NLO})}(p, E) = \frac{Z_{\{t,s\}} - 1}{2\gamma_{\{t,s\}}} \left(\gamma_{\{t,s\}} + \sqrt{\frac{3}{4}p^2 - M_N E} \right). \tag{23}$$

$Z_t = 1.6908$ ($Z_s = 0.9015$) is the residue at the $^3S_1$ ($^1S_0$) channel pole [43,45]. The residues Z_s and Z_t are equal in the Wigner-SU(4) symmetric limit. Expanding these equations in δ gives the $\mathcal{O}(r\delta^0)$ term[1]

$$\mathcal{G}_+^{(\text{NLO})}(p) = \mathcal{G}_+^{(0)}(p) R_+^{(\text{NLO})}(p, E) + \frac{4}{\pi} \int_0^{\Lambda} dq q^2 D(q, E) R^{(\text{LO})}(q, p, E) \mathcal{G}_+^{(\text{NLO})}(q), \tag{24}$$

---

[1] Note, when expanding in powers of δ the $\gamma_{\{t,s\}}$ in the expression $(Z_{\{t,s\}} - 1)/(2\gamma_{\{t,s\}})$ is not expanded, because this whole quantity is taken as the range correction.

 Springer

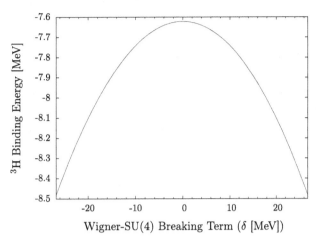

Fig. 2 Binding energy of the triton as a function of the Wigner-SU(4)-breaking parameter δ, where the three-body force is fit to the triton binding energy at the physical value of δ. The same three-body force is used for all other values of δ, and Wigner-SU(4) breaking is treated non-perturbatively

where $\mathcal{G}_+^{(NLO)}(p)$ is the NLO-in-range-but-LO-in-Wigner correction $(\mathcal{O}(r\delta^0))$ to $\mathcal{G}_+(p)$ and we have dropped the part of the range correction that breaks Wigner-SU(4) symmetry. The Wigner-SU(4)-symmetric part of the range correction involves the function $R_+^{(NLO)}(p, E)$, defined as

$$R_+^{(NLO)}(p, E) = \rho_{AVE}\left(\gamma_{AVE} + \sqrt{\frac{3}{4}p^2 - M_N E}\right),\tag{25}$$

where

$$\rho_{AVE} = \frac{1}{2}\left(\frac{Z_t - 1}{2\gamma_t} + \frac{Z_s - 1}{2\gamma_s}\right).\tag{26}$$

This means that, for the $\mathcal{O}(r)$ correction, in addition to expanding in powers of δ, we also expand in powers of

$$\delta_r = \frac{1}{2}\left(\frac{Z_t - 1}{2\gamma_t} - \frac{Z_s - 1}{2\gamma_s}\right),\tag{27}$$

and the equations derived here are $\mathcal{O}(\delta_r^0)$.

4 Binding Energy

To understand Wigner-SU(4) breaking in the three-body system we first investigate its effects on the triton binding energy. We do this at LO in the EFT($\not{\pi}$) expansion. Figure 2 plots the binding energy of the triton as a function of the Wigner-SU(4) breaking parameter δ, with Wigner-SU(4) breaking treated nonperturbatively. In this calculation we employ a three-body force that is independent of δ, and is fixed so as to reproduce the triton binding energy, $B_{^3H} = 8.48$ MeV, at the physical value of $\delta = 26.80$ MeV, which corresponds to the right edge of Fig. 2. The difference between the binding energy at the physical δ and in the Wigner-SU(4) limit, $\delta = 0$, is only 11%. The shape of the curve is essentially quadratic, demonstrating that the first Wigner-SU(4) correction to the binding energy comes in at $\mathcal{O}(\delta^2)$. This should come as no surprise: the vertex functions are SU(4) symmetric at leading order in the expansion in powers of δ, and so the insertion of an SU(4)-breaking correction between them must yield zero. This, indeed, is why the self energy $\Sigma(E)$, has no term of $\mathcal{O}(\delta)$.

Since in this paper we expand all observables around the Wigner-SU(4) limit, all our remaining calculations here are carried out with the binding energy chosen to have its $\delta = 0$ value, $B = 7.62$ MeV. This corresponds to using the same three-body force that was used to generate Fig. 2.

5 Charge and Matter Radii

5.1 Relations Between Radii Under Wigner-SU(4) Symmetry

In the absence of Coulomb, and assuming isospin is a conserved symmetry, ^{3}He is the isospin mirror of ^{3}H. Therefore, the proton radius of ^{3}He is the neutron radius of ^{3}H and vice versa. Using this fact it is straightforward to show that for the ^{3}H and ^{3}He wavefunctions

$$\left\langle ^3\text{H} \left| \sum_i \tau_3^{(i)} \vec{\mathbf{x}}_i^2 \right| ^3\text{H} \right\rangle = \langle r_{^3\text{H}}^2 \rangle - 2 \langle r_{^3\text{He}}^2 \rangle, \tag{28}$$

and

$$\left\langle ^3\text{He} \left| \sum_i \tau_3^{(i)} \vec{\mathbf{x}}_i^2 \right| ^3\text{He} \right\rangle = 2 \langle r_{^3\text{He}}^2 \rangle - \langle r_{^3\text{H}}^2 \rangle, \tag{29}$$

where $\langle r_{^3\text{H}}^2 \rangle$ and $\langle r_{^3\text{He}}^2 \rangle$ are the ^{3}H and ^{3}He point charge radii squared respectively and i sums over the nucleons. In the Wigner-SU(4) limit the wavefunction is spatially symmetric such that

$$\left\langle ^A Z \left| \sum_i \tau_3^{(i)} \vec{\mathbf{x}}_i^2 \right| ^A Z \right\rangle = \frac{1}{3} \left\langle ^A Z \left| 2T_3 \sum_i \vec{\mathbf{x}}_i^2 \right| ^A Z \right\rangle, \tag{30}$$

where $| ^A Z \rangle$ is either the ^{3}H or ^{3}He wavefunction, and T_3 the operator for isospin in the z-direction on these wavefunctions. (For a proof of this statment see Appendix.) Noting that

$$\left\langle ^3\text{H} \left| \sum_i \vec{\mathbf{x}}_i^2 \right| ^3\text{H} \right\rangle = 2 \langle r_{^3\text{He}}^2 \rangle + \langle r_{^3\text{H}}^2 \rangle, \tag{31}$$

and using Eqs. (30) and (28) we find

$$-\frac{1}{3} \left(2 \langle r_{^3\text{He}}^2 \rangle + \langle r_{^3\text{H}}^2 \rangle \right) = \langle r_{^3\text{H}}^2 \rangle - 2 \langle r_{^3\text{He}}^2 \rangle. \tag{32}$$

Solving this gives $\langle r_{^3\text{H}}^2 \rangle = \langle r_{^3\text{He}}^2 \rangle$, and therefore in the Wigner-SU(4) limit the charge radii of ^{3}H and ^{3}He are equivalent. In addition the point matter radii for ^{3}H and ^{3}He will be the same and equivalent to their point charge radii.

Assuming that Wigner-SU(4) corrections are kept to all orders Eq. (31) still holds. Therefore, considering $\mathcal{O}(\delta)$ corrections Eq (31) gives

$$\left\langle ^3\text{H} \left| \sum_i \vec{\mathbf{x}}_i^2 \right| \delta^3\text{H} \right\rangle + \left\langle \delta^3\text{H} \left| \sum_i \vec{\mathbf{x}}_i^2 \right| ^3\text{H} \right\rangle = 2 \langle r_{^3\text{He}}^2 \rangle_\delta + \langle r_{^3\text{H}}^2 \rangle_\delta, \tag{33}$$

where $|\delta^3\text{H}\rangle$ is the first order Wigner-correction to the ^{3}H wavefunction and $\langle r_{^3\text{H}}^2 \rangle_\delta$ and $\langle r_{^3\text{He}}^2 \rangle_\delta$ are the first order Wigner-corrections to the ^{3}H and ^{3}He charge radii squared respectively. This relationship is exactly the same for the ^{3}He wavefunctions. The quantity

$$\sum_Z \left\langle ^A Z \left| \sum_i \vec{\mathbf{x}}_i^2 \right| \delta^A Z \right\rangle = 0, \tag{34}$$

where the sum over "Z" simply sums both the ^{3}H and ^{3}He wavefunctions. Taking the sum over "Z" makes this quantity a Wigner-SU(4) scalar, but it has one insertion of an operator that breaks Wigner-SU(4) symmetry and therefore must be zero. Combining Eqs. (33) and (34) gives

$$4 \langle r_{^3\text{He}}^2 \rangle_\delta + 2 \langle r_{^3\text{H}}^2 \rangle_\delta = 0. \tag{35}$$

From this it follows that that the $\mathcal{O}(\delta)$ correction to the ^{3}H point charge radius squared is twice as large and has the opposite sign as the $\mathcal{O}(\delta)$ correction to the ^{3}He point charge radius squared. This relationship can also be proven using the identities in Ref. [33] and expanding them to $\mathcal{O}(\delta)$. However, this method is long and tedious.

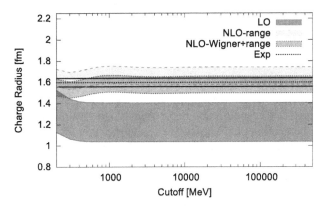

Fig. 3 Plot of cutoff dependence of LO, NLO, and $\mathcal{O}(r + \delta)$ prediction for the triton point charge radius. The *pink band* corresponds to a 15% error estimate about the LO central value, the *green band* to a 5% error about the NLO central value, and the *blue band* a 5% error estimate about the $\mathcal{O}(r + \delta)$ value. The *dotted black line* is the experimental value for the triton point charge radius of 1.5978 ± 0.040 fm and the *solid black lines* about it its error [1]

5.2 Results

To obtain the triton charge radius in the Wigner-SU(4) limit the results of Ref. [33] can simply be recalculated setting $\gamma_t = \gamma_s = \gamma_{\mathrm{AVE}}$ and $(Z_t - 1)/2\gamma_t = (Z_s - 1)/2\gamma_s = \rho_{\mathrm{AVE}}$. A second approach is to take the analytical expressions in Ref. [33] and expand them about the Wigner-limit to $\mathcal{O}(\delta)$. This allows calculation of the $\mathcal{O}(\delta)$ correction and the calculation of the triton charge radius in the Wigner-SU(4) limit using only the triton vertex functions $\mathcal{G}_+^{(0)}(p)$ and $\mathcal{G}_-^{(1)}(p)$. Both approaches give the same result in the limit $\delta = 0$.

We compute the triton point charge radius at LO ($\mathcal{O}(r^0\delta^0)$), NLO ($\mathcal{O}(r\delta^0)$), and $\mathcal{O}(r + \delta)$, where the last calculation involves the addition of both a single range insertion and a single Wigner-SU(4)-breaking insertion, but only considered separately, not in combination. Cutoff dependence of these three different results is displayed in Fig. 3.

All orders of the triton point charge radius considered here converge as a function of cutoff, and are therefore properly renormalized. The LO triton point charge radius is 1.22 fm, the NLO value 1.66 fm, and the $\mathcal{O}(r + \delta)$ value 1.58 fm. The experimental value for the triton point charge radius is 1.5978 ± 0.040 fm [1], which agrees well with our $\mathcal{O}(r + \delta)$ calculation. When Wigner-SU(4) breaking is included to all orders, i.e., the physical values of γ_s and γ_t, and the physical triton binding energy, $B_{3\mathrm{H}} = 8.48$ MeV, are employed, at LO (NLO) in EFT($\not{\pi}$) the triton point charge radius is 1.14 fm (1.59 fm) [33]. ^{3}He has an experimental point charge radius of 1.77527 ± 0.0054 fm [1]. This is about 7% away from the NLO-in-range-but-SU(4)-symmetric prediction of 1.66 fm. As already noted, the Wigner-SU(4)-breaking correction for the ^{3}He point charge radius squared is half that for the ^{3}H point charge radius squared and of opposite sign. Therefore, the $\mathcal{O}(r + \delta)$ ^{3}He point charge radius squared is 1.70 fm, about 4% away from the experimental value.

The error due to missing range corrections is about 10%. The dominant, SU(4)-symmetric, part of this correction will affect the ^{3}H and ^{3}He charge radii equally. In contrast, the effects of the Coulomb interaction, not included here, will affect only the charge radius of ^{3}He. We estimate this effect to be of order $\alpha M_N / \kappa_t$ (where $\kappa_t = \sqrt{M_N B_{3\mathrm{H}}}$ is the binding momentum of the triton), which is about 8%. Meanwhile, the uncertainty due to Wigner-SU(4) breaking in the NN effective ranges is naively 3% since

$$\frac{\delta_r}{a} \approx \left\{ \left(\frac{Z_t - 1}{2\gamma_t} - \frac{Z_s - 1}{2\gamma_s} \right) \Big/ \left(\frac{Z_t - 1}{2\gamma_t} + \frac{Z_s - 1}{2\gamma_s} \right) \right\} \left(\frac{\gamma_t}{m_\pi} \right) \sim 0.033. \tag{36}$$

Terms of $\mathcal{O}(r\delta)$ are also omitted. These could also be as large as a few per cent of the individual radii, since range corrections to those are large. Corrections that are Wigner-SU(4) odd (e.g. $\mathcal{O}(\delta)$, $\mathcal{O}(r\delta)$, and $\mathcal{O}(r\delta_r)$) will affect only the isovector combination of trinucleon charge radii,

$$\langle r_v^2 \rangle = \frac{1}{2} \left(2 \langle r_{3\mathrm{He}}^2 \rangle - \langle r_{3\mathrm{H}}^2 \rangle \right), \tag{37}$$

and give zero contribution to to the isoscalar combination:

$$\langle r_s^2 \rangle = \frac{1}{2} \left(2 \langle r_{3\mathrm{He}}^2 \rangle + \langle r_{3\mathrm{H}}^2 \rangle \right). \tag{38}$$

Finally, considering the convergence of the expansion in powers of δ, e. g., the ratio between $\mathcal{G}_+^{(0)}(p)$ and $\mathcal{G}_+^{(2)}(p)$, suggests that $\mathcal{O}(\delta^2)$ effects could have perhaps a 5% effect on the radii.

6 Convergence of the Wigner-SU(4) Expansion

In order to assess the efficacy of expanding about the Wigner-SU(4) limit we plot the relative error of the triton vertex function with the breaking of Wigner-SU(4) symmetry in the *NN* scattering lengths included to all orders, as compared with that obtained when this source of Wigner-SU(4) symmetry breaking is treated perturbatively order-by-order in δ.

Figure 4 shows the relative error of the cumulative sum in the expansion in powers of δ, $\sum_{m=0}^{n} \mathcal{G}_+^{(2m)}(q)\delta^{2m}$, compared to $\mathcal{G}_+^{(\mathrm{LO})}(q)$, up to $\mathcal{O}(\delta^6)$, over a range of momenta that essentially corresponds to the domain of validity of EFT($\not\pi$), $q = 0$–200 MeV. The data is chosen at the cutoff $\Lambda = 51{,}286$ MeV; by this cutoff all results have effectively converged as a function of Λ. Order-by-order convergence in the δ expansion can clearly be seen in the relative error.

In Fig. 5 the relative difference between $\mathcal{G}_-^{(\mathrm{LO})}(q)$ and the cumulative sum $\sum_{m=0}^{n} \mathcal{G}_-^{(2m+1)}(q)\delta^{2m+1}$ is shown for the same range of momenta and the same cutoff Λ, up to $\mathcal{O}(\delta^5)$. Again, order-by-order convergence is clearly observed. The dip at NLO merely corresponds to the fact that $\mathcal{G}_-^{(\mathrm{LO})}(q)$ and $\mathcal{G}_-^{(1)}(q)\delta$ cross each other at a momentum $\approx \gamma_t$, i.e., about 50 MeV. Finally, we compare the size of $\mathcal{G}_-^{(\mathrm{LO})}(q)$ and $\mathcal{G}_+^{(\mathrm{LO})}(q)$, see Fig. 6. $\mathcal{G}_-^{(\mathrm{LO})}(q)$ is at most 8% of $\mathcal{G}_+^{(\mathrm{LO})}(q)$ over the entire momentum region of interest.

These results support the claim that an expansion about the Wigner-SU(4) limit converges rapidly. We recognize, of course, that triton vertex functions are not observables. However, since the construction of any three-nucleon bound-state property in EFT($\not\pi$) will involve these non-perturbative objects, the fact that they converge rapidly in the δ expansion suggests that the expansion will generally be successful for three-nucleon bound-state observables.

7 Comments on the Unitary Limit

König et al. [48] have recently argued that the binding energies of nuclei up to $A = 4$ can be understood in an expansion about the unitary limit, where $\gamma_s = \gamma_t = 0$. The unitary limit is clearly a special case of Wigner-SU(4) symmetry; taking $\gamma_s = \gamma_t = 0$ enlarges the symmetry group still further, since the discrete scale invariance of EFT($\not\pi$) at LO then relates all the unitary-limit Efimov states by a fixed rescaling. In the case of finite scattering lengths the Efimov spectrum still possesses discrete scale invariance, but a particular

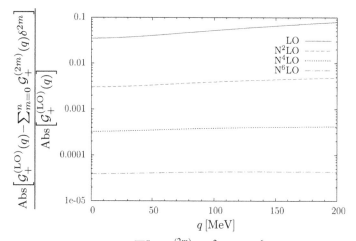

Fig. 4 Plot of relative difference between $\mathcal{G}_+(q)$ and $\sum_{m=0}^{n} \mathcal{G}_+^{(2m)}(q)\delta^{2m}$ to $\mathcal{O}(\delta^6)$. The relative error is plotted over the range $q = 0$–200 MeV and the data is for the cutoff $\Lambda = 51{,}286$ MeV. The LO ($\mathcal{O}(\delta^0)$) result is given by the *solid red curve*, the N^{2}LO ($\mathcal{O}(\delta^2)$) result by the *long-dashed green curve*, the N^{4}LO ($\mathcal{O}(\delta^4)$) result by the *short-dashed blue curve*, and the N^{6}LO ($\mathcal{O}(\delta^6)$) result by the *short-long-dashed purple curve*

Springer

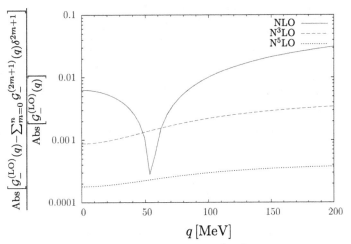

Fig. 5 Plot of relative error of difference between $\mathcal{G}_-^{(LO)}(q)$ and $\sum_{m=0}^n \mathcal{G}_-^{(2m+1)}(q)\delta^{2m+1}$ to $\mathcal{O}(\delta^5)$. The relative error is plotted over the range $q = 0\text{–}200$ MeV and the data is for the cutoff $\Lambda = 51,286$ MeV. The NLO ($\mathcal{O}(\delta)$) result is given by the *solid red curve*, the N^3LO ($\mathcal{O}(\delta^3)$) result by the *long-dashed green curve*, and the N^5LO ($\mathcal{O}(\delta^5)$) result given by the *short-dashed blue curve*

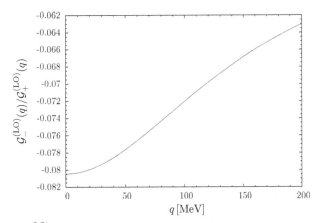

Fig. 6 The ratio of $\mathcal{G}_-^{(LO)}(q)$ and $\mathcal{G}_+^{(LO)}(q)$, again computed for a cutoff of $\Lambda = 51,286$ MeV

Efimov state is related to others at a different NN scattering length [2]. As we have done here, König et al. fix the size of the three-body force to reproduce the binding energy of the physical triton. They demonstrate that the binding-energy difference of ^{3}He and ^{3}H remains well predicted in the unitary limit (cf. Refs. [49,50]). They also show that the alpha particle, while overbound by about 10 MeV at exact unitarity, attains almost exactly its experimental binding energy (28.30 MeV) once first-order corrections in the expansion in γ_t are included.

It is straightforward for us to take the limit $\gamma_{AVE} \to 0$ in our results and so obtain point charge radii for three-nucleon bound states in the unitary limit. At leading order in the range expansion this gives $\langle r^2 \rangle_{pt}^{1/2} = 1.10$ fm (for $B = 7.62$ MeV), in accord with the analytic result [2][2]

$$\langle r^2 \rangle_{pt} M_N B = \frac{(1+s_0)^2}{9} \approx 0.224. \tag{39}$$

This lends support to the argument of König et al., since it is within 10% of either the Wigner-SU(4) limit result quoted above, or the full LO EFT($\not\pi$) answer of 1.14 fm [33].

Adding corrections of first order in the range of the NN interaction corrects the radius obtained from Eq. (39) by an amount $\sim r\kappa_t$—such effects are present even though $r/a = 0$ at unitarity. (Note, however, that

---

[2] In fact, Ref. [2] quotes this as the result for the matter radius in the unitary limit for three equal-mass particles. However, in that limit the symmetry of the three-body wave function leads to equal charge and matter radii.

Table 1 Anatomy of the point charge radii $\langle r^2 \rangle_{\text{pt}}^{1/2}$, of three-nucleon bound states

	Unitary limit	Wigner-SU(4) limit	$\mathcal{O}(\delta)$	Full Wigner-SU(4) breaking
LO EFT($\not{\pi}$) ($r = 0$)	1.10	1.22	1.08/1.19	1.14/1.26
$\mathcal{O}(r)$	1.42	1.66	1.58/1.70	1.59/1.72

When only one number is quoted the radii are equal for ^{3}H and ^{3}He, if two numbers are given the first is for the triton and the second for ^{3}He. All numbers are in fm. Note that the lower line, third entry in the table is the $\mathcal{O}(r + \delta)$ calculation of this paper. The "Full Wigner-SU(4) breaking" numbers treat Wigner-breaking in the scattering lengths and (in the second line) effective ranges nonperturbatively and use the physical triton binding energy [33,51]. The experimental evaluation of Ref. [1] quotes 1.598(40)/1.7753(54)

the factors Z_s and Z_t remain at their LO values of one as long as we consider $\gamma_s = \gamma_t = 0$ [3].) This shifts $\langle r^2 \rangle_{\text{pt}}^{1/2}$ to 1.42 fm, i.e., the size of the range correction at unitarity is about 70% of that when γ_{AVE} takes its physical value.

Table 1 summarizes the effect that different limits in the NN system have on the radii of the three-nucleon bound states. In the case of the triton we can compare to Ref. [33], which obtained 1.14 fm at LO in EFT($\not{\pi}$), but with Wigner-SU(4) breaking included to all orders in δ, and 1.59 fm in a calculation that was first order in the ranges (including SU(4) breaking therein), and again had the physical values of γ_s and γ_t. The proximity of our $\mathcal{O}(r + \delta)$ results to these is very striking.

The unitary limit seems a worse starting point—at least for radii—especially since the shift that results from range corrections is significantly underpredicted there. It may be that radii are more challenging for the expansion proposed in Ref. [48], since they are quite sensitive to infra-red physics, and the long-distance properties of the three-nucleon system in the unitary limit differ dramatically from reality: at both LO and NLO in the expansion of Ref. [48] infinite towers of bound Efimov excited states occur.

8 Conclusions

Working to $\mathcal{O}(r + \delta)$ in the range and δ expansion of EFT($\not{\pi}$) we obtain a point charge radius for ^{3}H of 1.58 fm, which agrees with the experimental number, 1.5978 ± 0.040 fm [1], within the experimental errors. It also agrees with the NLO result of 1.59 fm obtained using the physical values of the NN scattering lengths and triton binding energy [33], within theoretical errors. It follows that all higher-order corrections in δ and δ_r must conspire to give a total correction of only .01 fm to $\langle r_{3\text{H}}^2 \rangle_{\text{pt}}^{1/2}$. Naively $\mathcal{O}(\delta^2)$ corrections could give 5% of the LO Wigner SU(4)-symmetric charge radius 1.22 fm, i.e., they should be ≈ 0.06 fm. However, at $\mathcal{O}(\delta^2)$ there will be effects both from expanding the expressions of Ref. [33] out to $\mathcal{O}(\delta^2)$, and from the $\mathcal{O}(\delta^2)$ shift of the three-nucleon bound state energy from $B = 7.62$ MeV to the physical triton binding energy. The small overall result of a 0.01 fm shift is probably due to a cancellation between these two classes of $\mathcal{O}(\delta^2)$ corrections.

Working to first order in both Wigner-SU(4) breaking and the NN effective range produces a ^{3}He point charge radius of 1.70 fm, about 4% below the experimental value of 1.77527 ± 0.0054 fm [1]. The difference is mostly from missing Coulomb and higher-order range corrections, since the ^{3}He charge radius with Wigner SU(4)-breaking included to all orders (including breaking in ranges), but no Coulomb effects, is 1.72 fm at NLO in the range expansion [51].

In this (isospin symmetric) limit ^{3}H and ^{3}He have a common binding energy. It is thus an SU(4) scalar, and so receives no correction at $\mathcal{O}(\delta)$. We find that $\mathcal{O}(\delta^2)$ effects make the triton 11% less bound in the Wigner-SU(4) limit than it is at the physical value, $\delta = 27$ MeV.

δ is in fact larger than $\gamma_{\text{AVE}} = 19$ MeV, and so the rapid convergence of the expansion in powers of δ at first glance is somewhat mysterious. However, the expansion is really an expansion in powers of $\delta D(q, E)$, with $D(q, E)$ the EFT($\not{\pi}$) propagator for the NN system that appears in the three-body equations. This renders the expansion around the SU(4) limit one in $(\gamma_t - \gamma_s)/\kappa_t$, with $\kappa_t = 89$ MeV the binding momentum of the triton.

Examining both the three-nucleon binding energy and the relative size of the SU(4)-symmetric and SU(4)-anti-symmetric pieces of the three-nucleon vertex function, $\mathcal{G}_-^{(\text{LO})}(p)/\mathcal{G}_+^{(\text{LO})}(p)$, suggests that the error induced in observables through going to the Wigner-SU(4) limit will be at most 10%. This implies that an efficient way to account for Wigner-SU(4) breaking is to equate $\delta \sim r^2$, i.e., only compute one correction in Wigner-SU(4)

---

[3] In the unitary limit and Wigner-SU(4) limit $(Z_{\{t,s\}} - 1)/(2\gamma_{\{t,s\}}) \rightarrow \frac{1}{2}\rho$, where $\rho = \frac{1}{2}(\rho_t + \rho_s)$, with $\rho_t = 1.765$ fm ($\rho_s = 2.730$ fm) being the effective range about the ^{3}S$_1$ (^{1}S$_0$) pole.

breaking for every two orders in the range expansion. Unfortunately, an $\mathcal{O}(r^2)$ calculation requires a new three-body force that must be renormalized to a three-body datum [29,31]. Since the Wigner-SU(4) limit is not expected to work nearly as well for scattering observables that additional three-body force should be fit to a three-body bound state observable. We postpone this to future work.

Finally, we note that the Wigner-SU(4) symmetry which emerges in EFT($\not\pi$) is not obviously related to the contracted SU(4) of QCD in the limit of a large number of colors (N_C) [52]. In the large-N_C limit Wigner-SU(4) symmetry of nuclear forces naturally emerges [53–55], but this happens only at a renormalization scale $\sim \Lambda_{QCD}$, whereas the SU(4) in EFT($\not\pi$) emerges already for renormalization scales $\sim m_\pi$.

Acknowledgements We thank Shung-Ichi Ando for comments that helped us clarify the manuscript. We are grateful to the ExtreMe Matter Institute EMMI at the GSI Helmholtz Centre for Heavy Ion Research for support as part of the Rapid Reaction Task Force, "The systematic treatment of the Coulomb interaction in few-body systems". We acknowledge financial support by the US Department of Energy, Office of Science, Office of Nuclear Physics, under Award No. DE-FG02-93ER40756.

Appendix: Proof of Eq. (30)

In order to prove Eq. (30) the spatial permutation operator is defined as P_{ij}. This operator permutes the ith and jth particles in the spatial part of the wavefunction while leaving the spin and isospin parts of the wavefunction untouched. Noting $P_{ij}^2 = 1$ gives

$$
\left\langle {}^A Z \middle| \sum_i \tau_3^{(i)} \vec{\mathbf{x}}_i^2 \middle| {}^A Z \right\rangle
$$
$$
= \frac{1}{3} \left\langle {}^A Z \middle| \sum_i \tau_3^{(i)} \left(\vec{\mathbf{x}}_i^2 + P_{ij}^2 \vec{\mathbf{x}}_i^2 P_{ij}^2 + P_{ik}^2 \vec{\mathbf{x}}_i^2 P_{ik}^2 \right) \middle| {}^A Z \right\rangle, \tag{A1}
$$

where $i \neq j$, $i \neq k$, and $j \neq k$. In the Wigner SU(4)-limit the spatial part of the tri-nucleon wavefunction is spatially symmetric since it is equivalent to that of three bosons, and is thus invariant under any spatial permutation. Now, since the spatial permutation operator does not act on isospin it can be commuted with $\tau_3^{(i)}$, leading to

$$
\left\langle {}^A Z \middle| \sum_i \tau_3^{(i)} \vec{\mathbf{x}}_i^2 \middle| {}^A Z \right\rangle
$$
$$
= \frac{1}{3} \left\langle {}^A Z \middle| \sum_i \tau_3^{(i)} \left(\vec{\mathbf{x}}_i^2 + P_{ij} \vec{\mathbf{x}}_i^2 P_{ij}^\dagger + P_{ik} \vec{\mathbf{x}}_i^2 P_{ik}^\dagger \right) \middle| {}^A Z \right\rangle, \tag{A2}
$$

which reduces finally to

$$
\left\langle {}^A Z \middle| \sum_i \tau_3^{(i)} \vec{\mathbf{x}}_i^2 \middle| {}^A Z \right\rangle = \frac{1}{3} \left\langle {}^A Z \middle| \sum_i \tau_3^{(i)} \left(\vec{\mathbf{x}}_i^2 + \vec{\mathbf{x}}_j^2 + \vec{\mathbf{x}}_k^2 \right) \middle| {}^A Z \right\rangle
$$
$$
= \frac{1}{3} \left\langle {}^A Z \middle| 2 T_3 \sum_i \vec{\mathbf{x}}_i^2 \middle| {}^A Z \right\rangle. \tag{A3}
$$

References

1. I. Angeli, K. Marinova, Table of experimental nuclear ground state charge radii: An update. At. Data Nucl. Data Tables **99**(1), 69–95 (2013)
2. E. Braaten, H.-W. Hammer, Universality in few-body systems with large scattering length. Phys. Rep. **428**, 259–390 (2006). arXiv:cond-mat/0410417
3. V. Efimov, Energy levels arising form the resonant two-body forces in a three-body system. Phys. Lett. B **33**, 563–564 (1970)
4. V.N. Efimov, Weakly-bound states of 3 resonantly-interacting particles. Sov. J. Nucl. Phys. **12**, 589 (1971)

5. B. Huang, L.A. Sidorenkov, R. Grimm, J.M. Hutson, Observation of the second triatomic resonance in Efimov's scenario. Phys. Rev. Lett. **112**, 190401 (2014)
6. M. Kunitski et al., Observation of the Efimov state of the helium trimer. Science **348**, 551–555 (2015). arXiv:1512.02036
7. V. Efimov, Qualitative treatment of three-nucleon properties. Nucl. Phys. A **362**, 45–70 (1981)
8. S.K. Adhikari, A.C. Fonseca, L. Tomio, Fonseca, and Lauro Tomio. Method for resonances and virtual states: Efimov virtual states. Phys. Rev. C **26**, 77–82 (1982)
9. T. Frederico, I.D. Goldman, A. Delfino, Extension of the minimal three-nucleon model to the unphysical sheet of energy. Phys. Rev. C **37**, 497 (1988)
10. A. Kievsky, M. Gattobigio, Efimov physics with 1/2 spin-isospin fermions. Few Body Syst. **57**, 217 (2016). arXiv:1511.09184
11. U. van Kolck, Effective field theory of short range forces. Nucl. Phys. A **645**, 273 (1999). arXiv:nucl-th/9808007
12. D.B. Kaplan, M.J. Savage, M.B. Wise, A new expansion for nucleon-nucleon interactions. Phys. Lett. B **424**, 390 (1998). arXiv:nucl-th/9801034
13. D.B. Kaplan, M.J. Savage, M.B. Wise, Two-nucleon systems from effective field theory. Nucl. Phys. B **534**, 329 (1998). arXiv:nucl-th/9802075
14. M.C. Birse, J.A. McGovern, K.G. Richardson, A renormalization group treatment of two-body scattering. Phys. Lett. B **464**, 169 (1999). arXiv:hep-ph/9807302
15. J.-W. Chen, G. Rupak, M.J. Savage, Nucleon-nucleon effective field theory without pions. Nucl. Phys. A **653**, 386 (1999). arXiv:nucl-th/9902056
16. P.F. Bedaque, H.-W. Hammer, U. van Kolck, Effective theory of the triton. Nucl. Phys. A **676**, 357 (2000). arXiv:nucl-th/9906032
17. P.F. Bedaque, H.-W. Hammer, U. van Kolck, Renormalization of the three-body system with short range interactions. Phys. Rev. Lett. **82**, 463 (1999). arXiv:nucl-th/9809025
18. P.F. Bedaque, H.-W. Hammer, U. van Kolck, The three boson system with short range interactions. Nucl. Phys. A **646**, 444 (1999). arXiv:nucl-th/9811046
19. T. Mehen, I.W. Stewart, M.B. Wise, Wigner symmetry in the limit of large scattering lengths. Phys. Rev. Lett. **83**, 931 (1999). arXiv:hep-ph/9902370
20. G.V. Skornyakov, K.A. Ter-Martirosian, Three body Problem for short range forces. I. Scattering of Low Energy Neutrons by Deuterons. Sov. Phys. JETP **4**, 648 (1957)
21. G. Danilov, On the three-body problem in the case of short-range forces. Sov. Phys. JETP **13**, 349 (1961)
22. L.H. Thomas, The interaction between a neutron and a proton and the structure of ^{3}H. Phys. Rev. **47**, 903 (1935)
23. H.W. Grießhammer, Naive dimensional analysis for three-body forces without pions. Nucl. Phys. A **760**, 110 (2005). arXiv:nucl-th/0502039
24. T. Barford, M.C. Birse, Effective theories of scattering with an attractive inverse-square potential and the three-body problem. J. Phys. A **38**, 697 (2005). arXiv:nucl-th/0406008
25. M.C. Birse, Renormalisation-group analysis of repulsive three-body systems. J. Phys. A **39**, L49 (2006). arXiv:nucl-th/0509031
26. L. Platter, C. Ji, D.R. Phillips, Range corrections to three-body observables near a Feshbach resonance. Phys. Rev. A **79**, 022702 (2009). arXiv:0808.1230
27. C. Ji, D. Phillips, L. Platter, Beyond universality in three-body recombination:an effective field theory treatment. Europhys. Lett. **92**, 13003 (2010). arXiv:1005.1990
28. C. Ji, D.R. Phillips, L. Platter, The three-boson system at next-to-leading order in an effective field theory for systems with a large scattering length. Ann. Phys. **327**, 1803 (2012). arXiv:1106.3837
29. C. Ji, D.R. Phillips, Effective field theory analysis of three-boson systems at next-to-next-to-leading order. Few Body Syst. **54**, 2317 (2013). arXiv:1212.1845
30. H.-W. Hammer, T. Mehen, Range corrections to doublet S-wave neutron deuteron scattering. Phys. Lett. B **516**, 353 (2001). arXiv:nucl-th/0105072
31. P.F. Bedaque, G. Rupak, H.W. Griesshammer, H.-W. Hammer, Low-energy expansion in the three-body system to all orders and the triton channel. Nucl. Phys. A **714**, 589 (2003). arXiv:nucl-th/0207034
32. J. Vanasse, Fully perturbative calculation of nd scattering to next-to-next-to-leading order. Phys. Rev. C **88**, 044001 (2013). arXiv:1305.0283
33. J. Vanasse, Three-body bound states and the triton charge radius; perturbative corrections to next-to-next-to-leading order in pionless effective field theory. Phys. Rev. C (2017). arXiv:1512.03805 (to appear)
34. E. Wigner, On the consequences of the symmetry of the nuclear hamiltonian on the spectroscopy of nuclei. Phys. Rev. **51**, 106 (1937)
35. E.P. Wigner, On coupling conditions in light nuclei and the lifetimes of beta radioactivities. Phys. Rev. **56**, 519 (1939)
36. L.L. Foldy, J. Walecka, Muon capture in nuclei. Nuovo Cim. **34**, 1026 (1964)
37. G.E. Walker, Muon capture and supermultiplet symmetry breaking in ^{16}O. Phys. Rev. **151**, 745 (1966)
38. T. DeForest, Muon capture and inelastic electron scattering in ^{12}C and ^{16}O. Phys. Rev. **139**, B1217 (1965)
39. W. Detmold et al., Unitary limit of two-nucleon interactions in strong magnetic fields. Phys. Rev. Lett. **116**, 112301 (2016). arXiv:1508.05884
40. E. Braaten, H.-W. Hammer, An infrared renormalization group limit cycle in QCD. Phys. Rev. Lett. **91**, 102002 (2003). doi:10.1103/PhysRevLett.91.102002
41. D.L. Canham, H.-W. Hammer, Universal properties and structure of halo nuclei. Eur. Phys. J. **A37**, 367 (2008). arXiv:0807.3258
42. S. Weinberg, Effective chiral Lagrangians for nucleon-pion interactions and nuclear forces. Nucl. Phys. B **363**, 3 (1991)
43. H.W. Grießhammer, Improved convergence in the three-nucleon system at very low energies. Nucl. Phys. A **744**, 192 (2004). arXiv:nucl-th/0404073
44. S.R. Beane, P.F. Bedaque, W.C. Haxton, D.R. Phillips, M.J. Savage, in *At The Frontier of Particle Physics*, vol. 1, pp. 133–269, ed. by M. Shifman (2000). World Scientific, Singapore arXiv:nucl-th/0008064

45. D.R. Phillips, G. Rupak, M.J. Savage, Improving the convergence of NN effective field theory. Phys. Lett. B **473**, 209 (2000). arXiv:nucl-th/9908054
46. R. Amado, Theory of the triton wave function. Phys. Rev. **141**, 902 (1966)
47. C. Ji, C. Elster, D. Phillips, ^{6}He nucleus in halo effective field theory. Phys. Rev. C **90**, 044004 (2014). arXiv:1405.2394
48. S. König, H.W. Grießhammer, H.-W. Hammer, U. van Kolck, Nuclear physics around the unitarity limit. (2016). arXiv:1607.04623
49. J. Kirscher, D.R. Phillips, Constraining the neutron-neutron scattering length using the effective field theory without explicit pions. Phys. Rev. C **84**, 054004 (2011). arXiv:1106.3171
50. S. König, H.W. Grießhammer, H.-W. Hammer, U. van Kolck, Effective theory of ^{3}H and ^{3}He. J. Phys. **G43**, 055106 (2016). arXiv:1508.05085
51. J. Vanasse (in preparation) (2016)
52. R.F. Dashen, E.E. Jenkins, A.V. Manohar, The $1/N_c$ expansion for baryons. Phys. Rev. D **49**, 4713 (1994). arXiv:hep-ph/9310379 [Erratum: Phys. Rev. **D51**, 2489 (1995)]
53. D.B. Kaplan, M.J. Savage, The spin-flavor dependence of nuclear forces from large-N QCD. Phys. Lett. B **365**, 244 (1996). arXiv:hep-ph/9509371
54. D.B. Kaplan, A.V. Manohar, The Nucleon-nucleon potential in the $1/N_c$ expansion. Phys. Rev. C **56**, 76 (1997). arXiv:nucl-th/9612021
55. D.R. Phillips, C. Schat, Three-nucleon forces in the $1/N_c$ expansion. Phys. Rev. C **88**, 034002 (2013). arXiv:1307.6274

Few-Body Syst
DOI 10.1007/s00601-016-1180-3

E. Stephan · St. Kistryn · I. Skwira-Chalot · I. Ciepał ·
B. Kłos · A. Kozela · W. Parol · A. Rusnok · A. Wilczek ·
J. Zejma

Dynamics of Three-Nucleon System Studied in Deuteron–Proton Breakup Experiments

New Set of Invariant Coordinates.

Received: 19 September 2016 / Accepted: 7 December 2016
© Springer-Verlag Wien 2016

Abstract Systems composed of three nucleons have been a subject of precise experimental studies for many years. Recently, the database of observables for the deuteron breakup in collision with protons has been significantly extended at intermediate energies. In this region the comparison with exact theoretical calculations is possible, while the sensitivity to various aspects of the interaction, in particular to the subtle effects of the dynamics beyond the pairwise nucleon–nucleon force, is significant. The Coulomb interaction and relativistic effects show also their influence on the observables of the breakup reaction. All these effects vary with energy and appear with different strength in certain observables and phase-space regions, which calls for systematic investigations of a possibly rich set of observables determined in a wide range of energies. Moreover, a systematic comparison with theoretical predictions performed in coordinates related to the system dynamics in a possibly direct way is of importance. The examples of existing experimental data for the breakup reaction are briefly presented and the amenability of a set of invariant coordinates for that type of analysis is discussed.

1 Introduction

The understanding of the structure and dynamics of nuclei as systems of interacting protons and neutrons is one of the major goals of nuclear physics. The nucleon–nucleon potential is the leading part of the nuclear interaction and should be sufficient to describe basic properties of nuclei and main trends in observables for systems of few (and many) nucleons, if exact calculations are feasible. However, since the internal structure of nucleons is neglected, the question arises, how the suppressed degrees of freedom influence the observables in any system consisting of more than 2 nucleons. Such additional dynamics, called three-nucleon force (3NF), cannot be reduced to pairwise forces. It arises in the meson-exchange picture as an intermediate excitation of

This work was partially supported by Polish National Science Center from Grant DEC-2012/05/B/ST2/02556 and by the European Commission within the Seventh Framework Programme through IA-ENSAR (Contract No. RII3-CT-2010-262010).
This article belongs to the special issue "30th anniversary of Few-Body Systems".

E. Stephan (✉) · B. Kłos · A. Rusnok · A. Wilczek
Institute of Physics, University of Silesia, 40007 Katowice, Poland
E-mail: elzbieta.stephan@us.edu.pl

St. Kistryn · J. Zejma
Institute of Physics, Jagiellonian University, 30348 Kraków, Poland

I. Skwira-Chalot
Faculty of Physics, University of Warsaw, 02093 Warsaw, Poland

I. Ciepał · A. Kozela · W. Parol
Institute of Nuclear Physics, PAS, 31342 Kraków, Poland

a nucleon to a Δ isobar, or it appears fully naturally in Chiral Effective Field Theory at a certain order [1]. In the first case, 3NF models, like TM99 [2] or Urbana IX [3] forces, are combined with a given realistic nucleon–nucleon (2N) potential. Alternatively, the Δ isobar is included explicitly and the coupled-channel framework is applied [4]. Three-nucleon systems, as amenable to accurate ab-initio calculations, represent an excellent testing ground for 2N + 3NF interactions, constructed in any of the ways mentioned here.

The importance of 3NF contributions to the dynamics of systems of more than two nucleons was first established in binding energies of few-nucleon states [5–8]. Further verification of the role of the 3NF has been carried out on the basis of scattering experiments: the measurements of observables for elastic nucleon–deuteron scattering and for the breakup of a deuteron in its collision with a nucleon. The extensive discussions of the present status of our understanding of the $3N$ system dynamics, based on modern calculations and many precise and rich data sets, can be found in recent reviews [9–11]. The 3NF turned out to be very important for improving the description of the cross section for nucleon–deuteron elastic scattering. At beam energies above 100 MeV per nucleon certain discrepancies between data and calculations still persist, though significantly reduced as compared to predictions based on purely 2N potentials (cf. [12] and references therein). On the other hand, the precise experimental data demonstrate both the successes and the difficulties of the current models in describing analyzing powers, spin-transfer and spin-correlation coefficients for Nd elastic scattering.

Deuteron breakup in collision with a proton is another source of information on the three-nucleon system dynamics. When studied with a detector covering a big part of the phase space, it provides extensive information, and the effects of various dynamical ingredients can be studied as a function of a set of kinematic variables. Such an approach was applied in a series of experiments carried out at KVI, which comprised the studies of the ^{1}H($\mathbf{d}$,pp)n and ^{2}H($\mathbf{p}$,pp)n reactions with deuteron beams of energies 100, 130 and 160 MeV (50, 65, 80 MeV/nucleon) and proton beams with energies of 135 and 190 MeV. The detection systems of large angular acceptance, SALAD and BINA [11–13], were employed. Regarding differential cross sections, an experiment using the 4π WASA detector and deuteron beams of energies from 170 to 200 MeV/nucleon has recently been performed at the COSY ring of FZ-Jülich, while the investigations at lower (proton) beam energies between 108 and 160 MeV are currenly being carried out with the use of the BINA detector at the newly opened Cyclotron Center Bronowice (Cracow, Poland).

To a large extent, the general conclusions about the role of 3NF contributions for the description of the cross section for the breakup reaction are so far similar to those for elastic scattering. The significance of the 3NF for a correct description of the differential cross section has been confirmed [14,15]. On the other hand, certain discrepancies are observed for the tensor analyzing powers of the breakup reaction at the same beam energies, even if (or when) the 3NF is included in the calculations [16]. The experimental studies of the ^{2}H($\mathbf{p}$,pp)n reaction at 135 and 190 MeV [17,18] show a large (and growing with beam energy) discrepancy between the measured data and the theoretical predictions for the vector analyzing power A_y^p. It is located at small relative azimuthal angles of the two breakup protons, and even increases with the three-nucleon force included. The predicted relativistic effects do not explain this behaviour. These facts confirm a problem with the description of spin observables in 3N systems. The precise data sets for polarization observables of the ^{1}H($\mathbf{d}$,pp)n reaction at the beam energy of 270 MeV, collected at IUCF [19] and at RIKEN [20], also showed "a mixed picture" in the sector of spin observables. For example, A_y^d obtained in several angular configurations is described properly by the pure 2N force predictions, while the inclusion of the 3NFs leads to the deterioration of the agreement [20]. Such a behaviour is opposite to the one observed in elastic scattering at 270 MeV.

Elastic scattering is practically insensitive to the Coulomb interaction [21] and relativistic effects [22] over a wide range of energies. In contrast, the breakup reaction reveals, due to the variety of configurations of its final state, sensitivity to both of these effects, particularly enhanced at specific kinematics. In view of interplays of various effects in the breakup reaction, it is of great importance to compare experimental data to calculations which include all the dynamical ingredients. Significant progress in this respect has been achieved in recent years, and currently the calculations combining 3NF and Coulomb interactions are available [23] as well as relativistic calculations including 3NFs [22]. Coulomb effects turned out to be surprisingly large [15,24]. Therefore, an experiment probing the part of the phase space that is particularly sensitive to the Coulomb interaction was conducted at FZ-Jülich. A deuteron beam of 130 MeV and the GeWall detector [25] covering very forward angles were used in these studies. A strong influence of Coulomb effects and the general success of the theoretical calculations incorporating this long range interaction were confirmed [26]. The importance of relativistic calculations for describing the breakup reaction cross section has been shown at energies as low as 65 MeV ([27] using data from [28,29]). The calculations will be further tested by comparing with results of the breakup cross section measurements performed with the WASA detector at COSY [31] in the energy range where predicted relativistic effects exceed 30% [30].

2 Analysis Using Invariant Coordinates

The interplay of all dynamical effects in the breakup process requires a large database covering a wide range of energies, in order to verify the theoretical calculations. An optimal choice of kinematic coordinates is important for a systematic survey of such a database. In spite of the equivalence of many possible choices, the coordinates with possibly close relation to the dynamics of the process could be of particular interest. The relative momentum of a pair of outgoing particles is one example of this type, giving direct insight into the Final State Interaction (FSI) of this pair. For a 3-nucleon final state it is quite natural to define two such variables: for the pair of identical nucleons (pp in the case of pd breakup process) and for the np pair. Low relative momentum of the proton–neutron pair corresponds to configurations "similar" to elastic dp scattering, while for the pair of protons large effects of the Coulomb interaction are expected at the minimum value of such a variable. The motivation for the choice of two other coordinates is based on the important role of the four-momentum transfer in the two-body reaction dynamics. The differential cross section of elastic proton–deuteron scattering at small polar angles in the center-of-mass system (low momentum transfer) is dominated by the direct term in the two-nucleon potential and is rather well described by the calculations using only two-nucleon forces. At backward angles (large momentum transfer), the nucleon exchange term in the two-nucleon potential is responsible for the increase of the cross section. In these two regions the effects of three-nucleon forces are predicted to be mimimal. At intermediate angles, where the cross section has a minimum, it is expected that the effects of three-nucleon forces will be significant. In breakup, for the three-body exit channel, the 4-momentum transfer can be defined separately for individual nucleons. Intuitively, the less the momentum of the nucleon changes during the process, the less the nucleon is "involved" in the interaction, at the zero limit playing mainly the role of a spectator, which corresponds to Quasi-Free Scattering (QFS) of the other two nucleons. That is the reason why the so-called Chew-Low plot, combining invariant mass with 4-momentum transfer, was once used for breakup analyses [32,33] to identify events originating from quasi-elastic scattering in particular.

The coordinates can be constructed in analogy to the Mandelstam invariants. Denoting the four momenta in the exit channel by $p_p^{(1)}$, $p_p^{(2)}$ (for two protons) and p_n (for a neutron), and in the entrance channel by p_p for a proton and $p_d/2$ for any of the nucleons bound in a deuteron, we propose the following set of invariants (for further discussion see [11]):

- $s_{pp} = (p_p^{(1)} + p_p^{(2)})^2$ for a pair of protons in the exit channel,
- $s_{pn} \equiv s_{p^{(1)}n} = (p_p^{(1)} + p_n)^2$ for a proton–neutron pair - arbitrarily the pair with the lower s_{pn} value can be chosen,
- $t_n = (p_d/2 - p_n)^2$, corresponding to the 4-momentum transfer from a bound neutron in the entrance channel to a free neutron in the exit channel,
- $t_p = (p_p - p_p^{(2)})^2$, corresponding to the 4-momentum transfer from an unbound proton in the entrance channel to one of the two protons in the exit channel - arbitrarily, we choose the proton, which was not used in the calculation of s_{pn}.

One can easily recognize that the so-called star configurations are characterized by $s_{pp} = s_{pn}$ (with its value dependent on beam energy), while variation of the angle between the star plane and beam directions, α, can be translated to variation of t_n. This is only one out of many examples of a simple and clear description given in terms of the proposed coordinates, apart from the already mentioned FSI and QFS.

For convenience, the 4-momenta based coordinates can be redefined as energies, corresponding to the energy of the relative motion of two nucleons (pp, pn pairs), E_{rel}^{pp}, E_{rel}^{pn}, and the kinetic energy "transferred" to a particle:

$$
\begin{aligned}
E_{rel}^{pp} &= \sqrt{s_{pp}} - 2m_p, \\
E_{rel}^{pn} &= \sqrt{s_{pn}} - m_p - m_n, \\
E_{tr}^{p} &= \frac{-t_p}{2m_p}, \\
E_{tr}^{n} &= \frac{-t_n}{2m_n}.
\end{aligned}
\tag{1}
$$

The full dynamical information is maintained only on the 4-dimensional surface, while the integration over any number of coordinates leads to obscuring of parts of the information. However, if an effect is

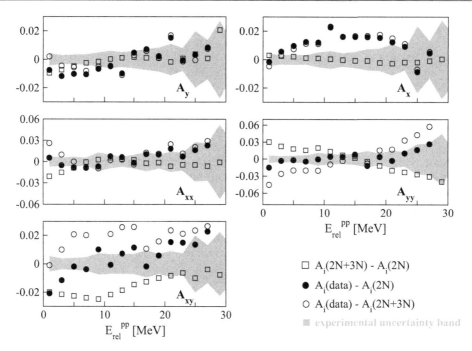

Fig. 1 Net effects of the 3NF on vector and tensor analyzing powers of the *dp* breakup reaction measured at the deuteron beam energy of 130 MeV. Open squares represent the predicted effects, i.e. the difference between the results of calculations including the TM99 3NF and those based on pure 2N potentials. Circles represent the difference between the measured and calculated observables, as explained in the legend, with the experimental uncertainty shown as *bands*. Theoretical calculations by Witała et al. For details see text

strongly correlated with a specific coordinate, it should not be smeared out by the integration and it stays clearly visible even in a one-dimensional spectrum. Such an analysis, performed in terms of E_{rel}^{pp}, turned out to be a very useful tool for tracing 3NFs and, in particular, Coulomb effects in the breakup cross section at 130 MeV (65 MeV/nucleon) [11]. The latter is consistent with the intuitive understanding of the relation between Coulomb effects and proton–proton FSI.

A similar analysis for vector and tensor analysing powers is presented here (a fraction of the results has already been discussed in [34]). The theoretical values were calculated for each vector and tensor analyzing power A_i at each experimental point, analysed conventionally on a grid of $\xi = (\theta_1, \theta_2, \phi_{12}, S)$ variables [11]. Then both data and theoretical predictions were sorted into E_{rel}^{pp} bins of 2 MeV. The predictions based on a set of two-nucleon (2N) realistic potentials (Argonne V18, Nijmegen I, II and CD-Bonn) were compared with the calculations including the TM99 3NF (2N + 3N) with the aim to analyse 3NF effects. The predictions obtained with the 2N potentials alone were considered together creating a band. The center of this band was then used for further comparisons with the data. The 2N + 3N calculations were treated in a similar way. Since the bands are relatively narrow (as compared to the effects under study) such an approach is justified. In order to magnify the effects, the theoretical predictions have been subtracted from the data. In addition, the difference between the two types of predictions has been calculated, see Fig. 1. Squares on the zero level represent the situation when the predicted 3NF effect is negligible. This is the case for A_x, A_y and A_{xx}. The predicted effects of the 3NF for A_{yy} and A_{xy} are sizeable and they exceed the experimental uncertainties, therefore the data are very suitable for testing them. For A_{yy} one can observe a systematic change of the predicted effect with E_{rel}^{pp}, signalling the importance of that coordinate in the description of the system dynamics. Full and open circles represent the difference between data and theory: the theoretical description is validated if the symbols are contained within the uncertainty bands. The most significant discrepancy is observed for A_{yy} and A_{xy} when the calculations with 3NFs (open circles) are applied. Apparently, the predicted effect is not confirmed with the data set.

It can be argued that Coulomb effects (not included in the calculations used in the above comparisons) may be the reason of the discrepancies. Therefore, the data were compared to calculations based on the Argonne V18 (AV18) potential combined with Urbana IX 3NF (UIX), with (AV18 + UIX + C) and without Coulomb interaction (AV18 + UIX) [23] in order to verify that hypothesis. As in the previous case, the theoretical

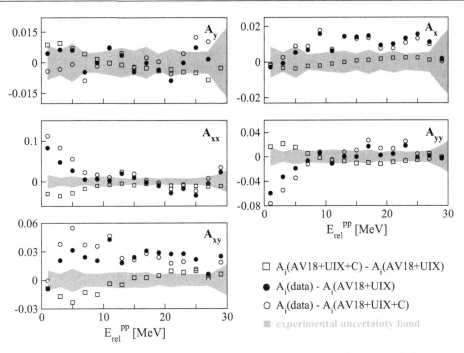

Fig. 2 Similar to Fig. 1, but for the net effects of the Coulomb interaction in vector and tensor analyzing powers. Theoretical calculations by Deltuva et al. For details see text

predictions have been individually subtracted from the data and the results are shown as circles in Fig. 2. The difference between the results of the two types of calculations i.e. the predicted net effects of the Coulomb interaction are shown as squares. In most cases the squares are grouped on the zero level, i.e. the predicted Coulomb effect is negligible. For tensor analyzing powers certain departures from zero are visible at low E_{rel}^{pp}. However, the predicted effects of the Coulomb interaction do not solve but rather exacerbate the problem of describing tensor analysing powers with currently available 2N + 3N potentials: Clearly, the open circles are even more distant from the zero level than the full ones.

Two-dimensional spectra based on two (out of four) invariants should provide us with deeper insight into the system dynamics. As the first step they are constructed for the differential cross section, which reveals interesting dependencies when studied as a function of E_{rel}^{pp} alone [11]. The ratio $\frac{\sigma_{2N+3N}-\sigma_{2N}}{\sigma_{2N}}$ was calculated in order to observe the relative influence of the given 3NF effect. σ_{2N+3N} denotes the differential cross section calculated with the 3NF included, while σ_{2N} are the results of calculations using only the 2N potential. The predicted effects of the TM99 3NF on the differential cross section of the breakup reaction at 130 MeV, presented as a function of E_{rel}^{pp} and E_{tr}^{n}, are shown in Fig. 3, left panel. The analysis has been performed for the region where data measured in the SALAD experiment exist [15], therefore the distributions are shown using the experimental thresholds and acceptance. Colours show changes in the magnitude of the effects. The 3NF effects are very local - the strongest on the very edge of the available phase space at large E_{rel}^{pp}. The experimental data σ_{exp} were compared to the calculations based on the CD-Bonn potential alone and the analogous ratio $\frac{\sigma_{exp}-\sigma_{2N}}{\sigma_{2N}}$ has been constructed. As it is shown in Fig. 3, right panel, the theoretical description based on the 2N interaction reveals a deficiency in the same region where 3NF effects are predicted and of very similar size. There is also an additional region in the data where a departure from the calculations is visible. It is located at the lowest E_{rel}^{pp} where a strong negative effect can be observed, i.e. the theory underestimates the data, and the effect changes rapidly to a positive one nearby. This region is, as we already know, particularly sensitive to the Coulomb force, which is not included in the calculations presented.

The region sensitive to the Coulomb repulsion can be investigated in detail on the basis of the results of the experiment mentioned earlier, devoted to study Coulomb effects in the forward angular region. The calculations using the AV18 potential and Urbana IX 3NF were used for comparisons, with ($\sigma_{2N+3N+C}$) and without (σ_{2N+3N}) the Coulomb interaction. The theoretically predicted effects $\frac{\sigma_{2N+3N+C}-\sigma_{2N+3N}}{\sigma_{2N+3N}}$ and the comparison of the data with the calculations without Coulomb $\frac{\sigma_{exp}-\sigma_{2N+3N}}{\sigma_{2N+3N}}$ are presented in Fig. 4. It becomes

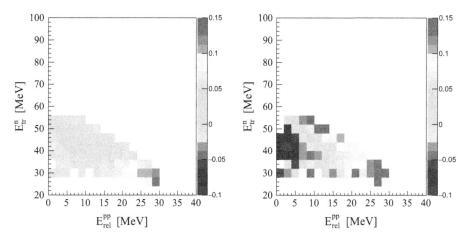

Fig. 3 Net effects of the 3NF on the differential cross section of the *dp* breakup at 130 MeV, presented as a function of two invariants. *Left panel* Difference of theoretical predictions (by Witała et al.) obtained for CD-Bonn potential with and without TM99 3NF, relatively normalized to calculations without 3NF. *Right panel* Difference between experimental data and calculations with CD-Bonn alone, normalized in the same way

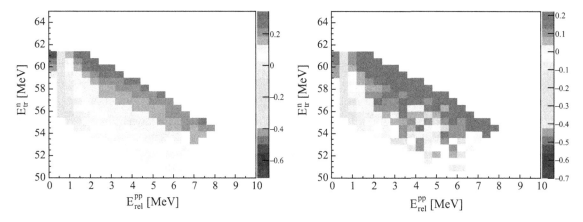

Fig. 4 Similar to Fig. 3, for the net effects of Coulomb interaction in the differential cross section of the *dp* breakup at 130 MeV. *Left panel* Difference of theoretical predictions (by Deltuva et al.) obtained for AV18 potential and Urbana IX 3NF, with and without Coulomb interaction, relatively normalized to the calculations without the Coulomb interaction. *Right panel* Difference between experimental data and calculations with AV18 potential and Urbana IX 3NF, normalized in the same way

immediately clear that the angular acceptance of the GeWall detector strongly "zooms in" at a certain region of the E_{tr}^n versus E_{rel}^{pp} plane, probing lower E_{rel}^{pp} and higher E_{tr}^n as compared to the case presented in Fig. 3. In this region, the theoretically predicted influence of the Coulomb force (left panel) changes quite dramatically from negative to positive values, which corresponds to "pushing out" of proton pairs from certain configurations to neighbouring ones. This effect amazingly well reproduces the deficiency of the Coulomb-less calculations in describing the data (right panel).

Extending the studies to the slightly higher deuteron beam energy of 160 MeV (80 MeV/nucleon), we have an opportunity to understand how the predicted and measured effects change when investigated in the coordinates proposed here. The experiment at 160 MeV was performed with the use of BINA. The BINA Wall detector represents a very similar angular acceptance, energy thresholds, angular and energy resolution as the SALAD detector used at 130 MeV. Therefore the influence of one parameter only, namely of the beam energy, can be studied. Due to the still ongoing analysis we have not shown the comparison to the data yet. The predicted effects are calculated, as in the previous cases, in the kinematic region where the experimental data were measured (excluding the lowest ϕ_{12}, corresponding to the lowest $E_{rel}^{p\bar{p}}$ values). The calculations using the AV18 2N potential with and without the Urbana IX 3NF have been compared. As it might be observed in Fig. 5, an increase of beam energy provides an access to the region of higher E_{tr}^n and E_{rel}^{pp}. The predicted 3NF effects are larger than the effects observed at 130 MeV (65 MeV/nucleon) and cumulated at the highest E_{rel}^{pp} values.

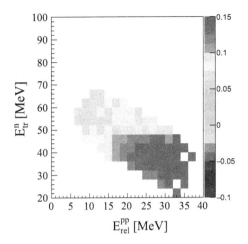

Fig. 5 Similar to Fig. 3, left panel, for the net effects of Urbama IX 3NF studied at the beam energy of 160 MeV. Theoretical predictions by Witała et al

3 Summary

Due to the recent progress in experimental techniques, the database for 3-nucleon system studies at intermediate energies has been significantly enriched and the progress in this area is continuing. The breakup reaction is studied in experiments employing detection systems covering large parts of the phase space, which allows for systematic analyses of observables as a function of kinematic variables. For the sake of comparing data obtained at different energies and learning more about the dynamics of the process, the analyses should rely on invariants, rather than on the classical kinematic variables related to geometrical configurations. The paper presents a pilot example of that kind of analysis as an indicator of its usefulness. Other combinations of invariants are also being considered and subsequently the analysis will be extended to other beam energies. New data sets will be collected in the near future for the breakup reaction. Thay will complement the basis for the tests of state-of-the-art and forthcoming theoretical calculations, which in turn will help to understand details of the few-nucleon system dynamics.

References

1. E. Epelbaum, H.W. Hammer, U.G. Meissner, Modern theory of nuclear forces. Rev. Mod. Phys. **81**, 1773 (2009)
2. S.A. Coon, H.K. Han, Reworking the Tucson–Melbourne three-nucleon potential. Few Body Syst. **30**, 131 (2001)
3. B.S. Pudliner, V.R. Pandharipande, J. Carlson, R.B. Wiringa, Quantum Monte Carlo calculations for $A \leq 6$ nuclei. Phys. Rev. Lett. **74**, 4397 (1995)
4. A. Deltuva, R. Machleidt, P.U. Sauer, Realistic two-baryon potential coupling two-nucleon and nucleon-Δ-isobar states: fit and applications to three-nucleon system. Phys. Rev. C **68**, 024005 (2003)
5. M. Viviani, A variational approach to three- and four-nucleon systems. Nucl. Phys. A **631**, 111c (1998)
6. A. Nogga, H. Kamada, W. Glöckle, Modern nuclear force predictions for the alpha particle. Phys. Rev. Lett. **85**, 944 (2000)
7. S.C. Pieper, R.B. Wiringa, Quantum Monte Carlo calculations of light nuclei. Ann. Rev. Nucl. Part. Sci. **51**, 53 (2001)
8. P. Navratil, V.G. Gueorguiev, J.P. Vary, W.E. Ormand, A. Nogga, Structure of $A = 10 - 13$ nuclei with two- plus three-nucleon interactions from chiral effective field theory. Phys. Rev. Lett. **99**, 042501 (2007)
9. K. Sagara, Experimental investigations of discrepancies in three-nucleon reactions. Few Body Syst. **48**, 59 (2010)
10. N. Kalantar-Nayestanaki, E. Epelbaum, J.G. Meschendorp, A. Nogga, Signatures of three-nucleon interactions in few-nucleon systems. Rep. Prog. Phys. **75**, 016301 (2012)
11. St Kistryn, E. Stephan, Deuteron–proton breakup at medium energies. J. Phys. G Nucl. Part. Phys. **40**, 063101 (2013)
12. A. Ramazani-Moghaddam-Arani et al., Elastic proton–deuteron scattering at intermediate energies. Phys. Rev. C **78**, 014006 (2008)
13. N. Kalantar-Nayestanaki et al., A small-angle large-acceptance detection system for hadrons. Nucl. Instrum. Methods A **444**, 591 (2000)
14. St Kistryn et al., Evidence of three-nucleon force effects from 130 MeV deuteron proton breakup cross section measurement. Phys. Rev. C **68**, 054004 (2003)
15. St Kistryn et al., Systematic study of three-nucleon force effects in the cross section of the deuteron–proton breakup at 130 MeV. Phys. Rev. C **72**, 044006 (2005)
16. E. Stephan et al., Vector and tensor analyzing powers in deuteron–proton breakup at 130 MeV. Phys. Rev. C **82**, 014003 (2010)

17. M. Eslami-Kalantari et al., Proton–deuteron break-up measurements with BINA at 135 MeV. Mod. Phys. Lett. A **24**, 839 (2009)
18. H. Mardanpour et al., Spin–isospin selectivity in three-nucleon forces. Phys. Lett. B **687**, 14 (2010)
19. H.O. Meyer et al., Axial observables in dp breakup and the three-nucleon force. Phys. Rev. Lett. **93**, 112502 (2004)
20. K. Sekiguchi et al., Three-nucleon force effects in the ^{1}H(d, pp)n reaction at 135 MeV/nucleon. Phys. Rev. C **79**, 054008 (2009)
21. A. Deltuva, A.C. Fonseca, P.U. Sauer, Momentum-space treatment of the Coulomb interaction in three-nucleon reactions with two protons. Phys. Rev. C **71**, 054005 (2005)
22. H. Witała et al., Three-nucleon force in relativistic three-nucleon Faddeev calculations. Phys. Rev. C **83**, 044001 (2011)
23. A. Deltuva, Momentum-space calculation of proton–deuteron scattering including Coulomb and irreducible three-nucleon forces. Phys. Rev. C **80**, 064002 (2009)
24. St Kistryn et al., Evidence of the Coulomb-force effects in the cross-sections of the deuteron–proton breakup at 130 MeV. Phys. Lett. B **641**, 23–27 (2006)
25. M. Betigeri et al., The germanium wall of the GEM detector system. Nucl. Instrum. Methods Phys. Res. A **421**, 447 (1999)
26. I. Ciepał et al., Investigation of the deuteron breakup on proton target in the forward angular region at 130 MeV. Few Body Syst. **56**, 665 (2015)
27. H. Witała, J. Golak, R. Skibiński, Selectivity of the nucleon-induced deuteron breakup and relativistic effects. Phys. Lett. B **634**, 374 (2006)
28. J. Zejma et al., Cross sections and analyzing powers Ay in the breakup reaction 2H(p, pp)n at 65 MeV: Star configurations. Phys. Rev. C **55**, 42 (1997)
29. M. Allet et al., Proton-induced deuteron breakup at $Elab^p = 65$ MeV in quasi-free scattering configurations. Few Body Syst. **20**, 27 (1996)
30. H. Witała, Private communication
31. B. Kłos et al., Systematic studies of the three-nucleon system dynamics in the cross section of the deuteron–proton breakup reaction. Few Body Syst. **55**, 721 (2014)
32. G. Baur, On the Chew-Low plot as a limiting case of the distorted wave theory of break-up reactions. Z. Phys. A **277**, 147 (1976)
33. J. Stepaniak, Pion production in 3He-proton interactions. Acta Phys. Pol. B **27**, 2971 (1996)
34. I. Ciepał, I. Skwira-Chalot, et al., Applications of polarized deuteron beams for studies of few-nucleon dynamics in d–p breakup, in *Proc. of the XVIth International Workshop in Polarized Sources, Targets, and Polarimetry (PSTP2015)* PoS 039

Few-Body Syst (2016) 57:1213–1225
DOI 10.1007/s00601-016-1156-3

H. Witała · J. Golak · R. Skibiński · K. Topolnicki ·
E. Epelbaum · K. Hebeler · H. Kamada · H. Krebs ·
U.-G. Meißner · A. Nogga

Role of the Total Isospin 3/2 Component in Three-Nucleon Reactions

Received: 6 May 2016 / Accepted: 6 September 2016 / Published online: 5 October 2016
© The Author(s) 2016. This article is published with open access at Springerlink.com

Abstract We discuss the role of the three-nucleon isospin $T = 3/2$ amplitude in elastic neutron–deuteron scattering and in the deuteron breakup reaction. The contribution of this amplitude originates from charge-independence breaking of the nucleon–nucleon potential and is driven by the difference between neutron–neutron (proton–proton) and neutron–proton forces. We study the magnitude of that contribution to the elastic scattering and breakup observables, taking the locally regularized chiral N^4LO nucleon–nucleon potential supplemented by the chiral N^2LO three-nucleon force. For comparison we employ also the Av18 nucleon–nucleon potential combined with the Urbana IX three-nucleon force. We find that the isospin $T = 3/2$ component is important for the breakup reaction and the proper treatment of charge-independence breaking in this case requires the inclusion of the $^1S_0$ state with isospin $T = 3/2$. For neutron–deuteron elastic scattering the $T = 3/2$ contributions are insignificant and charge-independence breaking can be accounted for by using the effective t-matrix generated with the so-called "$2/3 - 1/3$" rule.

This article belongs to the special issue "30th anniversary of Few-Body Systems".

H. Witała (✉) · J. Golak · R. Skibiński · K. Topolnicki
M. Smoluchowski Institute of Physics, Jagiellonian University, 30348 Kraków, Poland
E-mail: henryk.witala@uj.edu.pl

E. Epelbaum · H. Krebs
Institut für Theoretische Physik II, Ruhr-Universität Bochum, 44780 Bochum, Germany

K. Hebeler
Institut für Kernphysik, Technische Universität Darmstadt, 64289 Darmstadt, Germany

K. Hebeler
Extreme Matter Institute EMMI, GSI Helmholtzzentrum für Schwerionenforschung GmbH, 64291 Darmstadt, Germany

H. Kamada
Department of Physics, Faculty of Engineering, Kyushu Institute of Technology, Kitakyushu 804-8550, Japan

U.-G. Meißner
Helmholtz-Institut für Strahlen- und Kernphysik and Bethe Center for Theoretical Physics, Universität Bonn, 53115 Bonn, Germany

U.-G. Meißner
Institute for Advanced Simulation, Institut für Kernphysik, Jülich Center for Hadron Physics, and JARA - High Performance Computing, Forschungszentrum Jülich, 52425 Jülich, Germany

A. Nogga
Institut für Kernphysik, Institute for Advanced Simulation and Jülich Center for Hadron Physics, Forschungszentrum Jülich, 52425 Jülich, Germany

1 Introduction

Charge-independence breaking (CIB) is well established in the two-nucleon (2N) system in the $^1S_0$ state as evidenced by the values of the scattering lengths -23.75 ± 0.01, -17.3 ± 0.8, and -18.5 ± 0.3 fm [1,2] for the neutron–proton (np), proton–proton (pp) (with the Coulomb force subtracted), and neutron–neutron (nn) systems, respectively. That knowledge of CIB is incorporated into modern, high precision NN potentials, as exemplified by the standard semi-phenomenological models: Av18 [3], CD Bonn [4], or NijmI and NijmII [5], as well as by the chiral NN forces [6–8]. Treating neutrons and protons as identical particles requires that nuclear systems are described not only in terms of the momentum and spin but also isospin states. The general classification of the isospin dependence of the NN force is given in [9]. The isospin violating 2N forces induce an admixture of the total isospin $T = 3/2$ state to the dominant $T = 1/2$ state in the three-nucleon (3N) system. The CIB of the NN interaction thus affects 3N observables. The detailed treatment of the 3N system with CIB NN forces in the case of distinguishable or identical particles was formulated and described in [10]. We extend the investigation done in [10] by including a three-nucleon force (3NF). In the calculations performed with the standard semi-phenomenological potentials we use the UrbanaIX (UIX) [11] 3NF, while the chiral N^{2}LO 3N force [12] is used in addition to the recent and most accurate chiral NN interactions [13,14]. In this paper, based on such dynamics, we discuss the role of the amplitude with the total three-nucleon (3N) isospin $T = 3/2$ in elastic neutron–deuteron (nd) scattering and in the corresponding breakup reaction. In Sect. 2 we briefly describe the formalism of 3N continuum Faddeev calculations and the inclusion of CIB. The results are presented in Sect. 3. In Sect. 3.1 we discuss our results for elastic nd scattering and in Sect. 3.2 describe our findings for selected breakup configurations. We summarize and conclude in Sect. 4.

2 3N Scattering and Charge Independence Breaking

Neutron–deuteron scattering with nucleons interacting through a NN interaction v_{NN} and a 3NF V_{123}, is described in terms of a breakup operator T satisfying the Faddeev-type integral equation [15–17]

$$
T|\phi\rangle = tP|\phi\rangle + (1 + tG_0)V^{(1)}(1 + P)|\phi\rangle + tPG_0T|\phi\rangle \\
+ (1 + tG_0)V^{(1)}(1 + P)G_0T|\phi\rangle \,.
\tag{1}
$$

The two-nucleon t-matrix t is the solution of the Lippmann–Schwinger equation with the interaction v_{NN}. $V^{(1)}$ is the part of a 3NF which is symmetric under the interchange of nucleons 2 and 3: $V_{123} = V^{(1)}(1 + P)$. The permutation operator $P = P_{12}P_{23} + P_{13}P_{23}$ is given in terms of the transposition operators, P_{ij}, which interchange nucleons i and j. The incoming state $|\phi\rangle = |\mathbf{q}_0\rangle|\phi_d\rangle$ describes the free relative motion of the neutron and the deuteron with the relative momentum $\mathbf{q}_0$ and contains the internal deuteron state $|\phi_d\rangle$. Finally, G_0 is the resolvent of the three-body center-of-mass kinetic energy. The amplitude for elastic scattering leading to the corresponding two-body final state $|\phi'\rangle$ is then given by [16,17]

$$
\langle\phi'|U|\phi\rangle = \langle\phi'|PG_0^{-1}|\phi\rangle + \langle\phi'|PT|\phi\rangle + \langle\phi'|V^{(1)}(1 + P)|\phi\rangle \\
+ \langle\phi'|V^{(1)}(1 + P)G_0T|\phi\rangle,
\tag{2}
$$

while for the breakup reaction one has

$$
\langle\phi_0'|U_0|\phi\rangle = \langle\phi_0'|(1 + P)T|\phi\rangle,
\tag{3}
$$

where $|\phi_0'\rangle$ is the free three-body breakup channel state.

Solving Eq. (1) in the momentum-space partial wave basis, defined by the magnitudes of the 3N Jacobi momenta p and q [16] together with the angular momenta, spin and isospin quantum numbers α (β) [16], is performed by projecting Eq. (1) onto two types of basis states:

$$
|pq\alpha\rangle \equiv |pq \text{ angular momenta spins}\rangle \left| \left(t\frac{1}{2}\right) T = \frac{1}{2}M_T \right\rangle, \ (t = 0, 1),
\tag{4}
$$

and

$$
|pq\beta\rangle \equiv |pq \text{ angular momenta spins}\rangle \left| \left(t\frac{1}{2}\right) T = \frac{3}{2}M_T \right\rangle, \ (t = 1).
\tag{5}
$$

Assuming charge conservation and employing the notation where the neutron (proton) isospin projection is $\frac{1}{2}$ $(-\frac{1}{2})$, the 2N t-operator in the three-particle isospin space can be decomposed for the nd system as [10]:

$$\left\langle \left(t\frac{1}{2}\right) T M_T = \frac{1}{2} |t| \left(t'\frac{1}{2}\right) T' M_{T'} = \frac{1}{2}\right\rangle = \delta_{tt'}\delta_{TT'}\delta_{T1/2}\left[\delta_{t0}t_{np}^{t=0} + \delta_{t1}\left(\frac{2}{3}t_{nn}^{t=1} + \frac{1}{3}t_{np}^{t=1}\right)\right]$$

$$+ \delta_{tt'}\delta_{t1}(1-\delta_{TT'})\frac{\sqrt{2}}{3}\left(t_{nn}^{t=1} - t_{np}^{t=1}\right)$$

$$+ \delta_{tt'}\delta_{t1}\delta_{TT'}\delta_{T3/2}\left(\frac{1}{3}t_{nn}^{t=1} + \frac{2}{3}t_{np}^{t=1}\right), \tag{6}$$

where t_{nn} and t_{np} are solutions of the Lippman–Schwinger equations driven by the v_{nn} and v_{np} potentials, respectively.

As a result of solving Eq. (1) one gets the amplitudes $\langle pq\alpha|T|\phi\rangle$ and $\langle pq\beta|T|\phi\rangle$, which fulfill the following set of coupled integral equations:

$$\langle pq\alpha|T|\phi\rangle = \sum_{\alpha'}\int_{p'q'} \langle pq\alpha|t|p'q'\alpha'\rangle\langle p'q'\alpha'|P|\phi\rangle$$

$$+ \sum_{\alpha'}\int_{p'q'} \langle pq\alpha|V^{(1)}|p'q'\alpha'\rangle\langle p'q'\alpha'|(1+P)|\phi\rangle$$

$$+ \sum_{\alpha'}\int_{p'q'} \langle pq\alpha|t|p'q'\alpha'\rangle\langle p'q'\alpha'|G_0V^{(1)}(1+P)|\phi\rangle$$

$$+ \sum_{\alpha'}\int_{p'q'} \langle pq\alpha|t|p'q'\alpha'\rangle\langle p'q'\alpha'|PG_0T|\phi\rangle$$

$$+ \sum_{\beta'}\int_{p'q'} \langle pq\alpha|t|p'q'\beta'\rangle\langle p'q'\beta'|PG_0T|\phi\rangle$$

$$+ \sum_{\alpha'}\int_{p'q'} \langle pq\alpha|V^{(1)}|p'q'\alpha'\rangle\langle p'q'\alpha'|(1+P)G_0T|\phi\rangle$$

$$+ \sum_{\alpha'}\int_{p'q'}\sum_{\alpha''}\int_{p''q''} \langle pq\alpha|t|p'q'\alpha'\rangle\langle p'q'\alpha'|G_0V^{(1)}|p''q''\alpha''\rangle$$

$$\times\langle p''q''\alpha''|(1+P)G_0T|\phi\rangle$$

$$+ \sum_{\beta'}\int_{p'q'}\sum_{\beta''}\int_{p''q''} \langle pq\alpha|t|p'q'\beta'\rangle\langle p'q'\beta'|G_0V^{(1)}|p''q''\beta''\rangle$$

$$\times\langle p''q''\beta''|(1+P)G_0T|\phi\rangle$$

$$\langle pq\beta|T|\phi\rangle = \sum_{\alpha'}\int_{p'q'} \langle pq\beta|t|p'q'\alpha'\rangle\langle p'q'\alpha'|P|\phi\rangle$$

$$+ \sum_{\alpha'}\int_{p'q'} \langle pq\beta|t|p'q'\alpha'\rangle\langle p'q'\alpha'|G_0V^{(1)}(1+P)|\phi\rangle$$

$$+ \sum_{\alpha'}\int_{p'q'} \langle pq\beta|t|p'q'\alpha'\rangle\langle p'q'\alpha'|PG_0T|\phi\rangle$$

$$+ \sum_{\beta'}\int_{p'q'} \langle pq\beta|t|p'q'\beta'\rangle\langle p'q'\beta'|PG_0T|\phi\rangle$$

$$+ \sum_{\beta'}\int_{p'q'} \langle pq\beta|V^{(1)}|p'q'\beta'\rangle\langle p'q'\beta'|(1+P)G_0T|\phi\rangle$$

$$+ \sum_{\alpha'}\int_{p'q'}\sum_{\alpha''}\int_{p''q''} \langle pq\beta|t|p'q'\alpha'\rangle\langle p'q'\alpha'|G_0V^{(1)}|p''q''\alpha''\rangle$$

$$\times \langle p''q''\alpha''|(1+P)G_0 T|\phi\rangle$$
$$+ \sum_{\beta'} \int_{p'q'} \sum_{\beta''} \int_{p''q''} \langle pq\beta|t|p'q'\beta'\rangle\langle p'q'\beta'|G_0 V^{(1)}|p''q''\beta''\rangle$$
$$\times \langle p''q''\beta''|(1+P)G_0 T|\phi\rangle .$$

The form of the couplings in Eq. (7) follows from the fact that the incoming neutron–deuteron state $|\phi\rangle$ is a total isospin $T = 1/2$ state, the permutation operator P is diagonal in the total isospin, and the 3NF is assumed to conserve the total isospin T [18].

From Eq. (7) it is clear that only when the nn and np interactions differ in the same orbital and spin angular-momentum states with the 2N subsystem isospin $t = 1$ (CIB), then the amplitudes $\langle pq\beta|T|\phi\rangle$ will be nonzero. In addition, their magnitude is driven by the strength of the CIB as given by the difference of the corresponding t-matrices $\frac{\sqrt{2}}{3}(t_{nn}^{t=1} - t_{np}^{t=1})$ in Eq. (6). In such a case, not only the magnitude of CIB decides about the importance of the $\langle pq\beta|T|\phi\rangle$ contributions, but also the isospin $T = 3/2$ 3NF matrix elements, which participate in generating the $\langle pq\beta|T|\phi\rangle$ amplitudes. It is the set of equations Eq. (7) which we solve when we differentiate between nn and np interactions and include both $T = 1/2$ and $T = 3/2$ 3NF matrix elements.

In the case when the neglect of the $T = 3/2$ amplitudes $\langle pq\beta|T|\phi\rangle$ is justified, the CIB can be taken care of by using the effective two-body t-matrix generated with the so-called "$2/3 - 1/3$" rule, $t_{\text{eff}} = \frac{2}{3}t_{nn}^{t=1} + \frac{1}{3}t_{np}^{t=1}$, see Eq. (6), in the 2N subsystem isospin $t = 1$ states and restricting the treatment only to the amplitudes $\langle pq\alpha|T|\phi\rangle$.

Since the final state $|\phi'\rangle$ in elastic nd scattering also has the total 3N isospin $T = 1/2$, the amplitudes $\langle pq\beta|T|\phi\rangle$ do not contribute directly to this reaction. The $T = 3/2$ admixture enters in this case through a modification of the $T = 1/2$ amplitudes $\langle pq\alpha|T|\phi\rangle$ induced by the couplings given in Eq. (7). Contrary to that, for the nd breakup reaction both $T = 1/2$ ($\langle pq\alpha|T|\phi\rangle$) and $T = 3/2$ ($\langle pq\beta|T|\phi\rangle$) amplitudes contribute.

3 Results

In order to check the importance of the isospin $T = 3/2$ contributions we solved the 3N Faddeev equations Eq. (7) for four values of the incoming neutron laboratory energy: $E_{lab} = 13, 65, 135,$ and 250 MeV. As a NN potential we took the semi-locally regularized N^{4}LO chiral potential of Ref. [13,14,19] with the regulator $R = 0.9$ fm, alone or combined with the chiral N^{2}LO 3NF [12,20], regularized with the same regulator. We additionally regularized matrix elements of that 3NF by multiplying it with a nonlocal regulator $f(p, q) = \exp\{-(p^2 + \frac{3}{4}q^2)^3/\Lambda^6\}$ with large cut-off value $\Lambda = 1000$ MeV. This additional regulator is applied to the 3NF matrix elements only for technical reasons. The practical calculation of the local 3NFs involve the evaluation of convolution integrals whose calculation becomes numerically unstable at very large momenta. The value of the cutoff scale Λ is chosen sufficiently large so that low-energy physics is not affected by this additional regulator. In fact, we have checked explicitly that the effects of this regulator for the chosen cutoff value are negligible in three-body bound state and scattering calculations. As a nn force we took the pp version of that particular NN interaction (with the Coulomb force subtracted). The low-energy constants of the contact interactions in that 3NF were adjusted to the triton binding energy [12] and we used $c_D = 6.0$ for the one-pion exchange contribution and $c_E = -1.0943$ for the 3N contact term (we are using the notation of Ref. [12] with $\Lambda = 700$ MeV). That specific choice of the c_D and c_E values does not only reproduce the experimental triton binding energy when that N^{4}LO NN and N^{2}LO 3NF are combined but also improves description of the nucleon–deuteron elastic scattering cross section data at higher energies (see Fig. 4a) comparable to that obtained by combining realistic NN interactions with standard models of 3NF's [21,22]. In order to provide convergent predictions we solved Eq. (7) taking into account all partial wave states with the total 2N angular momenta up to $j_{max} = 5$ and 3N total angular momenta up to $J_{max} = 25/2$. The 3NF was included up to $J_{max} = 7/2$.

We present results for that particular combination of the chiral NN and 3N forces (of course, in the future a more consistent set of 2N and 3N forces needs to be employed, once the corresponding 3NFs are available). However, we checked using the example of the Av18 and Urbana IX 3NF combination that the conclusions remain unchanged when instead of the chiral forces so-called high precision realistic forces are used.

Since CIB effects are driven by the difference between np and nn t-matrices we display in Figs. 1 and 2 for the $^1S_0$ and $^3P_0$ NN partial waves, respectively, the np t-matrix $t_{np}(p, p'; E - \frac{3}{4}q^2)$ (a) and the difference

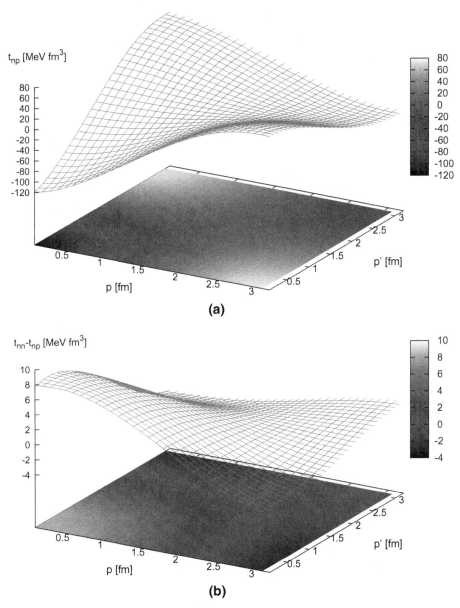

Fig. 1 (Color online) The np t-matrix $t_{np}(p, p'; E - \frac{3}{4}q^2)$ (**a**) and the difference of the nn and np t-matrices (**b**) at incoming neutron laboratory energy $E_{lab} = 13$ MeV for the $^1S_0$ partial wave, as a function of the relative NN momenta p and p' at a particular value of the spectator nucleon momentum $q = 0.528$ fm^{-1} at which the 2N subsystem energy is equal to the binding energy of the deuteron E_d: $E - \frac{3}{4}q^2 = E_d$

$t_{np}(p, p'; E - \frac{3}{4}q^2) - t_{nn}(p, p'; E - \frac{3}{4}q^2)$ (b), at the laboratory energy of the incoming neutron $E_{lab} = 13$ MeV. They are displayed as a function of the NN relative momenta p and p' for a chosen value of the spectator nucleon momentum $q = 0.528$ fm^{-1} at which the 2N subsystem energy is equal to the binding energy of the deuteron E_d: $E - \frac{3}{4}q^2 = E_d$. The behaviour of the t_{np} as well as of the difference $t_{np} - t_{nn}$ is similar at other energies. It is interesting to note that the difference between the np and nn t-matrices ranges up to $\approx 10\%$.

In Tables 1, 2, and 3 we show at the chosen energies the total cross section for the nd interaction, the total nd elastic scattering cross section, and the total nd breakup cross section, respectively, calculated with different underlying dynamics based on the chiral N^{4}LO NN or/and N^{2}LO 3NF force. Namely, the results in column 2 of those tables (no CIB, $V_{123} = 0$, $^1S_0$ np) were obtained with the NN force only, assuming no CIB and using in all $t = 1$ partial waves the effective t-matrix $t_{eff} = (2/3)t_{nn} + (1/3)t_{np}$, with the exception of the $^1S_0$ partial wave, where only the np force was taken.

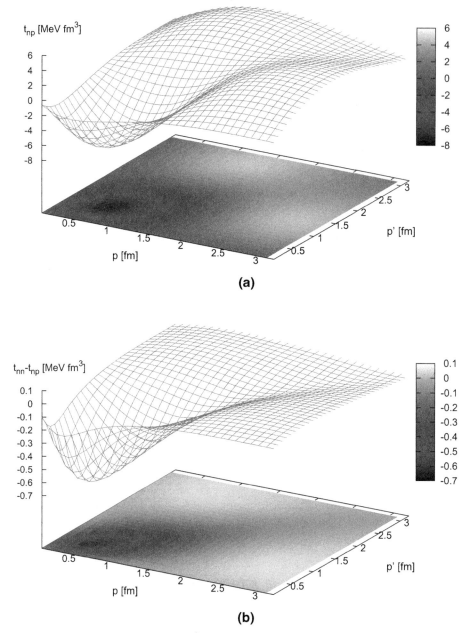

t_{np} [MeV fm^3]

(a)

t_{nn}-t_{np} [MeV fm^3]

(b)

Fig. 2 (Color online) The same as in Fig. 1 but for the $^3P_0$ partial wave

In column 3 instead of the np $^1S_0$ NN force a nn one was taken (no CIB, $V_{123} = 0$, $^1S_0$ nn). In column 4 in all $t = 1$ partial waves (including also $^1S_0$ one) only $T = 1/2$ was taken into account and the effective t-matrix $t_{eff} = (2/3)t_{nn} + (1/3)t_{np}$ was used (no CIB, $V_{123} = 0$, t_{eff}). In column 5 the NN interaction of column 4 was combined with 3NF (no CIB, V_{123}, t_{eff}). In column 6 a proper treatment of the CIB in the $^1S_0$ partial wave was performed by taking in that partial wave both np and nn interactions and keeping in addition to the total isospin $T = 1/2$ also $T = 3/2$. In all other $t = 1$ states the effective t-matrix t_{eff} was used and only states with $T = 1/2$ were kept. No 3NF was allowed ($^1S_0$ CIB, $V_{123} = 0$). Results, when in addition also the 3NF was active, are shown in column 7 ($^1S_0$ CIB, V_{123}). The proper treatment of CIB in all states with $t = 1$, when both np and nn interactions were used and both $T = 1/2$ and $T = 3/2$ states were kept, are shown in column 8 and 9 for the cases when NN interactions were used alone (CIB, $V_{123} = 0$) and combined with 3NF (CIB, V_{123}), respectively.

Table 1 The nd total cross section (in [mb]) at energies given in the first column. Dynamical models related to the particular columns are: in 2nd, 3rd, 4th, and 5th no CIB in any $t = 1$ states was assumed. In the $^1S_0$ state the t-matrix has been taken as t_{np} and t_{nn} for the 2nd and 3rd, and as $t_{\text{eff}} = (2/3)t_{nn} + (1/3)t_{np}$ for the 4th and 5th. In all other $t = 1$ states t_{eff} was used. Descriptions $V_{123} = 0$ and V_{123} means that the NN interaction was taken alone and combined with the 3NF, respectively. In the 6th and 7th columns CIB was exactly taken into account by using in $^1S_0$ state both t_{np} and t_{nn} t-matrices and state with isospin $T = 3/2$ was taken into account. In all other $t = 1$ states effective t-matrix t_{eff} was used. In the 8th and 9th columns for all $t = 1$ states CIB was treated exactly by taking in addition to $T = 1/2$ also $T = 3/2$ states and the corresponding t_{np} and t_{nn} t-matrices

1 E_{lab} [MeV]	2 no CIB $V_{123} = 0$ $^1S_0$ np	3 no CIB $V_{123} = 0$ $^1S_0$ nn	4 no CIB $V_{123} = 0$ t_{eff}	5 no CIB V_{123} t_{eff}	6 $^1S_0$ CIB $V_{123} = 0$	7 $^1S_0$ CIB V_{123}	8 CIB $V_{123} = 0$	9 CIB V_{123}
13.0	867.0	858.8	861.5	874.6	861.6	874.8	861.6	874.8
65.0	163.4	159.4	160.7	169.3	160.7	169.4	160.7	169.3
135.0	77.21	75.91	76.34	80.92	76.34	80.93	76.34	80.91
250.0	53.35	53.41	53.39	55.48	53.39	55.49	53.39	55.48

Table 2 The nd elastic scattering total cross section (in [mb]). For the description of underlying dynamics see Table 1

1 E_{lab} [MeV]	2 no CIB $V_{123} = 0$ $^1S_0$ np	3 no CIB $V_{123} = 0$ $^1S_0$ nn	4 no CIB $V_{123} = 0$ t_{eff}	5 no CIB V_{123} t_{eff}	6 $^1S_0$ CIB $V_{123} = 0$	7 $^1S_0$ CIB V_{123}	8 CIB $V_{123} = 0$	9 CIB V_{123}
13.0	699.8	699.3	699.4	711.8	699.5	712.0	699.5	712.0
65.0	71.43	69.53	70.15	75.68	70.15	75.70	70.15	75.67
135.0	20.80	20.31	20.46	22.63	20.46	22.64	20.46	22.62
250.0	8.769	8.774	8.769	9.472	8.769	9.475	8.769	9.471

Table 3 The nd breakup total cross section (in [mb]). For the description of underlying dynamics see Table 1

1 E_{lab} [MeV]	2 no CIB $V_{123} = 0$ $^1S_0$ np	3 no CIB $V_{123} = 0$ $^1S_0$ nn	4 no CIB $V_{123} = 0$ t_{eff}	5 no CIB V_{123} t_{eff}	6 $^1S_0$ CIB $V_{123} = 0$	7 $^1S_0$ CIB V_{123}	8 CIB $V_{123} = 0$	9 CIB V_{123}
13.0	167.2	159.8	162.1	162.9	162.1	162.8	162.1	162.9
65.0	91.92	89.89	90.58	93.69	90.58	93.69	90.58	93.61
135.0	56.42	55.60	55.88	58.29	55.88	58.29	55.88	58.28
250.0	44.58	44.64	44.62	46.01	44.62	46.01	44.62	46.01

From Tables 1, 2, and 3 it is clear that it is sufficient to use the effective t-matrix t_{eff} and to neglect $T = 3/2$ states completely to account exactly for CIB effects in all three total cross sections. That is true in both cases, when the 3NF is present or absent. In the case when the 3NF is not included, the total cross sections for the nd interaction, for elastic scattering, and for breakup, depend slightly on the $^1S_0$ t-matrix used in the calculations. Changing it from t_{np} to t_{nn} leads to differences of the cross section values up to $\approx 2\%$ (columns 2 and 4). Using in the $^1S_0$ channel the effective t-matrix $t_{\text{eff}} = (2/3)t_{nn} + (1/3)t_{np}$ accounts for all CIB effects exactly, without the necessity to introduce the total isospin $T = 3/2$ components in any of the $t = 1$ partial wave states. Namely, the exact treatment of CIB by using in all $t = 1$ states the t_{np} and t_{nn} t-matrices and both $T = 1/2$ and $T = 3/2$ partial wave states (column 8) gives the same value for all three total cross sections. Also restricting the exact treatment of CIB to the $^1S_0$ state only (column 6) provides the same values for the total cross sections. It shows that contribution of $T = 3/2$ states, as far as the total cross sections are concerned, can be neglected and all CIB effects properly taken into account by restricting to the total isospin $T = 1/2$ states only and using in all $t = 1$ channels the effective t-matrix generated according to the "2/3 − 1/3" rule.

The same is true when 3NF is included. In this case, in all three total cross sections, clear effects increasing with energy are seen. But again, all three treatments of CIB yield the same numbers (columns 5, 7, and 9). Since in the cases when $T = 3/2$ states were included (columns 7 and 9) also the corresponding $T = 3/2$ matrix elements of a 3NF were used, we conclude that their influence on the total cross sections is negligible.

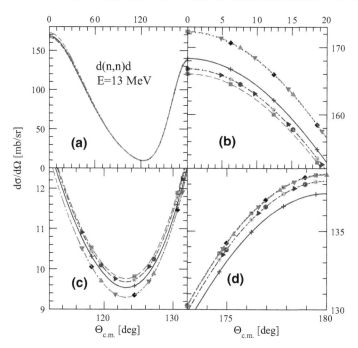

Fig. 3 (Color online) The nd elastic scattering cross section at 13 MeV of the incoming neutron laboratory energy. In **a** the full angular distribution is shown while in **b**, **c** and **d** the forward, intermediate and backward regions of angles are displayed. *Different lines* and *symbols* lying on them are predictions obtained with the locally regularized (regulator $R = 0.9$ fm) N^{4}LO NN potential alone or combined with the locally regularized N^{2}LO 3NF force for different underlying dynamics: *short-dashed* (*circles* indigo) *line*—NN potential alone and in all $t = 1$ states effective t-matrix t_{eff} was used, *solid* (*pluses: blue*) *line*—NN potential alone, in the $^1S_0$ state np force and in all other $t = 1$ states the effective t-matrix t_{eff} was used, *dashed* (*squares: red*) *line*—NN potential alone, in the $^1S_0$ state the nn force and in all other $t = 1$ states the effective t-matrix t_{eff} was used, *dotted* (*diamonds: black*) *line*—NN potential combined with 3NF and in all $t = 1$ states the effective t-matrix t_{eff} was used, *dashed-dotted* (*stars: orange*) *line*—NN potential alone, in the $^1S_0$ channels np and nn forces with isospin $T = 3/2$ included and in all other $t = 1$ states the effective t-matrix t_{eff} was used, *dotted* (*triangles right: maroon*) *line*—NN potential alone, in all $t = 1$ states np and nn forces used and isospin $T = 3/2$ included, *dashed-double-dotted* (*triangles down: green*) *line*—NN potential combined with 3NF, in $^1S_0$ states the np and nn forces used and isospin $T = 3/2$ included and in all other $t = 1$ states the effective t-matrix t_{eff} was used, *dotted-double-dashed* (*triangles up: magenta*) *line*—NN potential combined with 3NF, in all $t = 1$ states the np and nn forces used and isospin $T = 3/2$ states included

3.1 Elastic Scattering

In Figs. 3 and 4 we display results for the nd elastic scattering angular distributions obtained using different assumptions about the underlying dynamics and treatment of CIB. To immprove the readability of the figures we put on each line a unique symbol of the same color as the line. As for the total cross sections, in case when the 3NF is inactive, the three treatments of the CIB, namely using t_{eff} and no $T = 3/2$ states, $T = 3/2$ in the $^1S_0$ state with the t_{np} and t_{nn} t-matrices in that state, and $T = 3/2$ in all $t = 1$ states with the corresponding np and nn t-matrices, provide the same elastic scattering cross sections (short-dashed (circle: indigo), dashed-dotted (star: orange), and dotted (triangle right: maroon) lines, respectively). These lines overlap in Figs. 3 and 4 and the results are displayed in more detail in (b), (c), and (d) for particular ranges of angles). We checked that also for all nd elastic scattering spin observables, encompassing the neutron (vector) and deuteron (vector and tensor) analyzing powers, the spin correlation as well as spin transfer coefficients, the above three approaches lead to the same results. Thus again the contribution of all $T = 3/2$ states can be neglected and CIB in elastic nd scattering treated exactly by restricting only to $T = 1/2$ states and using in all $t = 1$ states the "$2/3 - 1/3$" rule to generate from t_{np} and t_{nn} the effective t_{eff} t-matrix.

Restricting to t_{np} or t_{nn} t-matrices in the $^1S_0$ channel and neglecting all $T = 3/2$ states changes the elastic scattering cross sections and spin observables by up to $\approx 1\%$ (see the solid (plus: blue) and dashed (square: red) lines, respectively, in Figs. 3 and 4).

Adding the 3NF changes the elastic scattering cross section. The 3NF effects grow with the projectile energy and are especially large in the region of intermediate and backward angles. But again the three approaches

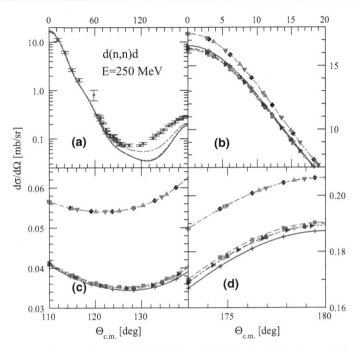

Fig. 4 (Color online) The same as in Fig. 3 but at 250 MeV. The *solid (blue) dots* in Fig. 4a are nd data from Ref. [22]

to CIB provide the same cross sections and spin observables (they are displayed for cross sections in Figs. 3 and 4 by overlapping lines: dotted (diamond: black), dashed-double-dotted (triangle down: green), and dotted-double-dashed (triangle up: magenta)). That again supports the conclusion that $T = 3/2$ states can be neglected together with the $T = 3/2$ 3NF matrix elements for all nd elastic scattering observables.

3.2 Breakup Reaction

In the final breakup state of three free nucleons both $T = 1/2$ and $T = 3/2$ total isospin components are allowed. Thus one would expect that here the influence of the $T = 3/2$ components will be better visible than in elastic scattering.

We show in Figs. 5, 6, 7 and 8 that this indeed is the case for the example of three kinematically complete breakup configurations: final-state-interaction (FSI), quasi-free-scattering (QFS), and symmetrical-space-star (SST).

In the FSI configuration under the exact FSI condition, the two outgoing nucleons have equal momenta. Their strong interaction in the $^1S_0$ state leads to a characteristic cross section maximum occurring at the exact FSI condition, the magnitude of which is sensitive to the $^1S_0$ scattering length. Since largest CIB effects are seen in the difference between np and nn (pp) $^1S_0$ scattering lengths, the region of the FSI peak should reveal largest CIB effects. That is clearly demonstrated in Fig. 5a for the nn and in Fig. 5b for the np FSI configuration, where the solid (plus: blue) and dashed (square: red) lines display FSI cross sections obtained with the $^1S_0$ np and nn t-matrices, respectively, using in all other $t = 1$ states the effective t-matrix t_{eff}. Only $T = 1/2$ states were used and the 3NF was omitted. Using also in the $^1S_0$ state the effective t-matrix leads to the short-dashed (circle: indigo) line, which changes to the dotted (diamond: black) line when, keeping the rest unchanged, the 3NF is also included. Most interesting is the effect of treating the CIB exactly in the state $^1S_0$ by including the $T = 3/2$ component and using both t_{np} and t_{nn} t-matrices: dashed-dotted (star: orange) and dashed-double-dotted (triangle down: green) lines in case when 3NF is omitted and included, respectively. As expected, the inclusion of the isospin $T = 3/2$ in the $^1S_0$ state brings the predictions, in the case when 3NF is omitted, close to the nn prediction (dashed (square: red) line in Fig. 5a) for the nn FSI, and close to the np prediction (solid (plus: blue) line in Fig. 5b) for the np FSI. However, a significant difference exists between that result and the pure nn or np ones as well as when compared to results obtained with the effective t-matrix t_{eff} (short-dashed (circle: indigo) line in Fig. 5a, b). This shows that the proper treatment of CIB in the FSI configurations of the nd breakup requires the inclusion of the total isospin $T = 3/2$ component and using

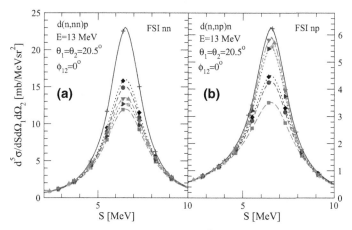

Fig. 5 (Color online) The nd complete breakup d(n, nn)p cross section $d^5\sigma/d\Omega_1 d\Omega_2 dS$ for the nn (**a**) and np (**b**) FSI configuration at 13 MeV of the incoming neutron laboratory energy and laboratory angles of detected outgoing nucleons $\theta_1 = \theta_2 = 20.5^o$ and $\phi_{12} = 0^o$, as a function of the arc-length of the S-curve. For the description of *lines* and *symbols* see Fig. 3

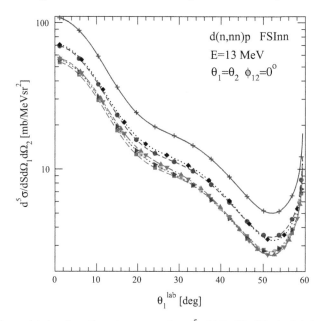

Fig. 6 (Color online) The nd complete breakup d(n, nn)p cross section $d^5\sigma/d\Omega_1 d\Omega_2 dS$ calculated exactly at the neutron–neutron final state interaction condition (maximum of the cross section along the S-curve) at 13 MeV of the incoming neutron laboratory energy as a function of the laboratory production angle of the outgoing final-state-interacting neutrons $\theta_1^{lab} = \theta_2^{lab}$ and $\phi_{12} = 0^o$. For the description of *lines* and *symbols* see Fig. 3

both t_{np} and t_{nn} t-matrices in the $^1S_0$ state. This is also sufficient for the exact treatment of CIB as shown by the results of the full CIB treatment, where in all $t = 1$ partial waves also isospin $T = 3/2$ states are taken into account and corresponding t_{np} and t_{nn} t-matrices are used, as shown in Fig. 5a, b by the dotted (triangle right: maroon) line for the NN interaction acting alone and dotted-double-dashed (triangle up: magenta) line when combined with the 3NF, respectively. These lines overlap with the lines corresponding to the case when $T = 3/2$ is included for the state $^1S_0$. It is interesting to note that the $T = 3/2$ component in the $^1S_0$ state is important and provides FSI cross sections which are different from the results obtained with particular NN $^1S_0$ interactions only (np for np FSI and nn for nn FSI). It proves the importance of including $T = 3/2$ in the $^1S_0$ state and shows that both np and nn interactions have to be employed when the FSI peaks are analyzed to extract the value of the corresponding scattering length.

In order to see how the magnitude of the effects induced by the $T = 3/2$ $^1S_0$ component depends on the particular FSI configuration, we present in Fig. 6 the cross section in the maximum of the FSI peak as a function of the laboratory production angle of the final-state interacting pair. Again it is clearly seen that restricting to

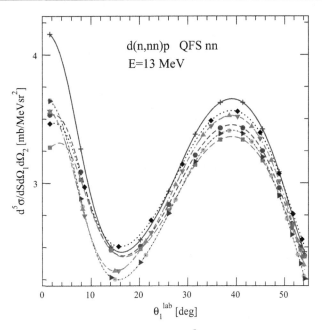

Fig. 7 (Color online) The nd complete breakup d(n,nn)p cross section $d^5\sigma/d\Omega_1 d\Omega_2 dS$ calculated exactly at the neutron–neutron quasi-free-scattering condition (maximum of the cross section along the S-curve at $E_3^{lab} = 0$ and $\phi_{12} = 180^o$) at 13 MeV of the incoming neutron laboratory energy as a function of the laboratory angle of the outgoing neutron 1. For the description of *lines* and *symbols* see Fig. 3

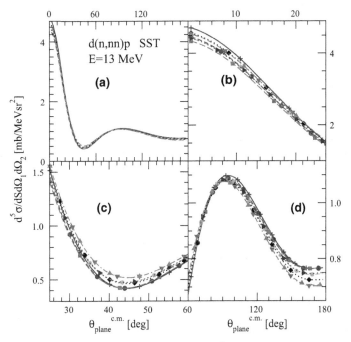

Fig. 8 (color online) The nd complete breakup d(n,nn)p cross section $d^5\sigma/d\Omega_1 d\Omega_2 dS$ exactly at the symmetrical-space-star condition (in the 3N c.m. system the momenta of three outgoing nucleons are equal and form a symmetric star in a plane inclined at an angle $\theta_{plane}^{c.m.}$ with respect to the incoming neutron momentum) at 13 MeV of the incoming neutron laboratory energy, as a function of the angle $\theta_{plane}^{c.m.}$. For the description of lines and symbols see Fig. 3

t_{eff} only and neglecting $T = 3/2$ components is insufficient to include all CIB effects. Inclusion of $T = 3/2$ component only in the $^1S_0$ state is, however, sufficient to fully account for the CIB effects. The importance of that component depends on the production angle. At the angles in the region around $\approx 45^o$ the contribution of that component is tiny but becomes significant at smaller and larger production angles.

For the QFS and SST configurations the picture is similar. Again in order to fully account for the CIB effects it is necessary and sufficient to include the total isospin $T = 3/2$ component in the $^1S_0$ state. We exemplify this in Fig. 7 for the nn QFS and in Fig. 8 for the SST configurations. Again there is an angle around which the contribution of that component is minimized. For the nn QFS it occurs around $\theta_1^{lab} \approx 28^o$ and for the SST around $\theta_{plane}^{c.m.} \approx 90^o$.

4 Summary

We investigated the importance of the scattering amplitude components with the total 3N isospin $T = 3/2$ in two 3N reactions. The inclusion of these components is required to account for CIB effects of the NN interaction. The difference between np and nn (pp) forces leads to a situation in which also the matrix elements of the 3NF between $T = 3/2$ states contribute to the considered 3N reactions. The modern NN interactions, which describe existing pp and np data with high precision, provide pp and np t-matrices which differ up to $\approx 10\%$. Such a magnitude of CIB requires that the isospin $T = 3/2$ components are included in the calculation of the breakup reaction, especially for the regions of the breakup phase-space close to the FSI condition. However, in order to account for all CIB effects it is sufficient to restrict the inclusion of $T = 3/2$ to the $^1S_0$ partial wave state only instead of doing it in all $t = 1$ states. For elastic scattering we found that the $T = 3/2$ components can be neglected completely and all CIB effects are accounted for by restricting oneself to total 3N isospin $T = 1/2$ partial waves only and using the effective t-matrix generated with the "2/3 − 1/3" rule $t_{\text{eff}} = (2/3)t_{nn} + (1/3)t_{np}$. These results allow one to reduce significantly the number of partial waves in time-consuming 3N calculations. This is of particular importance in view of the necessity to fix the parameters of the higher-order chiral 3NF components by fitting them to 3N scattering observables.

The presented results show that in 3N reactions the $T = 3/2$ components are overshadowed by the dominant $T = 1/2$ contributions. It will be interesting to investigate reactions with three nucleons in which only $T = 3/2$ components contribute in the final state such as e.g. $^3H + \pi^- \rightarrow n + n + n$. That will allow one to study the properties and the importance of 3NFs in the $T = 3/2$ states.

Acknowledgments This work was performed by the LENPIC collaboration with support from the Polish National Science Center under Grant No. DEC-2013/10/M/ST2/00420 and PRELUDIUM DEC-2013/11/N/ST2/03733, BMBF (Contract No. 05P2015 - NUSTAR R&D), and ERC Grant No. 307986 STRONGINT. The numerical calculations have been performed on the supercomputer cluster of the JSC, Jülich, Germany.

References

1. Schori, O., Gabioud, B., Joseph, C., Perroud, J.P., Rüegger, D., Tran, M.T., Truöl, P., Winkelmann, E., Dahme, W.: Measurement of the neutron–neutron scattering length a_{nn} with the reaction $\pi d \rightarrow nn\gamma$ in complete kinematics. Phys. Rev. C **35**, 2252–2257 (1987)
2. de Téramond, G.F., Gabioud, B.: Charge asymmetry of the nuclear interaction and neutron–neutron scattering parameters. Phys. Rev. C **36**, 691–701 (1987)
3. Wiringa, R.B., Stoks, V.G.J., Schiavilla, R.: Accurate nucleon–nucleon potential with charge-independence breaking. Phys. Rev. C **51**, 38–51 (1995)
4. Machleidt, R., Sammarruca, F., Song, Y.: Nonlocal nature of the nuclear force and its impact on nuclear structure. Phys. Rev. C **53**, R1483–1487 (1996)
5. Stoks, V.G.J., Klomp, R.A.M., Terheggen, C.P.F., de Swart, J.J.: Construction of high-quality NN potential models. Phys. Rev. C **49**, 2950–2963 (1994)
6. Epelbaum, E.: Few nucleon forces and systems in chiral effective field theory. Prog. Part. Nuclear Phys. **57**, 654–741 (2006)
7. Epelbaum, E., Hammer, H.W., Meißner, U.-G.: Modern theory of nuclear forces. Rev. Mod. Phys. **81**, 1773–1825 (2009)
8. Machleidt, R., Entem, D.R.: Chiral effective field theory and nuclear forces. Phys. Rep. **503**, 1–75 (2011)
9. Henley, E.M., Miller, G.A.: in Mesons and Nuclei, Rho M. and Brown G. E., eds. (North-Holland, Amsterdam 1979). Vol. I p. 405
10. Witała, H., Glöckle, W., Kamada, H.: Charge-independence breaking in the three-nucleon system. Phys. Rev. C **43**, 1619–1629 (1991)
11. Pudliner, B.S., Pandharipande, V.R., Carlson, J.: Pieper, Steven C., Wiringa, R.B.: Quantum Monte Carlo calculations of nuclei with $A < 7$. Phys. Rev. C **56**, 1720–1750 (1997)

12. Epelbaum, E., Nogga, A., Glöckle, W., Kamada, H., Meißner, Ulf-G, Witała, H.: Three-nucleon forces from chiral effective field theory. Phys. Rev. C **66**, 064001–064017 (2002)
13. Epelbaum, E., Krebs, H., Meißner, U.-G.: Improved chiral nucleon–nucleon potential up to next-to-next-to-next-to-leading order. Eur. Phys. J. A **51**, 53–81 (2015)
14. Epelbaum, E., Krebs, H., Meißner, U.-G.: Precision nucleon–nucleon potential at fifth order in the chiral expansion. Phys. Rev. Lett. **115**, 122301-1-5 (2015)
15. Witała, H., Cornelius, T., Glöckle, W.: Elastic scattering and break-up processes in the n–d system. Few Body Syst. **3**, 123–134 (1988)
16. Glöckle, W., Witała, H., Hüber, D., Kamada, H., Golak, J.: The three-nucleon continuum: achievements, challenges and applications. Phys. Rep. **274**, 107–285 (1996)
17. Hüber, D., Kamada, H., Witała, H., Glöckle, W.: How to include a three-nucleon force into Faddeev equations for the 3N continuum: a new form. Acta Phys. Pol. B **28**, 1677–1685 (1997)
18. Epelbaum, E., Meißner, U.-G., Palomar, J.E.: Isospin dependence of the three-nucleon force. Phys. Rev. C **71**, 024001-1-11 (2005)
19. Binder, S., et al.: Few-nucleon systems with state-of-the-art chiral nucleon–nucleon forces. Phys. Rev. C **93**, 044002–044006 (2016)
20. Hebeler, K., Krebs, H., Epelbaum, E., Golak, J., Skibiński, R.: Efficient calculation of chiral three-nucleon forces up to N^3LO for ab initio studies. Phys. Rev. C **91**, 044001–044009 (2015)
21. Witała, H., Glöckle, W., Golak, J., Nogga, A., Kamada, H., Skibiński, R., Kuroś-Żołnierczuk, J.: Nd elastic scattering as a tool to probe properties of 3N forces. Phys. Rev. C **63**, 024007–024012 (2001)
22. Maeda, Y., et al.: Differential cross section and analyzing power measurements for $\vec{n}\,d$ elastic scattering at 248 MeV. Phys. Rev. C **76**, 014004–13 (2007)

Few-Body Syst (2017) 58:46
DOI 10.1007/s00601-017-1211-8

A. C. Fonseca · A. Deltuva

Numerical Exact Ab Initio Four-Nucleon Scattering Calculations: from Dream to Reality

Received: 14 October 2016 / Accepted: 30 November 2016 / Published online: 19 January 2017
© Springer-Verlag Wien 2017

Abstract In the present manuscript we review the work of the last ten years on the pursuit to obtain numerical exact solutions of the four-nucleon scattering problem using the most advanced force models that fit two nucleon data up to pion production threshold with a χ^2 per data point approximately one, together with the Coulomb interaction between protons; three- and four-nucleon forces are also included in the framework of a meson exchange potential model where NN couples to NΔ. Failure to describe the world data on four-nucleon scattering observables in the framework of a non relativistic scattering approach falls necessarily on the force models one uses. Four-nucleon observables pose very clear challenges, particular in the low energy region where there are a number of resonances whose position and width needs to be dynamically generated by the nucleon–nucleon (NN) interactions one uses. In addition, our calculations constitute the most advance piece of work where observables for all four-nucleon reactions involving isospin $I = 0$, $I = 0$ coupled to $I = 1$ and isospin $I = 1$ initial states are calculated at energies both below and above breakup threshold. We also present a very extensive comparison between calculated results and data for cross sections and spin observables. Therefore the present work reveals both the shortcomings and successes of some of the present NN force models in describing four-nucleon data and serve as a benchmark for future developments.

1 Introduction

Almost 40 years ago one of us (ACF) published the first four-nucleon calculation [1] based on the numerically exact solution of a momentum space integral equation that was derived from a field theoretical model [2] inspired on Bronzan's extended Lee model [3]. The work involved the simplest extension to the four-body sector of the three-body work of Aaron, Amado and Young for the three-nucleon problem [4]. At the time, all pair interactions were described by separable s-wave couplings, and particle-pair subsystem amplitudes were also expressed in a separable form using only s-waves. States involving two non-interacting pairs were treated through the convolution of the respective propagators. The equations satisfied unitarity up to the three-body breakup threshold and could be solved by standard contour rotation methods and matrix inversion. Years later Haberzettl and Sandhas [5] were able to show that our model four-nucleon equations could be derived from the Alt, Grassberger and Sandhas (AGS) equations [6,7] once s-wave separable interactions are used between pairs and the underlying three-body subamplitudes are also represented in a separable form.

This article belongs to the Topical Collection "30th anniversary of Few-Body Systems".

A. C. Fonseca (✉)
Centro de Física Nuclear da Universidade de Lisboa, 1649-003 Lisbon, Portugal
E-mail: acsafonseca29@gmail.com

A. Deltuva
Institute of Theoretical Physics and Astronomy, Vilnius University, 10222 Vilnius, Lithuania

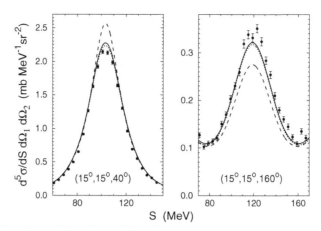

Fig. 1 Differential cross section for $p - d$ breakup at 130 MeV deuteron lab energy. Results including Δ-isobar excitation and the Coulomb interaction (*solid curves*) are compared to results without Coulomb (*dashed curves*). In order to appreciate the size of the Δ-isobar effect, the purely nucleonic results including Coulomb are also shown (*dotted curves*). The experimental data are from Ref. [19]

Over the years improvements were made to include higher number of partial waves and sophisticated one term separable representations of realistic interactions and multi-term separable representations of (3+1) subsystem amplitudes, but progress was slow [8–10] and conditioned to the size and speed of available computers, given the increased dimensionality of the four-nucleon problem that, unlike the three-nucleon system, shows in the low energy region a complex structure of resonances [11] and several open two-body channels that are coupled. Nevertheless there was one additional major obstacle: The inability at the time to include the Coulomb repulsion between protons in the solution of momentum space scattering equations. In the absence of the Coulomb force a second exited 0+ state in ^{4}He may emerge, depending on the interaction one uses, and the ^{3}H and ^{3}He binding energy difference is zero leading to a 0- resonance [12] in neutron-^{3}He (n-^{3}He) scattering that gets shifted to the point where n-^{3}He phases and observables develop a strange behavior at low energy.

It wasn't until 2005 that a viable procedure was developed [13–15] to handle the Coulomb repulsion in the solution of momentum space three-nucleon scattering calculations that could be extended to the four-nucleon problem. The method we developed involved: (a) The screening of the Coulomb potential between protons using a screening function $e^{-(r/R)^n}$, where R is the screening radius with n ranging from 4 to 8; (b) The numerical solution of the Faddeev/AGS equations for three-/four-nucleons with nuclear plus screened Coulomb potentials; (c) The use of a two-potential formula to separate the long range Coulomb part of the amplitude from the Coulomb modified short range part; (d) The renormalization method to obtain the correct limit when R goes to infinity.

This approach leads to the fully converged results for proton-deuteron (p-d) elastic scattering and breakup over the energy range from 1 to 150 MeV in the center of mass (c.m.); breakup results are still unique in the literature and have been widely used by experimentalists to interpret data in a multitude of final state configurations, namely the ones shown in Fig. 1 where the importance of Coulomb is shown to exceed the effect of including a three-nucleon force. Elastic results for p-d scattering had been obtained earlier using configuration space equations [16,17] and appropriate boundary conditions. Benchmark calculations indicate that the two approaches lead to the same results for elastic observables in the 3–65 MeV proton laboratory energy regime [18].

2 Four-Nucleon Results in Review

After the progress achieved in the solution of the three-nucleon problem with the Coulomb repulsion included, in 2007 we presented numerically exact ab initio results for neutron-triton elastic scattering (n-^{3}H) [20] below breakup threshold that were followed by calculations for p-^{3}He observables [21] using the method of screening and renormalization used before for p-d scattering. Although our work in this energy region succeeded some of the precise ab initio coordinate space calculations by the Pisa [22–26] and Grenoble-Strasbourg [27–31]

Table 1 Binding energies for ^{3}H, ^{3}He, and ^{4}He derived from the potentials CD Bonn and CD Bonn + Δ and the corresponding experimental values are given in the first three rows

	^{3}H	^{3}He	^{4}He
CD Bonn	8.00	7.26	26.18
CD Bonn + Δ	8.28	7.53	27.10
exp	8.48	7.72	28.30
ΔE_2	−0.51	−0.48	−2.80
ΔE_3^{FM}	0.50	0.48	2.25
$\Delta E_3^{\text{h.o.}}$	0.29	0.27	1.30
ΔE_4			0.17

The last four rows split the complete Δ effect up into $2N$ dispersion ΔE_2, Fujita-Miyazawa type $3N$ force effect ΔE_3^{FM}, higher order $3N$ force effect $\Delta E_3^{\text{h.o.}}$, and $4N$ force effect ΔE_4 for ^{4}He. All results are given in MeV

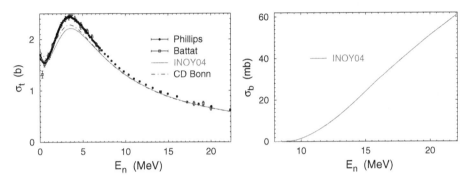

Fig. 2 (Color online) Total (*left*) and breakup (*right*) cross sections for n-^{3}H scattering as a function of the neutron lab energy. Results obtained with INOY04 (*solid curves*) and CD Bonn (*dashed-dotted curves*) potentials are compared with the experimental data from Refs. [37,38]

groups, there were a number of issues to be settled given the complexity of the problem and the lack of benchmarking.

The calculations we present were done using realistic NN interactions based on meson exchange models (CD Bonn [32], AV18 [33], INOY04 [34]) or derived from Effective Field Theory (N3LO [35])) and were fully converged vis-a-vis the number of 2N, 3N and 4N partial waves that were included, as well as number of mesh points in the three Jacobi momentum variables or size of the screening radius. Below breakup threshold we used real axis integration together with subtraction methods and spline interpolation; the amplitudes are obtained by summing the corresponding Neumann series using the Pade method. In order to test the importance of three-nucleon and four-nucleon forces in 4N observables, we use the CD Bonn+ Δ [36] potential where NN couples to NΔ, leading to effective 3N and 4N forces and currents that are compatible with the underlying 2N forces. In Table 1 we show the contribution of 3N and 4N forces to the binding energy of ^{3}H, ^{3}He and ^{4}He.

Because two-body dispersion effects are repulsive and relatively large in ^{3}H, ^{3}He, and ^{4}He, the overall gain in binding energy resulting from adding three- and four-nucleon forces is relatively small and of the order of 3.5% in all three nuclei and insufficient to obtain the respective experimental binding energies. Nevertheless the calculation shows for the first time a realistic estimate of the contribution of the four-nucleon force to the ^{4}He binding energy.

The simplest four-nucleon reactions one can think of are n-^{3}H and p-^{3}He reactions which are dominated by isospin $I = 1$. Results are shown in Figs. 2, 3, 4 and 5 but already extended to energies well above breakup threshold [42,43] where we still use real axis integration and spline interpolation, but solve instead the equations with complex energy $E + i\epsilon$ for four to six different ϵ and extrapolate the resulting amplitudes to $\epsilon = 0$. In addition a special subtraction method was developed to cope with the integration over the new singularities that emerge above breakup threshold. In Fig. 2 (left) we show the total cross section for n-^{3}H scattering calculated with INOY04 and CD Bonn potentials and compare with the experimental data for the total cross section. Since there is a strong correlation between the triton binding energy and n-^{3}H singlet and triplet scattering length [22], the zero energy cross section is well reproduced by the INOY04 potential curve, but as the energy rises the calculation fails to describe the maximum of the total cross section. Using CD Bonn or any other potential does not solve this issue [20] but improvement is observed if a specific three-nucleon

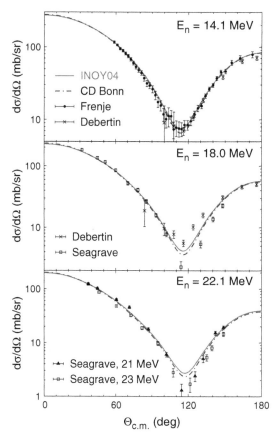

Fig. 3 (Color online) Differential cross section for elastic n-^{3}H scattering at 14.1, 18.0, and 22.1 MeV neutron energy. Results obtained with INOY04 (*solid curves*) and CD Bonn (*dashed-dotted curves*) potentials are compared with the experimental data from Refs. [39–41]

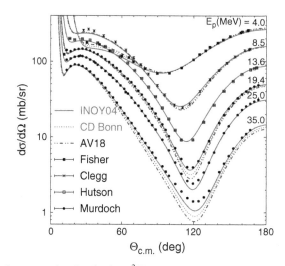

Fig. 4 (Color online) Differential cross section for elastic p-^{3}He scattering at 4.0, 8.52, 13.6, 19.4, 25.0, and 35.0 MeV proton energy as function of the c.m. scattering angle. Results obtained with INOY04 (*solid curves*), and at selected energies, with CD Bonn (*dashed-dotted curves*) and AV18 (*dotted curves*) potentials are compared with the experimental data from Refs. [44–46]

force is added [26]. Nevertheless as the energy rises both calculations describe the data reasonably well. Above three-body breakup threshold we show for the first time in Fig. 2 (right) predictions for the total breakup cross section using INOY04 potential. As discussed in Ref. [20], the calculated differential cross sections for elastic scattering below breakup threshold describe the data except for small deviations at backward and forward

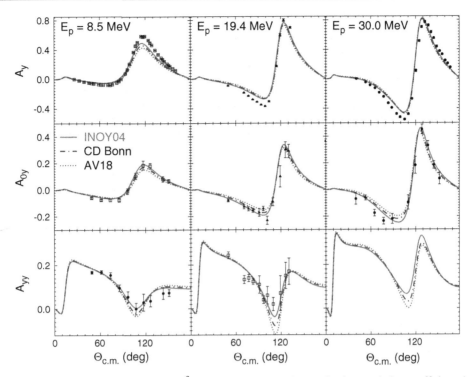

Fig. 5 (Color online) Proton analyzing power A_y, ^{3}He analyzing power A_{0y}, and spin correlation coefficient A_{yy} for elastic p-^{3}He scattering at 8.52, 19.4, and 30.0 MeV proton energy. See Ref. [43] for details on theoretical calculations and additional experimental data

angles; it is not obvious to us that the disagreement with the total cross section data is compatible with the discrepancies we observe in the differential cross section data. Above breakup threshold [42], as shown in Fig. 3, the disagreement between calculated cross sections and data is limited to the overestimation of the cross section at the minimum, as the energy rises beyond $E_n = 18$ MeV. Nevertheless the different sets of data in this energy region are not compatible with each other which raises doubts about the quality of the data.

Since p-^{3}He scattering data is much more abundant and covers the measurement of different polarization observables, we show in Fig. 4 the differential cross section as a function of the c.m. scattering angle at energies ranging from $E_p = 4$ to 35 MeV and in Fig. 5 results for A_y, A_{0y} and A_{yy} for $E_p = 8.5$ to 30 MeV. Calculations are done for INOY04, CD Bonn and AV18 pair potentials. The cross section decreases rapidly with energy and changes shape. Calculations describe the angular dependence fairly well with a slight under prediction at forward angles for $E_p < 15$ MeV much like what has been reported in calculations below breakup threshold [21]. At the minimum, the $d\sigma/d\Omega$ predictions scale with the ^{3}He binding energy; the weaker the ^{3}He binding the lower the minimum of $d\sigma/d\Omega$. The scaling is more pronounced at higher E_p. For the INOY04 potential that fits ^{3}He binding energy, one gets a good agreement in the whole angular region up to $E_p = 20$ MeV [43], but as the energy increases, the calculated cross section under predicts the data at the minimum much like what happens in n-d and p-d elastic scattering, but at a much higher center of mass energy. This under prediction may be conjectured as a sign for the need to include a three-nucleon force. As for the polarization observables shown in Fig. 5 we find that the calculations describe the data considerably well over the shown energy range except at the peak of the A_y for $E_p < 15$ MeV and at the minimum for $E_p > 6$ MeV. Surprisingly A_{0y} is well described at the maximum but shows similar behavior at the minimum for $E_p > 6$ MeV. As shown both here and in Ref. [43] A_{yy} and A_{xx} are well described by the calculations; all polarization observables are sensitive to the choice of NN force in the angular region where extremes take place (maxima or minima). Although data for spin transfer coefficients are sparse, show large error bars, and are limited to narrow angular regions, the calculations seem to follow the trends of the data [43]. Below breakup threshold recent benchmark calculations [47] indicate that both n-^{3}H and p-^{3}He observables can be accurately calculated using different numerical methods based on different equations, but above breakup mostly the solutions of the AGS equations in momentum space have so far produced what we believe to be the first ab initio predictions for a wide range of

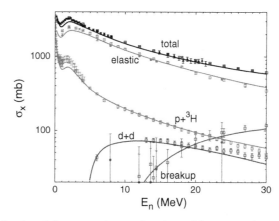

Fig. 6 (Color online) n-^{3}He total and partial cross sections as functions of the neutron beam energy calculated using INOY04 potential. The data are from Refs. [48,49] (*open square*), [50] (*times symbol*), [51] (*filled circle*), [37] (*filled triangle*), [52] (*plus symbol*)

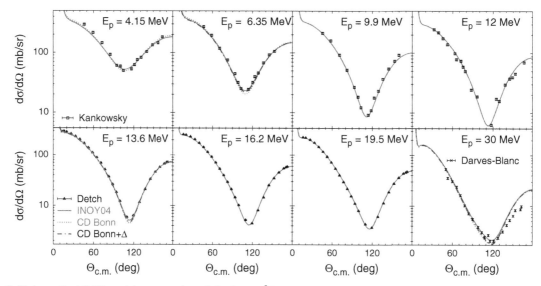

Fig. 7 (Color online) Differential cross section of elastic $p + {}^3$H scattering at proton energy between 4.15 and 30 MeV. Results obtained with potentials INOY04 (*solid curves*), CD Bonn (*dotted curves*), and CD Bonn + Δ (*dashed-dotted curves*) are compared with data from Refs. [53–56]

observables at energies up to 35 MeV. The work in Ref. [31] is another step in this direction, using coordinate space framework with complex scaling, but so far limited to n-^{3}H scattering.

In the case of reactions initiated by n-^{3}He, p-^{3}H and d+d we face a much complex reaction dynamics because all three reactions are coupled and involve both isospin $I = 0$ and $I = 1$ contributions [58–62]. One major outcome of these calculations is shown in Fig. 6 where we plot the total cross sections for all possible reactions initiated by n-^{3}He and include the first prediction ever for the total breakup cross section as a function of energy. Given the complicated structure of 4N resonances at low energy, the calculations do not quite follow the data up to 10 MeV, but thereafter the results are very impressive, showing that above $E_n > 23$ MeV breakup becomes the largest reaction process after elastic scattering. Calculations were done with the INOY04 potential which fits both ^{3}He and ^{3}H binding energies. Other results for n-^{3}He elastic observables, $d\sigma/d\Omega$ and neutron A_y, are shown in Refs. [60,63]. The calculated differential cross sections also under predict the data at the minimum but to a much lesser extent than in p-^{3}He. Likewise neutron A_y seems to be closer to the data both at the minima and maxima.

Next we show in Figs. 7 and 8 results for elastic p-^{3}H $d\sigma/d\Omega$ and proton A_y for energies ranging from $E_p = 4.15$ to 30 MeV. Calculations are done with INOY04 potential but at specific energies with CD Bonn and CD Bonn + Δ interactions in order to show the sensitivity of the observables to NN force models. Except for

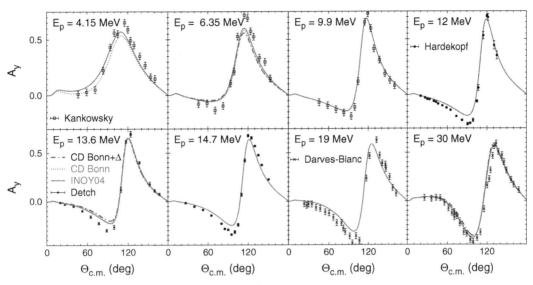

Fig. 8 (Color online) Proton analyzing power of elastic $p + {}^3$H scattering at proton energy between 4.15 and 30 MeV. Curves are as in Fig. 7. Data are from Refs. [53,55–57]

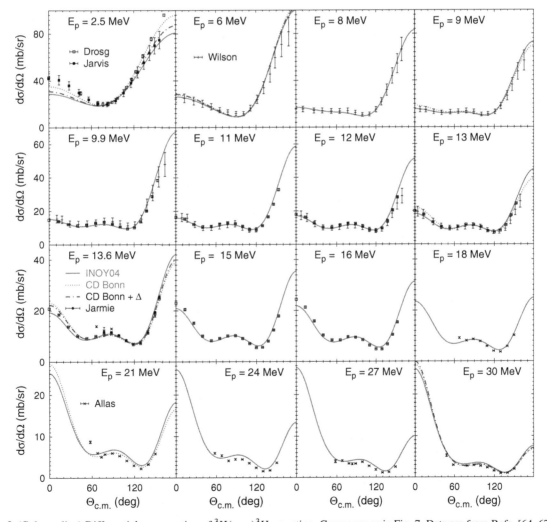

Fig. 9 (Color online) Differential cross section of ^{3}H$(p, n)^3$He reaction. Curves are as in Fig. 7. Data are from Refs. [64–67]

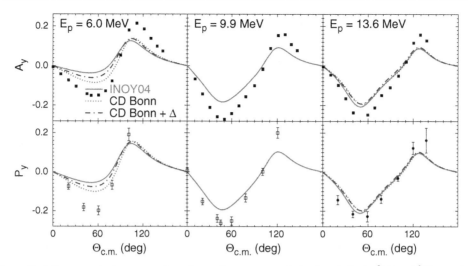

Fig. 10 (Color online) Proton analyzing power A_y and outgoing neutron polarization P_y in the ^{3}H$(p, n)^3$He reaction at 6.0, 9.9, and 13.6 MeV proton energy. Curves are as in Fig. 7. The data are from Ref. [68] for A_y and from Refs. [69] (*open square*) and [70] (*filled circle*) for P_y

a small under prediction of the data at forward angles the calculated elastic differential cross sections follow the data over the whole energy range, including the region of the minima. As for proton A_y there is a very consistent over prediction of the data at the minima but a smaller under prediction at the maxima compared to p-^{3}He, The most interesting, and in some way impressive results, emerge in Fig. 9 for the charge exchange reaction p-^{3}H to n-^{3}He differential cross sections in the energy range $E_p = 2.5$–30 MeV. The data has a complicated energy evolution starting at low energy by having a shallow minimum followed by a backward peak; as the energy rises it develops a second minimum while slowly developing a forward peak that rises past the backward peak that decreases in magnitude. The calculations follow the energy behavior of the data in a

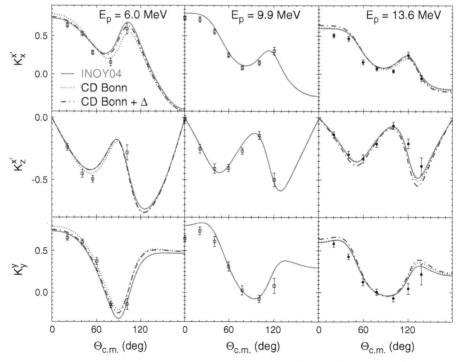

Fig. 11 (Color online) Proton-to-neutron polarization transfer coefficients of ^{3}H$(p, n)^3$He reaction at 6.0, 9.9, and 13.6 MeV proton energy. Curves are as in Fig. 7. The data are from Refs. [69] (*open square*) and [70] (*filled circle*)

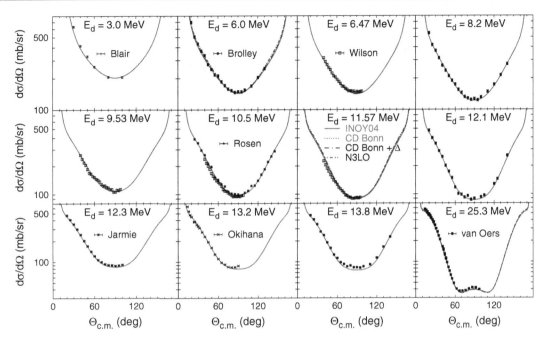

Fig. 12 (Color online) Differential cross section of $d + d$ elastic scattering as a function of c.m. scattering angle at deuteron beam energies ranging from 3 to 25.3 MeV. Results are obtained using INOY04 potential (*solid curves*), and, at 6.0, 11.57, and 25.3 MeVm also CD Bonn + Δ (*dashed-dotted curves*), CD Bonn (*dotted curves*), and N3LO (*double-dotted-dashed curves*) potentials. The experimental data are from Refs. [54,72–77]

very consistent way except for a few discrepancies at low energy at the forward angles and the region between the two minima as E_p rises beyond 20 MeV. Sensitivity to the NN force model is mostly around the forward and backward angles. In Fig. 10 we show the charge exchange proton A_y and outgoing neutron polarization P_y for three different proton laboratory energies E_p. The charge exchange proton A_y and P_y constitute the worst disagreement between calculations and data in this energy region which is surprising given the good overall agreement between proton-to-neutron polarization transfer coefficients K_{ij} shown in Fig. 11.

Finally in Figs. 12 and 13 we show the differential cross section for d-d elastic scattering between $E_d = 3$ MeV and $E_d = 25.3$ MeV and the tensor analyzing powers for the reactions dd-n^{3}He and dd-p^{3}H at $E_d = 10$ MeV lab energy. Given the identity of both particles, elastic $d\sigma/d\Omega$ is symmetric around 90° in the c.m. and has a minimum at 90°, except at 25.3 MeV where it shows a maximum at 90° placed in between two minima [71]. The calculations follow the data over the whole energy range except for a small discrepancy at the minimum at 13.8 MeV which may need further validation in case new data is measured around that energy. Given the large size of the deuteron, agreement with data was naively expected at low energy where the interaction between the two deuterons is mostly peripheral. On the contrary tensor analyzing power data for the transfer reactions dd-n^{3}He and dd-p^{3}H constitute a challenge for theory considering the complicated structure of the data. Therefore the excellent agreement with data displayed in Fig. 13, except for iT_{11} where a large sensitivity to the NN force model is observed, is an indication of the quality of the chosen dynamics and force models used in the calculation. More results for dd-n^{3}He and dd-p^{3}H reactions at higher energies will be forthcoming.

3 Conclusions

In the present manuscript we review the last 10 years of relentless progress in the solution of numerical exact ab initio calculations of scattering observables involving all four-nucleon reactions both below and above breakup threshold. The work involves the momentum-space solution of the AGS equations, including realistic NN force models, the Coulomb interaction between protons, and effective three- and four-nucleon forces through NN to NΔ coupling. Except for the calculation of breakup observables for which there are, at present, very few data to compare with, these calculations constitute the most advanced and complete study of four-nucleon scattering observables covering all the available data in the field we know of. Overall one could say that the force models

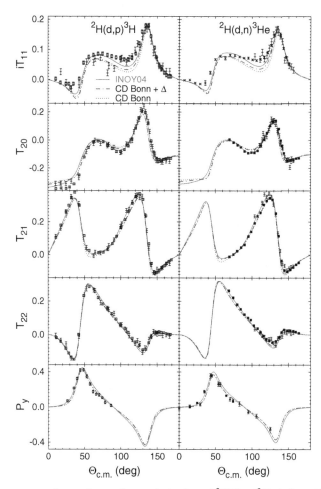

Fig. 13 Deuteron analyzing powers and outgoing nucleon polarization of ^{2}H($\mathbf{d}$, p)^{3}H (*left*) and ^{2}H($\mathbf{d}$, n)^{3}He (*right*) transfer reactions at 10 MeV deuteron energy. Curves as in Fig. 7. The data are from Refs. [78] (*plus symbol*), [79] (*open square*), [80] (*times symbol*), [81] (*filled square*), [82] (*open circle*), [83] (*filled circle*), and [84] (*filled triangle*)

we use provide a very satisfactory explanation of the cross sections and spin observables over a broad energy region up to about 20 MeV in the c.m., including the proper energy dependence and the change in shape of the observables as the energy rises. Agreement with data can be in some cases paradigmatic, given the complex structure of some observables that display several maxima and minima, a feature that is not often observed in three-nucleon data. Nevertheless there are some notable failures in describing four-nucleon observables:

(a) The total n-^{3}H cross section at peak, using force models that fit the triton binding energy in order to reproduce the magnitude of the total cross section at threshold; (b) The total cross section for n-^{3}He in the energy region up to 5 MeV; (c) The peak of the angular distribution of proton A_y in p-^{3}He scattering for energies below $E_p < 15$ MeV, much like what has been observed in p-d proton A_y, as well as the minimum for $E_p > 6$ MeV; (d) The under prediction of the minimum in p-^{3}He differential cross section for $E_p > 20$ MeV; (e) A_y and P_y in the ^{3}H(p,n)^{3}He reaction, assuming there is no problem with this data.

As pointed out above, there are other small discrepancies between calculations and data but these are, in our opinion, the ones that offer greater sensitivity to force model investigation, particular in the region of isospin $I = 1$ physics which cannot be so easily constrained by three-nucleon studies. In addition four-nucleon observables display a number of maxima and minima that are force model sensitive, which by itself is a considerable advantage when testing different interactions between nucleons.

Although the four-nucleon scattering problem is much more difficult to manage from the computational view point than three-nucleon scattering, it has the advantage of providing highly sensitive discrepancies with data at low energy that involve two-cluster reactions alone. Therefore we expect in the future that four-nucleon calculations may become the preferred theoretical laboratory to study more advanced force models, given the richer structure of resonances of the 4N system, the diversity of reactions in different Isospin channels and complex structure of observables.

References

1. A.C. Fonseca, Four-body model of the four-nucleon system. Phys. Rev. C **19**, 1711 (1979)
2. A.C. Fonseca, P.E. Shanley, Soluble model involving four identical particles. Phys. Rev. C **14**, 1343 (1976)
3. J.B. Bronzan, Soluble model field theory with vertex function. Phys. Rev. **139**, B751 (1965)
4. R. Aaron, R.D. Amado, Y.Y. Yam, Calculations of neutron–deuteron scattering. Phys. Rev. **140**, B1291 (1965)
5. H. Haberzettl, W. Sandhas, Effective two-body equations for the four-body problem with exact treatment of (2+2)-subsystem contributions. Phys. Rev. C **24**, 359 (1981)
6. P. Grassberger, W. Sandhas, Systematical treatment of the non-relativistic n-particle scattering problem. Nucl. Phys. **B2**, 181 (1967)
7. E.O. Alt, P. Grassberger, W. Sandhas: JINR report No. E4-6688 (1972)
8. A.C. Fonseca, Contribution of p-wave (3) + 1 subamplitudes to ^{4}H binding energy and scattering observables below four-body breakup threshold. Few-Body Syst. **1**, 69 (1986)
9. A.C. Fonseca, ^{4}H binding energy calculation including full tensor-force effects. Phys. Rev. C **40**, 1390 (1989)
10. A.C. Fonseca, Contribution of nucleon–nucleon P waves to nt-nt, dd-pt, and dd-dd scattering observables. Phys. Rev. Lett. **83**, 4021 (1999)
11. D.R. Tilley, H. Weller, G.M. Hale, $A = 4$ energy levels. Nucl. Phys. A **541**, 1 (1992)
12. A.C. Fonseca, G. Hale, J. Haidenbauer, Contribution of nucleon–nucleon P waves to nt-nt, dd-pt, and dd-dd scattering observables. Few-Body Syst. **31**, 139 (2002)
13. A. Deltuva, A.C. Fonseca, P.U. Sauer, Momentum-space treatment of Coulomb interaction in three-nucleon reactions with two protons. Phys. Rev. C **71**, 054005 (2005)
14. A. Deltuva, A.C. Fonseca, P.U. Sauer, Momentum-space description of three-nucleon breakup reactions including the Coulomb interaction. Phys. Rev. C **72**, 054004 (2005)
15. A. Deltuva, A.C. Fonseca, P.U. Sauer, Nuclear many-body scattering calculations with the Coulomb interaction. Annu. Rev. Nucl. Part. Sci. **58**, 27 (2008)
16. A. Kievsky, M. Viviani, S. Rosati, Polarization observables in p-d scattering below 30 MeV. Phys. Rev. C **64**, 024002 (2001)
17. A. Kievsky, M. Viviani, L.E. Marcucci, N-d scattering including electromagnetic forces. Phys. Rev. C **69**, 014002 (2004)
18. A. Deltuva, A.C. Fonseca, A. Kievsky, S. Rosati, P.U. Sauer, M. Viviani, Benchmark calculation for proton-deuteron elastic scattering observables including Coulomb. Phys. Rev. C **71**, 064003 (2005)
19. S. Kistryn et al., Proton-deuteron breakup. Phys. Rev. C **72**, 044006 (2005)
20. A. Deltuva, A.C. Fonseca, Four-nucleon scattering: ab initio calculations in momentum space. Phys. Rev. C **75**, 014005 (2007)
21. A. Deltuva, A.C. Fonseca, Four-body calculation of proton-^{3}He scattering. Phys. Rev. Lett. **98**, 162502 (2007)
22. M. Viviani, S. Rosati, A. Kievsky, Neutron-^{3}H and proton-^{3}H zero energy scattering. Phys. Rev. Lett. **81**, 1580 (1998)
23. M. Viviani, A. Kievsky, S. Rosati, E.A. George, L.D. Knutson, The A_y problem for $p-^3He$ elastic scattering. Phys. Rev. Lett. **86**, 3739 (2001)
24. A. Kievsky, S. Rosati, M. Viviani, L.E. Marcucci, L. Girlanda, A high-precision variational approach to three- and four-nucleon bound and zero-energy scattering states. J. Phys. G **35**, 063101 (2008)
25. M. Viviani, R. Schiavilla, L. Girlanda, A. Kievsky, L.E. Marcucci, Parity-violating asymmetry in the ^{3}He($\mathbf{n}$, p)^{3}H reaction. Phys. Rev. C **82**, 044001 (2010)
26. M. Viviani, L. Girlanda, A. Kievsky, L.E. Marcucci, Effect of three-nucleon interactions in p-He3 elastic scattering. Phys. Rev. Lett. **111**, 172302 (2013)
27. R. Lazauskas: Four-nucleon problem. Ph.D. thesis, Université Joseph Fourier, Grenoble (2003), http://tel.ccsd.cnrs.fr/documents/archives0/00/00/41/78/
28. R. Lazauskas, J. Carbonell, Testing nonlocal nucleon-nucleon interactions in four-nucleon systems. Phys. Rev. C **70**, 044002 (2004)
29. R. Lazauskas, Elastic proton scattering on tritium below the n-^{3}He threshold. Phys. Rev. C **79**, 054007 (2009)
30. R. Lazauskas, Application of the complex-scaling method to four-nucleon scattering above break-up threshold. Phys. Rev. C **86**, 044002 (2012)
31. R. Lazauskas, Modern nuclear force predictions for $n-^3$H scattering above the three- and four-nucleon breakup thresholds. Phys. Rev. C **91**, 041001 (2015)
32. R. Machleidt, High-precision, charge-dependent Bonn nucleon–nucleon potential. Phys. Rev. C **63**, 024001 (2001)
33. R.B. Wiringa, V.G.J. Stoks, R. Schiavilla, Accurate nucleon–nucleon potential with charge-independence breaking. Phys. Rev. C **51**, 38 (1995)
34. P. Doleschall, Influence of the short range nonlocal nucleon-nucleon interaction on the elastic n-d scattering: below 30 MeV. Phys. Rev. C **69**, 054001 (2004)
35. D.R. Entem, R. Machleidt, Accurate charge-dependent nucleon-nucleon potential at fourth order of chiral perturbation theory. Phys. Rev. C **68**, 041001(R) (2003)
36. A. Deltuva, R. Machleidt, P.U. Sauer, Realistic two-baryon potential coupling two-nucleon and nucleon-Δ-isobar states: fit and applications to three-nucleon system. Phys. Rev. C **68**, 024005 (2003)
37. M.E. Battat et al., Total neutron cross sections of the hydrogen and helium isotopes. Nucl. Phys. **12**, 291 (1959)

38. T.W. Phillips, B.L. Berman, J.D. Seagrave, Neutron total cross section for tritium. Phys. Rev. C **22**, 384 (1980)
39. J.A. Frenje, C.K. Li, F.H. Seguin, D.T. Casey, R.D. Petrasso, D.P. McNabb, P. Navratil, S. Quaglioni, T.C. Sangster, V.Y. Glebov, D.D. Meyerhofer, Measurements of the differential cross sections for the elastic n-^{3}H and n-^{2}H scattering at 14.1 MeV by using an inertial confinement fusion facility. Phys. Rev. Lett. **107**, 122502 (2011)
40. K. Debertin, E. Roessle, J. Schott, n+t scattering data, in *EXFOR Database* (NNDC, Brookhaven, 1967)
41. J. Seagrave, J. Hopkins, D. Dixon Jr., P.K. Kerr, E. Niiler, A. Sherman, R. Walter, Elastic scattering and polarization of fast neutrons by liquid deuterium and tritium. Ann. Phys. **74**, 250 (1972)
42. A. Deltuva, A.C. Fonseca, Neutron-^{3}H scattering above the four-nucleon breakup threshold. Phys. Rev. C **86**, 011001(R) (2012)
43. A. Deltuva, A.C. Fonseca, Calculation of proton-^{3}He elastic scattering between 7 and 35 MeV. Phys. Rev. C **87**, 054002 (2013)
44. T.B. Clegg, A.C.L. Barnard, J.B. Swint, J.L. Weil, The elastic scattering of protons from 3He from 4.5 to 11.5 MeV. Nucl. Phys. **50**, 621 (1964)
45. R.L. Hutson, N. Jarmie, J.L. Detch, J.H. Jett, $p - ^3$He elastic scattering from 13 to 20 MeV. Phys. Rev. C **4**, 17 (1971)
46. B.T. Murdoch, D.K. Hasell, A.M. Sourkes, W.T.H. van Oers, P.J.T. Verheijen, R.E. Brown, ^{3}He(p, p)^{3}He scattering in the energy range 19 to 48 MeV. Phys. Rev. C **29**, 2001 (1984)
47. M. Viviani, A. Deltuva, R. Lazauskas, J. Carbonell, A.C. Fonseca, A. Kievsky, L.E. Marcucci, S. Rosati, Benchmark calculation of n-^{3}H and p-^{3}He scattering. Phys. Rev. C **84**, 054010 (2011)
48. B. Haesner, W. Heeringa, H.O. Klages, H. Dobiasch, G. Schmalz, P. Schwarz, J. Wilczynski, B. Zeitnitz, F. Käppeler, *EXFOR Database* (NNDC, Brookhaven, 1982)
49. B. Haesner, W. Heeringa, H.O. Klages, H. Dobiasch, G. Schmalz, P. Schwarz, J. Wilczynski, B. Zeitnitz, F. Käppeler, Measurement of the He3 and He4 total neutron cross sections up to 40 MeV. Phys. Rev. C **28**, 995 (1983)
50. J.D. Seagrave, L. Cranberg, J.E. Simmons, Elastic scattering of fast neutrons by tritium and He^3. Phys. Rev. **119**, 1981 (1960)
51. M. Drosg, D.K. McDaniels, J.C. Hopkins, J.D. Seagrave, R.H. Sherman, E.C. Kerr, Elastic scattering of neutrons from He between 7.9 and 23.7 MeV. Phys. Rev. C **9**, 179 (1974)
52. J.H. Gibbons, R.L. Macklin, Total neutron yields from light elements under proton and alpha bombardment. Phys. Rev. **114**, 571 (1959)
53. R. Kankowsky, J.C. Fritz, K. Kilian, A. Neufert, D. Fick, Elastic scattering of polarized protons on tritons between 4 and 12 MeV. Nucl. Phys. A **263**, 29 (1976)
54. J.E. Brolley, T.M. Putnam, L. Rosen, L. Stewart, Hydrogen-helium isotope elastic scattering processes at intermediate energies. Phys. Rev. **117**, 1307 (1960)
55. J.L. Detch, R.L. Hutson, N. Jarmie, J.H. Jett, Accurate measurements of the nuclear processes T(p, p)T, T(p,^{3}He)n, T(p, d)D, and T(**p**, p)T from 13 to 20 MeV. Phys. Rev. C **4**, 52 (1971)
56. R. Darves-Blanc, N. Sen, J. Arvieux, A. Fiore, J. Gondrand, G. Perrin, Elastic scattering of polarized protons by3H between 19 and 57 MeV. Lett. Nuovo Cimento **4**, 16 (1972)
57. R. Hardekopf, P. Lisowski, T. Rhea, R. Walter, T. Clegg, Four-nucleon studies: (IV). Analyzing powers in the elastic scattering of protons from tritium. Nucl. Phys. A **191**, 481 (1972)
58. A. Deltuva, A.C. Fonseca, Ab initio four-body calculation of n-^{3}He, p-^{3}H, and d-d scattering. Phys. Rev. C **76**, 021001(R) (2007)
59. A. Deltuva, A.C. Fonseca, Calculation of multichannel reactions in the four-nucleon system above breakup threshold. Phys. Rev. Lett. **113**, 102502 (2014)
60. A. Deltuva, A.C. Fonseca, Calculation of neutron-^{3}He scattering up to 30 MeV. Phys. Rev. C **90**, 044002 (2014)
61. A. Deltuva, A. Fonseca, Deuteron-deuteron scattering above four-nucleon breakup threshold. Phys. Lett. B **742**, 285 (2015)
62. A. Deltuva, A.C. Fonseca, Proton-^{3}H scattering calculation: elastic and charge-exchange reactions up to 30 MeV. Phys. Rev. C **91**, 034001 (2015)
63. J. Esterline, W. Tornow, A. Deltuva, A.C. Fonseca, Analyzing power $A_y(\theta)$ of **n**-^{3}He elastic scattering between 1.60 and 5.54 MeV. Phys. Rev. Lett. **110**, 152503 (2013)
64. W.E. Wilson, R.L. Walter, D.B. Fossan, Differential cross sections for the T(p, n)3He reaction. Nucl. Phys. **27**, 421 (1961)
65. M. Drosg, Unified absolute differential cross sections for neutron production by the hydrogen isotopes for charged-particle energies between 6 and 17 MeV. Nucl. Sci. Eng. **67**, 190 (1978), in *EXFOR Database* (NNDC, Brookhaven, 1978)
66. N. Jarmie, J.H. Jett, Neutron source reaction cross sections. Phys. Rev. C **16**, 15 (1977)
67. R.G. Allas, L.A. Beach, R.O. Bondelid, E.M. Diener, E.L. Petersen, J.M. Lambert, P.A. Treado, I. Slaus, Mechanisms in the reactions 3H+3He and 3H+2H and and measurement of the reaction 3H(p, n)3He from 13.6 to 32.8 MeV. Phys. Rev. C **9**, 787 (1974)
68. J.J. Jarmer, R.C. Haight, J.E. Simmons, J.C. Martin, T.R. Donoghue, Analyzing-power measurements in the ^{3}H(p, n)^{3}He reaction. Phys. Rev. C **9**, 1292 (1974)
69. J.J. Jarmer, R.C. Haight, J.C. Martin, J.E. Simmons, Polarization and polarization transfer in the reaction 3H(p, n)3He at 5.97 and 9.9 MeV. Phys. Rev. C **10**, 494 (1974)
70. R.C. Haight, J.E. Simmons, T.R. Donoghue, Polarization and polarization transfer in the reaction T(p, n)3He. Phys. Rev. C **5**, 1826 (1972)
71. A. Deltuva, A.C. Fonseca, Four-body calculation of elastic deuteron–deuteron scattering. Phys. Rev. C **92**, 024001 (2015)
72. A. Wilson, M. Taylor, J. Legg, G. Phillips, The elastic scattering of deuterons by deuterium. Nucl. Phys. A **126**, 193 (1969)
73. L. Rosen, J.C. Allred, d-d Elastic scattering at 10.5 Mev. Phys. Rev. **88**, 431 (1952)
74. N. Jarmie, J.H. Jett, Various cross sections for A $\leq$ 3 nuclei. Phys. Rev. C **10**, 54 (1974)
75. A. Okihana et al., The D(d,d)D, D(d,p)T and D(d,3He)N reactions at 13.2 MeV. Jpn. Phys. J. **46**, 707 (1979)
76. H. Itoh, Phenomenological potential for d-d elastic scattering at 14.2 MeV. Prog. Theor. Phys. **39**, 1361 (1968)

77. W.T.H. Van Oers, H. Arnold, K.W. Brockman Jr., Elastic scattering of deuterons by deuterons at 25.3 MeV. Nucl. Phys. **46**, 611 (1963)
78. W. Grüebler, V. König, P.A. Schmelzbach, R. Risler, R.E. White, P. Marmier, Investigation of excited states of 4He via the 2H(d, p)3H and 2H(d, n)3He reactions using a polarized deuteron beam. Nucl. Phys. A **193**, 129 (1972)
79. W. Grüebler, V. König, P.A. Schmelzbach, B. Jenny, J. Vybiral, New highly excited 4He levels found by the 2H(d, p)3H reaction. Nucl. Phys. A **369**, 381 (1981)
80. P. Guss, K. Murphy, R. Byrd, C. Floyd, S. Wender, R. Walter, T. Clegg, W. Wylie, The analyzing power for the 2H(d, n)3He reaction from 5.5 to 11.5 MeV. Nucl. Phys. A **395**, 1 (1983)
81. V. König, W. Grüebler, R.A. Hardekopf, B. Jenny, R. Risler, H. Brgi, P. Schmelzbach, R. White, Investigation of charge symmetry violation in the mirror reactions 2H(d, p)3H and 2H(d, n)3He. Nucl. Phys. A **331**, 1 (1979)
82. R. Hardekopf, P. Lisowski, T. Rhea, R. Walter, T. Clegg, Four-nucleon studies: (III). Measurement of the 2H(d, p)3H polarization via the 3H(p, d)2H reaction and comparison to polarization in 2H(d, n)3He. Nucl. Phys. A **191**, 468 (1972)
83. G. Spalek, R. Hardekopf Jr., J.T. Stammbach, R. Walter, Four-nucleon studies: (I). A redetermination of the neutron polarization for the 2H(d, n)3He reaction from 6 to 14 MeV. Nucl. Phys. A **191**, 449 (1972)
84. G. Salzman, G.G. Ohlsen, J. Martin, J. Jarmer, T. Donoghue, Polarization and polarization transfer in the 2H(d, n)3He reaction at 10 MeV. Nucl. Phys. A **222**, 512 (1974)

Few-Body Syst (2017) 58:1–12
DOI 10.1007/s00601-016-1167-0

CrossMark

Sergio Deflorian · Victor D. Efros · Winfried Leidemann

Calculation of the Astrophysical S-Factor S_{12} with the Lorentz Integral Transform

Received: 29 July 2016 / Accepted: 13 October 2016 / Published online: 9 December 2016
© Springer-Verlag Wien 2016

Abstract The LIT approach is tested for the calculation of astrophysical S-factors. As an example the S-factor of the reaction $^2\mathrm{H}(p, \gamma)^3\mathrm{He}$ is considered. It is discussed that a sufficiently high density of LIT states at low energies is necessary for a precise determination of S-factors. In particular it is shown that the hyperspherical basis is not very well suited for such a calculation and that a different basis system is much more advantageous. A comparison of LIT results with calculations, where continuum wave functions are explicitly used, shows that the LIT approach leads to reliable results. It is also shown how an error estimate of the LIT inversion can be obtained.

1 Introduction

The study of stellar nucleosynthesis is one of the central issues of nuclear astrophysics. In order to understand the details of this process it is necessary to have a precise determination of a large number of reaction cross sections at relatively low energies. Considering for example the solar proton–proton cycle and taking into account that the temperature of the core of the sun is about 1.5×10^7 K one finds that the relevant energies are below 100 keV [1]. At such low energies cross sections can become extremely small, in particular in presence of a Coulomb barrier between the reacting particles. In many cases data have been obtained only at higher energies, which makes extrapolations to lower energies necessary. Therefore it is very helpful to have additional input from the theory side, especially calculations with ab initio methods [2,3] employing modern realistic nuclear forces can help to reduce error estimates for cross sections.

Among the relevant nuclear reactions of astrophysical interest there are many electroweak processes. Concerning such kind of reactions the Lorentz integral transform (LIT) [4] is a particularly interesting ab initio method, since it reduces a continuum-state problem to a much simpler to solve bound-state like problem, however, involves an inversion of the transform [5–7]. In the past the LIT was applied to quite a number of

---

This article belongs to the special issue "30th anniversary of Few-Body Systems".

S. Deflorian · W. Leidemann (✉)
Dipartimento di Fisica, Università di Trento, 38123 Trento, Italy
E-mail: leideman@science.unitn.it

S. Deflorian · W. Leidemann
Istituto Nazionale di Fisica Nucleare, TIFPA, 38123 Trento, Italy

V. D. Efros
National Research Centre "Kurchatov Institute", 123182 Moscow, Russia

V. D. Efros
National Research Nuclear University MEPhI (Moscow Engineering Physics Institute), Moscow, Russia

reactions [8,9], where in most cases the bound-state methods of choice were expansions in hyperspherical harmonics (HH). Up to today the LIT was never applied to calculations of cross sections relevant in stellar nucleosynthesis. In fact extremely small low-energy cross sections are a challenge for the method because of the above mentioned LIT inversion. In such a scenario one needs a rather high density of LIT states in the low-energy region in order to have a sufficient resolution of the LIT. That such a request can be problematic became evident in recent LIT calculations for the ^{4}He isoscalar monopole resonance [10,11], where the effective interaction HH expansion technique [12–14] was applied. On the one hand the resonance strength was successfully determined, on the other hand the resonance width could not be computed since the density of LIT states was much too low in the resonance region. In [15] it was then shown that with a four-body hybrid basis, consisting of a three-body HH basis plus a single-particle basis, one obtains a much higher density of LIT states in the ^{4}He isoscalar monopole resonance region, which is located below the three-body breakup threshold.

The aim of the present paper is to check whether the LIT method succeeds to reliably determine the low-energy cross section in presence of a Coulomb barrier. To this end we have chosen to calculate the S-factor S_{12} of the reaction ^{2}H$(p, \gamma)^3$He. A positive outcome of the check would allow to apply the LIT method also for the calculation of S-factors involving a higher number of nucleons. The calculation is carried out in two different ways: (i) via the LIT method and (ii) with the explicit calculation of the d-p continuum wave function. For this check it is not necessary to use a realistic nuclear force, therefore we take the central MT-I/III potential [16] as NN interaction, however we would like to mention that S_{12} was calculated in rather complete ab initio calculations [17,18].

The paper is organized as follows. After the definition of the S-factor S_{12} in Sect. 2, in Sect. 2.1 the LIT approach for the calculation of the S-factor is described. Since we want to determine the S-factor also in the conventional way, in Sect. 2.2 we discuss the calculation of continuum states with the Kohn variational principle. Section 3 contains a detailed study of the LIT method. It is shown that the density of LIT states in the low-energy region depends significantly on the basis system chosen for the solution of the LIT equation. The section closes with a comparison of LIT and conventional results for the low-energy ^{3}He photodisintegration cross section and S-factor S_{12} and with a brief summary.

2 Calculation of the S-Factor S_{12}

The S-factor S_{12} is defined as follows

$$S_{12}(E) = \sigma_{\mathrm{cap}}\, E\, \exp(2\pi\eta)\,, \tag{1}$$

where σ_{cap} is the cross section of the reaction $d + p \rightarrow {}^3\mathrm{He} + \gamma$, E denotes the relative energy of the deuteron-proton pair, and $\exp(2\pi\eta)$ is the Gamow factor taking into account the effect of the Coulomb barrier with

$$\eta = \sqrt{\frac{\mu c^2}{2E}}\,\alpha\,, \tag{2}$$

where μ is the reduced mass of the deuteron-proton pair and α is the fine structure constant.

We determine σ_{cap} by first calculating the cross section σ_γ of the inverse reaction ^{3}He$+\gamma \rightarrow d + p$ and then using the relation

$$\sigma_{\mathrm{cap}}(E) = \frac{2E_\gamma^2}{3k^2}\sigma_\gamma(E_\gamma)\,, \tag{3}$$

where E_γ is the photon energy and k denotes the relative momentum of the deuteron-proton pair. The photodisintegration cross section of ^{3}He is calculated in unretarded dipole approximation,

$$\sigma_\gamma(E_\gamma) = 4\pi^2\alpha E_\gamma R(E_\gamma)\,, \tag{4}$$

where

$$R(E_\gamma) = \int df\, |\langle f|D_z|0\rangle|^2 \delta\left(E_f - E_0 - E_\gamma\right) \tag{5}$$

is the dipole response function. In Eq. (5) $|0\rangle$ and $|f\rangle$ are the ^{3}He ground state and the deuteron-proton final state, respectively, while E_0 and E_f are the corresponding eigenenergies. Finally, D_z is the third component of the nuclear dipole operator.

As mentioned in the introduction we calculate $R(E_\gamma)$ in two different ways: (i) with the LIT approach, where bound-state methods can be used, and (ii) with the explicit calculation of the continuum state $|f\rangle$. Both methods are described briefly in the following two subsections.

2.1 Calculation with LIT Approach

The LIT of the response function $R(E_\gamma)$ is defined as follows

$$L(\sigma) = \int dE_\gamma\, \mathcal{L}(E_\gamma, \sigma)\, R(E_\gamma)\,, \tag{6}$$

where the kernel $\mathcal{L}$ is a Lorentzian with a width of $2\sigma_I$, which is located at $E_\gamma = \sigma_R$:

$$\mathcal{L}\left(E_\gamma, \sigma = \sigma_R + i\sigma_I\right) = \frac{1}{(E_\gamma - \sigma_R)^2 + \sigma_I^2}\,. \tag{7}$$

In fact the width can in principle be adjusted to resolve the detailed structure of $R(E_\gamma)$ and due to the variable width the LIT is a transform with a controlled resolution. However, an increase of the resolution by a reduction of σ_I does not come for free and it requires in general an increase of the precision of the calculation.

The LIT $L(\sigma)$ is calculated by solving the following equation

$$(H - E_0 - \sigma)\, |\tilde{\Psi}(\sigma)\rangle = D_z|0\rangle\,, \tag{8}$$

where H is the Hamiltonian of the particle system under consideration. The solution $\tilde{\Psi}(\sigma)$ is localized, since the rhs of Eq. (8) is asymptotically vanishing. Therefore one can determine $\tilde{\Psi}(\sigma)$ using bound-state methods. The solution directly leads to the transform:

$$L(\sigma) = \langle\tilde{\Psi}(\sigma)|\tilde{\Psi}(\sigma)\rangle\,. \tag{9}$$

Finally, the response function $R(E_\gamma)$ is obtained from the inversion of the transform (for details see [5–8]).

Here we solve the LIT Eq. (8) via an expansion on a complete basis, where the number of basis functions N is increased up to a sufficient convergence. One can understand such an expansion for the solution of the LIT equation as follows. The spectrum of the Hamiltonian for the basis is determined, thus one has N eigenstates ϕ_n with eigenenergies E_n. The LIT solution assigns to any eigenenergy E_n a LIT state, which is a Lorentzian with strength S_n and width $2\sigma_I$. The strength S_n depends on the source term on the rhs of the LIT equation:

$$S_n = |\langle\phi_n|D_z|0\rangle|^2\,. \tag{10}$$

The LIT result is then just given by the the sum over the N LIT states:

$$L(\sigma) = \sum_{n=1}^{N} \frac{S_n}{(\sigma_R - (E_n - E_0))^2 + \sigma_I^2}\,. \tag{11}$$

From the equation above it is evident that at a given resolution of the LIT, which is characterized by the value of σ_I, one needs a sufficient density of LIT states as discussed in detail in Ref. [15]. There it is illustrated that the density of LIT states is not only correlated to the number of basis functions N, but depends also on the specific basis. For example, for the electromagnetic ^{4}He breakup it was discussed that it is very difficult to increase the density of LIT states below the three-body breakup for a hyperspherical harmonics (HH) basis. As is discussed in the following section a similar problems occurs at use of the HH basis also in the three-body case considered in the present work. At this point we would like to emphasize that the LIT contains in general for a generic electroweak reaction the full response function R with all breakup channels and one may use any complete localized A-body basis set for the calculation of the LIT. On the other hand one has to have in mind that in a given energy range one basis set can be more advantageous than another one.

In order to take into account the findings of [15] we use for the LIT calculation two different basis systems. A HH basis with two-body correlations of the Jastrow type as was done for the same NN potential in [19].

For the second basis we use the two Jacobi coordinates of the three-body system in an explicit way, therefore this basis will be called Jacobi basis. The spatial part of this basis starts from the following definition

$$
\psi_{n_1, n_2, l_1, l_2} = \sum_{m_1 m_2} \mathcal{R}_{n_1}^{[1]}(\eta_1) Y_{l_1}^{m_1}(\theta_1, \phi_1)
$$
$$
\times \mathcal{R}_{n_2}^{[2]}(\eta_2) Y_{l_2}^{m_2}(\theta_2, \phi_2) \langle L = 1 \, M | l_1 m_1 l_2 m_2 \rangle, \tag{12}
$$

where $\boldsymbol{\eta}_1 = (\eta_1, \theta_1, \phi_1)$ is the relative ("pair") coordinate of particles 1 and 2, $\boldsymbol{\eta}_2 = (\eta_2, \theta_2, \phi_2)$ is the single-particle coordinate of the third particle with respect to the center of mass of particles 1 and 2, the $Y_l^m(\theta, \phi)$ are spherical harmonics and $\langle L = 1 \, M | l_1 m_1 l_2 m_2 \rangle$ denotes a Clebsch–Gordan coefficient (note that because of the dipole operator in Eq. (5) one needs only basis states with angular momentum $L = 1$). The radial functions $\mathcal{R}_n^{[1,2]}(\eta)$ are defined as follows

$$
\mathcal{R}_{n_1}^{[1]}(\eta_1) = \sqrt{\frac{n_1!}{(n_1 + 2)!}} L_{n_1}^{(2)}\left(\frac{\eta_1}{b_1}\right) e^{-\frac{\eta_1}{2b_1}} b_1^{-\frac{3}{2}} \tag{13}
$$

$$
\mathcal{R}_{n_2}^{[2]}(\eta_2) = \sqrt{\frac{n_2!}{(n_2 + 2)!}} L_{n_2}^{(2)}\left(\frac{\eta_2}{b_2}\right) e^{-\frac{\eta_2}{2b_2}} b_2^{-\frac{3}{2}} , \tag{14}
$$

where $L_{n_i}^{(2)}$ is a Laguerre polynomial of order n_i ($n_i \in \{0, 1, 2, \ldots, N_i - 1\}$) with parameter b_i. This is very similar to our expansions of the HH hyperradial function R_n, in fact, in this case we have

$$
\mathcal{R}_n(\rho) = \sqrt{\frac{n!}{(n + 5)!}} L_n^{(5)}\left(\frac{\rho}{b}\right) e^{-\frac{\rho}{2b}} b^{-\frac{3}{2}} , \tag{15}
$$

where ρ is the hyperradius and $n \in \{0, 1, 2, \ldots, N - 1\}$.

Including the spin-isospin part to $\psi_{n_1, n_2, l_1, l_2}$ of Eq. (12) one has

$$
\phi_{n_1, n_2, l_1, l_2, s_{12}, t_{12}} = \psi_{n_1, n_2, l_1, l_2} \chi^S(s_{12}) \chi^T(t_{12}), \tag{16}
$$

where the spin and isospin functions $\chi^S(s_{12})$ and $\chi^T(t_{12})$ are defined to have spin s_{12} and isospin t_{12} equal to 1 or 0 for the first two particles and total spin and isospin $S = \frac{1}{2}$ and $T = \frac{1}{2}$. A totally antisymmetric basis state is given by

$$
\Phi_{n_1, n_2, l_1, l_2, s_{12}, t_{12}} = \mathcal{A} \, \phi_{n_1, n_2, l_1, l_2, s_{12}, t_{12}}, \tag{17}
$$

where $\mathcal{A}$ is a proper antisymmetrization operator.

2.2 Explicit Calculation of the Continuum States

To obtain the deuteron–proton final states entering Eq. (5) we apply the version [20,21] of the general trial function approach which employs the HH expansion. The continuum wave function is written as $\Psi_f = X + Y$ where at large distances the Y component represents the two–body asymptotics of Ψ_f. The X component is an expansion over HH. At energies below the three–body breakup threshold it vanishes at large distances and above the threshold it reproduces the three–body breakup asymptotics in the absence of the Coulomb interaction.

Our calculation refers to the former case. One sets

$$
X = \sum_{i=1}^{i_{max}} c_i \psi_i \tag{18}
$$

where ψ_i are basis functions. They are the sums that are antisymmetric with respect to nucleon permutations of products of correlated hyperspherical harmonics mentioned above and spin–isospin functions, times the Laguerre type hyperradial basis functions (15). The c_i expansion coefficients are to be determined.

The Y component is of the form $Y_R + Y_I \tan \delta$ where δ is the trial scattering phase shift. The functions $Y_{R,I}$ are of the form $\mathcal{A}\varphi(12, 3)$ where 1, 2, and 3 are the nucleon numbers, and $\mathcal{A}$ is the antisymmetrization operator. The function $\varphi(12, 3)$ is the product of a channel function with a given spin and isospin of a system and a relative motion function pertaining to a given orbital momentum L. The channel function is obtained by coupling the deuteron wave function of the nucleons 1 and 2 and the spin–isospin function of the nucleon 3. The relative motion function is the product of the spherical harmonics and a radial function.

The radial function is $(kr)^{-1}F_L(kr, \eta)$ in the case of Y_R, and $g_L(r)(kr)^{-1}G_L(kr, \eta)$ in the case of Y_I. Here F_L and G_L are the regular and irregular Coulomb functions, and $g_L(r)$ is a correction factor. It is to be taken such that $g_L(r)$ turns to unity beyond the interaction region and $g_L G_L$ is regular and behaves e.g. like F_L at $r \to 0$. In our $L = 1$ case we used $g_1(r) = [1 - \exp(-r/r_0)]^3$, r_0 being a scale parameter, which is of the same form as in [17]. The results vary little in a broad range of r_0 values when convergence is achieved.

The above trial wave function may be written as

$$\Psi_f = \sum_{i=0}^{i_{max}} c_i \psi_i + Y_R \tag{19}$$

where $\psi_0 = Y_I$ and $c_0 = \tan \delta$. The system of equations

$$\sum_{i=0}^{i_{max}} \langle \psi_j | H - E | \psi_i \rangle c_i = -\langle \psi_j | (H - E) Y_R \rangle \tag{20}$$

with $j = 0, \ldots, i_{max}$ was used to obtain the c_i coefficients. These equations emerge in particular from the requirement for the Kohn functional to be stationary.

At a given i_{max} value, the quality of the wave function thus obtained apparently deteriorates when the energy approaches the eigenvalues of the $\langle \psi_j | H | \psi_i \rangle$ matrix. Corresponding vicinities of the eigenvalues in which results are unsatisfactory normally are narrow as compared to distances between the eigenvalues [22]. The least–square method involving in addition to Eq. (20) the equations of the same form with j exceeding i_{max} may cure the deficiency [23]. In our low–energy case, Eq. (20) did not lead to problems in the range of i_{max} considered so that the convergence trends of the results do not seem to depend on energy.

Let us denote $\delta \Psi_f$ the deviation of the approximate Ψ_f wave function from the exact one. The difference between the exact $\tan \delta$ and its value that pertains to an approximate Ψ_f may be represented as an integral with the integrand containing $\delta \Psi_f$ linearly. In the difference to this, the deviation of the Kohn functional value from the exact $\tan \delta$ is quadratic with respect to $\delta \Psi_f$. Thus the value of the Kohn functional is more accurate than obtained directly from Eq. (20) when a calculation is close to convergence. We replaced $c_0 \equiv \tan \delta$ with the value of the Kohn functional in the Eq. (20) with $j \geq 1$ to get the rest $c_1, \ldots, c_{i_{max}}$ coefficients such that the equations are satisfied. However, these coefficients are not necessarily more accurate than those obtained directly from Eq. (20).

For checking purposes we compared our P wave phase shifts with those obtained with the same MT-I/III potential by the Pisa group [24]. Their scattering calculations are known to be of a high precision [25]. The differences found between the Kohn functional values of the phase shifts are about 0.5% or less [26].

3 Discussion of Results

We start the discussion illustrating first results, where the HH basis is used for the calculation of the LIT of the ^{3}He photodisintegration. We consider only the final state in the isospin $T = 1/2$ channel, since the $T = 3/2$ channel corresponds exclusively to a three-body breakup. In Fig. 1 we show results for various values of σ_I. One sees that with $\sigma_I = 20\,\text{MeV}$ a smooth transform is obtained, then with an increase of the resolution to $\sigma_I = 2.5\,\text{MeV}$ the transform starts to have an oscillating behaviour beyond $20\,\text{MeV}$, and a still further increase of the resolution to $\sigma_I = 0.5\,\text{MeV}$ exhibits the underlying structure of the single LIT states [see Eq. (11)]. From the last result one can conclude that the resolving power of the LIT is certainly not just given by the chosen σ_I value.

For a higher degree of resolution one has to increase the density of LIT states, which can be achieved in two ways, namely by increasing the number of basis functions and by enhancing the b parameter of the hyperradial wave function of Eq. (15). Both measures are taken for the results shown in Fig. 2, where we illustrate the low-energy part of the LIT for rather small σ_I values. It is evident that the density of LIT states grows as expected. In Fig. 2d one observes a rather high LIT state density and one could easily further increase the density. However, one readily sees that there is not a single LIT state below the three-body breakup threshold

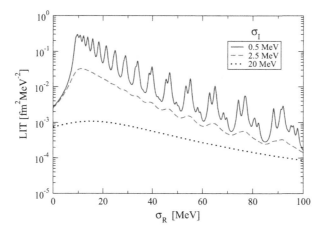

Fig. 1 LIT of the ^{3}He dipole response function for $T = 1/2$ with $\sigma_I = 0.5$, 2.5 and 20 MeV calculated with an HH basis of 30 hyperspherical and 31 hyperradial states for a total of 930 basis states ($b = 0.3$ fm)

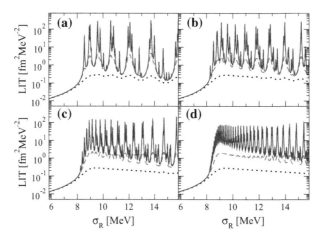

Fig. 2 Low-energy part of the LIT of Fig. 1 with $\sigma_I = 0.01$ (*full*), 0.1 (*dashed*) and 0.5 MeV (*dotted*), results with different HH basis systems: **a** same basis as in Fig. 1, **b** 40 hyperspherical and 51 hyperradial states ($b = 0.3$ fm), **c** as in (**b**), but with $b = 0.5$ fm, **d** 40 hyperspherical and 76 hyperradial states ($b = 1$ fm)

at about 8 MeV (^{3}He binding energy with MT-I/III potential). On the other hand the calculation of the S-factor S_{12} requests energies just beyond the two-body breakup threshold at 5.8 MeV (difference of binding energies of ^{2}H and ^{3}He for the MT-I/III potential). Thus one cannot expect that an inversion of the LITs of Fig. 2 leads to a high-precision result in the region of astrophysical relevance. Here a comment is in order concerning the use of a more realistic nuclear force for a calculation of the LIT with the HH basis. In this case one can find a few LIT states below the three-body breakup threshold, but also there one encounters the problem of further increasing the density of LIT states in a systematic way in order to obtain a smooth LIT with a sufficiently small σ_I [10,11,15].

Now we turn to the results with the Jacobi basis. Since in principle we are only interested in the cross section just above the two-body breakup threshold we only consider S-wave interaction for the pair coordinate, this means that in Eq. (12) only basis states with $l_1 = 0$ and $l_2 = 1$ are taken into account (as already mentioned for the dipole response we have $L = 1$). For the radial parts of the pair and single-particle wave functions we choose $b_1 = 0.75$ fm and $b_2 = 0.5$ fm, respectively.

In Fig. 3 we show the LIT for the cases that the pair in ϕ of Eq. (16) is solely in a $^3S_1$-state and the additional effect when also $^1S_0$-states are allowed. One sees that the contribution due to the $^1S_0$-states is quite tiny. In fact a rather small number of basis states with the pair in the $^1S_0$ state ($N_1 = 5$, $N_2 = 19$) is sufficient in order to obtain convergence. As shown in Fig. 4 the convergence of the main LIT contribution due to the $^3S_1$-states is not as rapid as in case of the $^1S_0$ states. On the one hand one needs only a rather moderate value for N_1 of about 20 to obtain a sufficient convergence in the pair coordinate as shown in Fig. 4b (note a result with $N_1=24$

 Springer

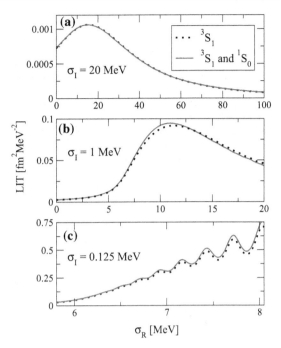

Fig. 3 LIT of the ^{3}He dipole response function for $T = 1/2$ with $\sigma_I = 20$, 1 and 0.125 MeV calculated with the Jacobi basis taking into account only $^3S_1$ states in the pair coordinate of ϕ (*dotted*) and in addition also $^1S_0$-states (*full*)

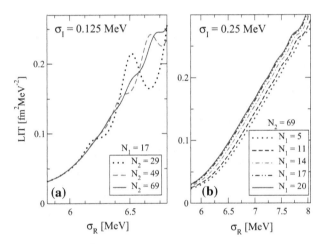

Fig. 4 Convergence pattern for the LIT of the ^{3}He dipole response function for $T = 1/2$ calculated with the Jacobi basis: **a** $N_1 = 17$ and various N_2 values ($\sigma_I = 0.125$ MeV) and **b** $N_2 = 69$ and various N_1 values ($\sigma_I = 0.25$ MeV)

could not be distinguished in the figure from the $N_1 = 19$ result). On the other hand the situation is different for the single-particle coordinate (see Fig. 4a). In order to have a sufficiently convergent LIT in the region just above the two-body breakup threshold with a small σ_I value of 0.125 MeV one has to go up to an N_2 of about 70. In fact for our calculation of the *S*-factor S_{12} we use $N_2 = 79$.

It is interesting to observe the different effects of an increase of N_1 and N_2. The enhancement of basis states for the pair coordinate in Fig. 4b shifts the transform to lower energies without changing the shape of the LIT. This corresponds to an energy shift of the low-energy LIT states to lower energies without a notable change of the density. On the contrary the increase of basis states for the single-particle coordinate (Fig. 4a) leads to a smoother result of the transform due to an increased density of LIT states.

In Fig. 5 we compare the low-energy LIT calculated with HH and Jacobi basis systems for various σ_I values. Note that different from the case with the Jacobi basis, where only *S*-wave interaction is taken into account, for the HH basis also interaction in higher partial waves is considered, however, the contribution of

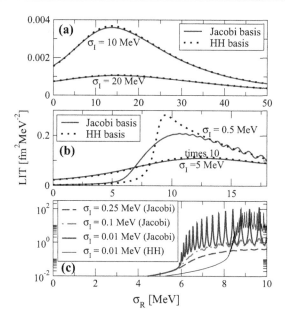

Fig. 5 Comparison of the LIT calculated with HH and Jacobi basis systems with various σ_I values as indicated in the figure

the latter should be quite small. In fact for large σ_I (see Fig. 5a) one can hardly find any difference between both results. Even for $\sigma_I = 5$ MeV, shown in Fig. 5b, the results are rather similar, whereas a decrease of σ_I to 0.5 MeV, also shown in Fig. 5b, exhibits quite some difference: the peak of the LIT of the HH basis is considerably more pronounced than that of the Jacobi basis. To a large extent the difference is caused by the missing LIT states at low energy for the HH basis and not by the additional interaction in higher partial waves. Thus one may conclude that the lack of low-energy LIT states leads to a shift of low-energy strength to the peak region region just above the two-body breakup threshold.

The energy distribution of low-energy LIT states for both basis systems is nicely illustrated in Fig. 5c for $\sigma_I = 0.01$ MeV. Only for the Jacobi basis one finds LIT states directly above the two-body breakup threshold. The LIT state density is so high that one obtains a smooth LIT in the very threshold region even with $\sigma_I = 0.1$ MeV and up to the three-body breakup threshold with $\sigma_I = 0.25$ MeV.

In order to determine the cross section σ_γ one has to invert the calculated transforms. With regard to the aim to determine the S-factor it is evident that close to the threshold region one wants to work with a high resolution, however, one has to take into account that with a small σ_I value one does not obtain a smooth LIT at higher energies because the density of LIT states decreases with growing energy. In fact it is better to work with an energy dependent σ_I. Therefore we divide the σ_R range in various intervals $[E_j, E_{j+1}]$ ($j = 1, 2, 3, \ldots, J$) and take in this interval $\sigma_I = \sigma_{I,j}$. Considering that we have calculated the LIT for a certain number of σ_R points $\sigma_{R,k}$ ($k = 1, 2, 3, \ldots, K$) we rescale the LIT for all $\sigma_{R,k} \geq E_{j+1}$ by the factor

$$f(j+1) = \frac{L\left(\sigma_{k_2(j)}, \sigma_{I,j}\right)}{L\left(\sigma_{k_1(j+1)}, \sigma_{I,j+1}\right)}, \tag{21}$$

where $\sigma_{k_1(j)}$ is the lowest and $\sigma_{k_2(j)}$ the highest σ_R value in interval $[E_j, E_{j+1}]$. Note that this is made in a cumulative way, thus for the LIT in the last interval ($\sigma_R \in [E_{J-1}, E_J]$) we have the total factor $F = f(2)f(3)\ldots f(J)$. The values we have chosen for E_j and $\sigma_{I,j}$ are given in Table 1. The application of Eq. (21) and the definitions given in Table 1 define a new transform $\mathcal{L}$.

In Fig. 6 we show the newly defined transform $\mathcal{L}(\sigma)$, where we use a Jacobi basis with ($N_1 = 24$, $N_2 = 79$) and ($N_1 = 5$, $N_2 = 19$) for the $^3S_1$ states and the $^1S_0$ states, respectively. The dashed curve in the figure shows the LITs L for the various energy intervals without any additional factor, whereas the continuous curve corresponds to the result when the additional factors of Eq. (21) are introduced. Note that according to the definition of the $f(j)$ the derivative of $\mathcal{L}$ seems to be not continuous, but actually this is not the case since the transform is only defined pointwise in K σ_R points. In principle one could also work with the transform described by the dashed curve in Fig. 6, but this would mean that the impact of the transform is reduced with growing energy. The rescaling simulates the case where the transform is calculated with a single σ_I.

Table 1 E_j and $\sigma_{I,j}$ values for the definition of the new transform $\mathcal{L}$

j	E_j (MeV)	$\sigma_{I,j}$ (MeV)
1	5.7	0.125
2	6.15	0.175
3	6.55	0.35
4	8.05	0.7
5	10.55	1.1
6	13.05	1.5
7	23.05	5
8	58.05	10
9	108.05	20
10	308.05	–

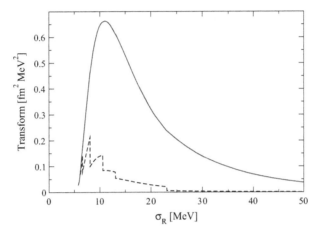

Fig. 6 The LITs L with $\sigma_I = \sigma_{I,j}$ in the various energy intervals $[E_j, E_{j+1}]$ (*dashed*) and the new transform $\mathcal{L}$ (*full*) as described in the text

For the inversion we use our standard method, where the response function R is expanded as follows

$$R\left(E_\gamma = E + E_{thr}\right) = \sum_{n=1}^{N} c_n g_n(E), \qquad (22)$$

where E is defined as in Eq. (1) and E_{thr} is the energy of the two-body breakup threshold. In order to consider the effect of the Coulomb barrier we include the Gamow factor of Eq. (1) taking

$$g_n(E) = \exp(-2\pi\eta) \exp[(-\alpha E)/n], \qquad (23)$$

where α is a non-linear parameter. The various $g_n(E)$ are then transformed numerically to the σ-space according to the LIT transformation given in Eq. (6) for the response function. Note that in case of the transform $\mathcal{L}$ the factors $f(j)$ of Eq. (21) have to be taken properly into account. In this way one obtains a set of functions $\tilde{g}_n(\sigma)$ which are then used for the expansion of the transform, here given for the case of $\mathcal{L}$,

$$\mathcal{L}(\sigma) = \sum_{n=1}^{N} c_n \tilde{g}_n(\sigma). \qquad (24)$$

For given values of N and α of Eqs. (22) and (23) a best fit to the calculated $\mathcal{L}$ is made, which determines the coefficients c_n. Varying then only the non-linear parameter α over a wide range values one obtains the absolute best fit for a specific N. Then one repeats the procedure increasing N by one. A stable inversion result should be obtained in a range $N_A \leq N \leq N_B$.

In Fig. 7 we show inversion results of $L(\sigma)$ for the HH basis and of $\mathcal{L}$ for the Jacobi basis. The parameters for the HH basis are the same as defined in caption of Fig. 2d, for the Jacobi basis we use the new transform $\mathcal{L}$ with the setting ($N_1 = 24$, $N_2 = 79$) and ($N_1 = 5$, $N_2 = 19$) for $^3S_1$- and $^1S_0$-states in the pair coordinate, respectively. Note that for the HH basis we take $\sigma_I = 20\,\text{MeV}$. We do not choose a higher resolution otherwise

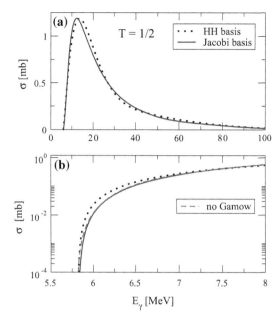

Fig. 7 The ^{3}He cross section σ_γ of Eq. (4) obtained from inversions of $L(\sigma_R, \sigma_I = 20\,\text{MeV})$ with HH basis (*dotted*) and of $\mathcal{L}(\sigma)$ with Jacobi basis (*full*); in **b** also shown inversion of $\mathcal{L}(\sigma)$ with Jacobi basis with factor $E^{3/2}$ instead of Gamow factor in functions g of Eq. (23) (*dashed*)

the inversion could be hampered too much by the fact that the low-energy strength is shifted to the peak region. Due to this misplaced strength one cannot expect that the two inversion results are extremely close to each other. On the other hand, as Fig. 7a shows, differences remain rather small. The peak heights are almost identical, but the peak of the HH basis is shifted somewhat to higher energies.

It is a bit surprising that the low-energy cross sections are not completely different (see Fig. 7b), but this is due to the correct implementation of the Gamow factor in the set of functions g for the inversion. It is interesting to check the effect of an inversion of $\mathcal{L}$, where the Gamow factor in Eq. (23) is replaced by the factor $E^{3/2}$ (correct threshold behaviour without Coulomb barrier). Although we have a high-precision transform for the Jacobi basis the inversion without the Gamow factor does not lead to the correct threshold behaviour, but at least coincides with the proper inversion result above about 6 MeV.

In Fig. 8 we show a comparison of the LIT result with that of a calculation with explicit continuum wave functions. In the upper panel the ^{3}He photodisintegration cross section σ_γ is depicted. It is evident that there is an excellent agreement between both results. However, because of the strong fall-off of σ_γ close to the breakup threshold it is difficult to understand the level of agreement in this energy range. This can be estimated much better for the S-factor since the Gamow factor is divided out. In Fig. 8b one finds also in this case a very good agreement between both calculations. It is worthwhile to mention that we find quite stable inversion results $11 \leq N \leq 18$, where N is the number of basis function used for the inversion [see Eqs. (22) and (23)]. This enables us to make the following error estimate for the LIT inversion. We take the inversions for $N = 11$ ($F_{\text{inv},11}(E)$) up to $N = 18$ ($F_{\text{inv},18}(E)$) and first determined an average inversion result $\bar{F}_{\text{inv}}(E) = \sum_{i=11}^{18} F_{\text{inv},i}(E)/8$, which is described by the full cure in Fig. 8b. In addition we have calculated the energy dependent standard deviation $\sigma_{\text{std}}(E)$ and the dashed curves correspond to $\bar{F}_{\text{inv}}(E) \pm \sigma_{\text{std}}(E)$. As one sees the inversion error is rather small, but grows towards lower energies. One could further improve the inversions by making an even more precise LIT calculation. In our specific case it would probably be better to change the b_i parameters of the radial basis function a bit rather than to increase the number of basis functions.

We summarize our work as follows. We have tested the LIT method for a calculation of the S-factor of the reaction ^{2}H$(p, \gamma)^3$He using a simple central NN interaction (MT-I/III potential). The calculation is performed by first computing the cross section of the inverse reaction in unretarted dipole approximation and then using the law of detailed balances in order to determine the deuteron-proton capture cross section which then leads to the determination of the S-factor. For a precise application of the LIT method it is necessary to have a sufficient density of LIT states in the energy region of interest. Considering our specific case this corresponds for the ^{3}He photodisintegration to the energy region between the two- and three-body breakup thresholds. We

 ✎ Springer

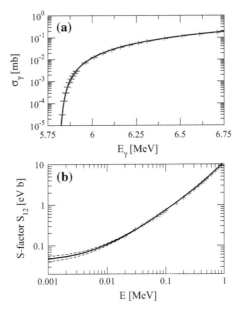

Fig. 8 a Full curve same as in Fig. 7b and in addition results from the direct calculation with explicit continuum wave function (*plus signs*); **b** same results as in (**a**) but rescaled in order to determine the *S*-factor [see Eqs. (1) and (3)], inversion error shown by dashed lines (see text)

have found that a solution of the LIT equation with the MT-I/III potential via an expansion in hyperspherical harmonics does not yield a single LIT-state below the three-body breakup threshold, even though using a rather high number of basis functions of a rather large spatial extension. With a more realistic nuclear force the picture does not change essentially as can be deduced from another low-energy observable, namely the ^{4}He isoscalar monopole resonance [10,11,15]. As pointed out in [15] for an increase of the LIT state density in the low-energy region one needs to use a basis where the relevant dynamical variable, namely the single-particle coordinate (vector pointing from the center of mass of the (A-1) particle system to the A-th particle), appears explicitly. Therefore we have taken a basis which is a product of expansions of two basis systems, each of them depending either on the single-particle coordinate or on the pair coordinate. We could show that using such a basis one can systematically increase the low-energy LIT state density. Furthermore, we show that in order to take into account that the LIT states become less dense with increasing energy it is advantageous to use different σ_I-values in different energy intervals.

In addition to the LIT approach we have carried out the calculation with explicit continuum wave functions. They have been determined via solving the Schrödinger equation with the help of an expansion over a proper basis set. A comparison of results from both methods shows a very good agreement. For the LIT method we have also included an estimate of the inversion error.

References

1. E.G. Adelberger et al., Rev. Mod. Phys. **83**, 195 (2011)
2. W. Leidemann, G. Orlandini, Prog. Part. Nucl. Phys. **68**, 158 (2013)
3. J. Carbonell, A. Deltuva, A.C. Fonseca, R. Lazauskas, Prog. Part. Nucl. Phys. **74**, 55 (2014)
4. V.D. Efros, W. Leidemann, G. Orlandini, Phys. Lett. B **338**, 130 (1994)
5. N. Barnea, V.D. Efros, W. Leidemann, G. Orlandini, Few Body Syst. **47**, 201 (2010)
6. D. Andreasi, W. Leidemann, Ch. Reiss, M. Schwamb, Eur. Phys. J. A **24**, 361 (2005)
7. W. Leidemann, Few Body Syst. **42**, 139 (2008)
8. V.D. Efros, W. Leidemann, G. Orlandini, N. Barnea, J. Phys. G **34**, R459 (2007)
9. S. Bacca, S. Pastore, J. Phys. G **41**, 123002 (2014)
10. S. Bacca, N. Barnea, W. Leidemann, G. Orlandini, Phys. Rev. Lett. **110**, 042503 (2013)
11. S. Bacca, N. Barnea, W. Leidemann, G. Orlandini, Phys. Rev. C **91**, 024303 (2015)
12. N. Barnea, W. Leidemann, G. Orlandini, Phys. Rev. C **61**, 054001 (2000)
13. N. Barnea, W. Leidemann, G. Orlandini, Nucl. Phys. A **693**, 565 (2001)
14. N. Barnea, V.D. Efros, W. Leidemann, G. Orlandini, Few Body Syst. **35**, 155 (2004)

15. W. Leidemann, Phys. Rev. C **91**, 054001 (2015)
16. R.A. Malfliet, J.A. Tjon, Nucl. Phys. A **127**, 161 (1969)
17. L.E. Marcucci, M. Viviani, R. Schiavilla, A. Kievsky, S. Rosati, Phys. Rev. C **72**, 014001 (2005)
18. L.E. Marcucci, G. Mangano, A. Kievsky, M. Viviani, Phys. Rev. Lett. **116**, 102501 (2016)
19. V.D. Efros, W. Leidemann, G. Orlandini, Phys. Lett. B **408**, 1 (1997)
20. B.N. Zakhariev, V.V. Pustovalov, V.D. Efros, Sov. J. Nucl. Phys. **8**, 234 (1968)
21. V.P. Permjakov, V.V. Pustovalov, YuI Fenin, V.D. Efros, Sov. J. Nucl. Phys. **14**, 317 (1972)
22. C. Schwartz, Ann. Phys. **16**, 36 (1961)
23. E.W. Schmid, K.H. Hoffmann, Nucl. Phys. A **175**, 443 (1971)
24. A. Kievsky, priv. comm. (2016)
25. A. Deltuva, A.C. Fonseca, A. Kievsky, S. Rosati, P.U. Sauer, M. Viviani, Phys. Rev. C **71**, 064003 (2005)
26. S. Deflorian, Ph.D. thesis, University of Trento (2016)

Few-Body Syst (2017) 58:1–13
DOI 10.1007/s00601-016-1183-0

D. V. Fedorov

Analytic Matrix Elements and Gradients with Shifted Correlated Gaussians

Received: 11 November 2016 / Accepted: 9 December 2016 / Published online: 30 December 2016
© Springer-Verlag Wien 2016

Abstract Matrix elements between shifted correlated Gaussians of various potentials with several form-factors are shown to be analytic. Their gradients with respect to the non-linear parameters of the Gaussians are also analytic. Analytic matrix elements are of importance for the correlated Gaussian method in quantum few-body physics.

1 Introduction

Correlated Gaussian method is a popular variational method to solve quantum-mechanical few-body problems in molecular, atomic, and nuclear physics [1,2]. One of the important advantages of the correlated Gaussian method is the ease of computing the matrix elements. In some cases the matrix elements and even their gradients with respect to optimization parameters are fully analytic [2–5]. This enables extensive numerical optimizations to be carried out leading to accurate results [6–8] despite the incorrect functional form of the Gaussians in certain asymptotic regions of the configuration space.

Although a number of analytic matrix elements have been calculated for different potentials and different forms of correlated Gaussians, one combination—short-range potentials with shifted correlated Gaussians—is still missing [1]. Indeed historically the shifted Gaussians have been applied more often to Coulombic systems rather than to atomic and nuclear systems where the short-range interactions are most important.

In this paper several types of short-range potentials are considered in a search for the form-factors that produce analytic matrix elements with shifted correlated Gaussians. A brief introduction to the correlated Gaussian method is given first and then the analytic matrix elements and their gradients are introduced.

2 Correlated Gaussian Method

Correlated Gaussian method is a variational method where the coordinate part of the wave-function of a quantum few-body system is expanded in terms of correlated Gaussians. Various forms of correlated Gaussians have been concieved [1], one of them being the shifted correlated Gaussian, $|g\rangle$, which for a system of N particles with coordinates $\vec{r}_i|_{i=1...N}$ has the form

$$\langle \mathbf{r}|g\rangle = \exp\left(-\sum_{i,j=1}^{N} A_{ij}\, \vec{r}_i \cdot \vec{r}_j + \sum_{i=1}^{N} \vec{s}_i \cdot \vec{r}_j\right) \equiv e^{-\mathbf{r}^\mathsf{T} A\mathbf{r} + \mathbf{s}^\mathsf{T}\mathbf{r}}, \tag{1}$$

This article belongs to the special issue "30th anniversary of Few-Body Systems".

D. V. Fedorov (✉)
Institute of Physics and Astronomy, Aarhus University, Ny Munkegade 120, 8000 Aarhus C, Denmark
E-mail: fedorov@phys.au.dk

where $\mathbf{r}$ is size-N column of particle coordinates $\vec{r}_i$, $A = \{A_{ij}\}$ is a size-N square symmetric positive-definite *correlation matrix*, $\mathbf{s}$ is size-N column of *shift vectors* $\vec{s}_i$, and where the following notation has been introduced,

$$\mathbf{r}^{\mathsf{T}} A \mathbf{r} \equiv \sum_{i,j=1}^{N} A_{ij} \vec{r}_i \cdot \vec{r}_j, \quad \mathbf{s}^{\mathsf{T}} \mathbf{r} \equiv \sum_{i=1}^{N} \vec{s}_i \cdot \vec{r}_j, \tag{2}$$

where "·" denotes the dot-product of two vectors. The elements of the correlation matrix and the shift vectors are the non-linear parameters of the Gaussians.

The coordinate part of the few-body wave-function $|\psi\rangle$ is represented as a linear combination of several Gaussians,

$$|\psi\rangle = \sum_{i=1}^{n_g} c_i |g_i\rangle, \tag{3}$$

where the coefficients c_i are the linear parameters, and where n_g is the number of Gaussians. Inserting this representation into the Schrdinger equation,

$$\hat{H}|\psi\rangle = E|\psi\rangle, \tag{4}$$

where $\hat{H}$ is the Hamiltonian of the few-body system, and multiplying from the left with $\langle g_j|$ leads to the generalized matrix eigenvalue equation,

$$\mathcal{H}c = E\mathcal{N}c, \tag{5}$$

where $c = \{c_i\}$ is the column of the linear parameters, and where the Hamilton matrix $\mathcal{H} = \{\mathcal{H}_{ij}\}$ and the overlap matrix $\mathcal{N} = \{\mathcal{N}_{ij}\}$ are given as

$$\mathcal{H}_{ij} = \langle g_i|\hat{H}|g_j\rangle, \quad \mathcal{N}_{ij} = \langle g_i|g_j\rangle. \tag{6}$$

The linear parameters together with the energy spectrum are found by solving the generalized eigenvalue problem (5) numerically using the standard linear algebra methods [9]. The non-linear parameters are optimized by one of many optimization methods which typically involve elements of stochastic-evolutionary [10] and direct optimization algorithms [7,8]. The direct optimization algorithms often employ gradients of the matrix elements with respect to the optimization parameters.

These optimization techniques involve numerous evaluations of the Hamiltonian matrix elements and their gradients. Therefore the analytica matrix elements are of particular importance for the method.

3 Matrix Elements

3.1 Overlap

The overlap $\langle g'|g\rangle$ between a shifted Gaussian $|g\rangle$ with parameters A, $\mathbf{s}$ and a shifted Gaussian $|g'\rangle$ with parameters A', $\mathbf{s}'$ is given as[1]

$$\langle g'|g\rangle = e^{\frac{1}{4}\mathbf{v}^{\mathsf{T}} B^{-1}\mathbf{v}} \left(\frac{\pi^N}{\det(B)}\right)^{3/2} \equiv M, \tag{8}$$

where $B = A' + A$, $\mathbf{v} = \mathbf{s}' + \mathbf{s}$.

---

[1] The overlap can be evaluated by an orthogonal coordinate transformation, $\mathbf{r} = Q\mathbf{x}$, to the basis where the matrix B is diagonal: $B = QDQ^{\mathsf{T}}$ where $Q^{\mathsf{T}}Q = QQ^{\mathsf{T}} = 1$ and D is a diagonal matrix,

$$\langle g'|g\rangle = \int d^3\vec{r}_1 \dots d^3\vec{r}_N \exp\left(-\mathbf{r}^{\mathsf{T}} B\mathbf{r} + \mathbf{v}^{\mathsf{T}}\mathbf{r}\right) = \int d^3\vec{x}_1 \dots d^3\vec{x}_N \exp\left(-\sum_{i=1}^{N} \vec{x}_i \cdot D_{ii}\vec{x}_i + \sum_{i=1}^{N} \vec{v}_i \cdot \vec{x}_i\right) \tag{7}$$

$$= \prod_{i=1}^{N} \int d^3\vec{x}_i \exp\left(-\vec{x}_i \cdot D_{ii}\vec{x}_i + \vec{v}_i \cdot \vec{x}_i\right) = \prod_{i=1}^{N} \exp\left(\frac{1}{4D_{ii}}\vec{v}_i^2\right)\left(\frac{\pi}{D_{ii}}\right)^{3/2} = e^{\frac{1}{4}\mathbf{v}^{\mathsf{T}} B^{-1}\mathbf{v}}\left(\frac{\pi^N}{\det(B)}\right)^{3/2} .$$

3.2 Kinetic Energy

The non-relativistic kinetic energy operator $\hat{K}$ for an N-body system of particles with coordinates $\vec{r}_i$ and masses m_i is given as

$$\hat{K} = -\sum_{i=1}^{N} \frac{\hbar^2}{2m_i} \frac{\partial^2}{\partial \vec{r}_i^2}. \tag{9}$$

For completeness we shall consider a more general form of the kinetic energy operator,

$$\hat{K} = -\sum_{i,j=1}^{N} \frac{\partial}{\partial \vec{r}_i} \Lambda_{ij} \frac{\partial}{\partial \vec{r}_j} \equiv -\frac{\partial}{\partial \mathbf{r}} \Lambda \frac{\partial}{\partial \mathbf{r}^{\mathsf{T}}}, \tag{10}$$

where Λ is a symmetric positive-definite matrix. The matrix element of this operator is given as[2]

$$\left\langle g' \left| -\frac{\partial}{\partial \mathbf{r}} \Lambda \frac{\partial}{\partial \mathbf{r}^{\mathsf{T}}} \right| g \right\rangle = \left\langle g' \left| (\mathbf{s}'^{\mathsf{T}} - 2\mathbf{r}^{\mathsf{T}} A') \Lambda (\mathbf{s} - 2A\mathbf{r}) \right| g \right\rangle$$

$$= \left(6 \operatorname{trace}(A' \Lambda A B^{-1}) + (\mathbf{s}' - 2A'\mathbf{u})^{\mathsf{T}} \Lambda (\mathbf{s} - 2A\mathbf{u}) \right) M, \tag{14}$$

where $\mathbf{u} \doteq \frac{1}{2} B^{-1} \mathbf{v}$, and M is the overlap.

3.3 Potential Energy

3.3.1 Central Potential

A one-body central potential, $V(\vec{r}_i)$, and a two-body central potential, $V(\vec{r}_i - \vec{r}_j)$, can be written in a convenient general form, $V(w^{\mathsf{T}}\mathbf{r})$, where w is a size-N column of numbers with all components equal zero except for $w_i = 1$ for the one-body potential and $w_i = -w_j = 1$ for the two body potential.

Gaussian form-factor For the Gaussian form-factor, $V(w^{\mathsf{T}}\mathbf{r}) \propto e^{-\gamma \mathbf{r}^{\mathsf{T}} w w^{\mathsf{T}} \mathbf{r}}$, the matrix element directly follows from the overlap integral,

$$\left\langle g' \left| e^{-\gamma \mathbf{r}^{\mathsf{T}} w w^{\mathsf{T}} \mathbf{r}} \right| g \right\rangle = e^{\frac{1}{4} \mathbf{v}^{\mathsf{T}} B'^{-1} \mathbf{v}} \left(\frac{\pi^N}{\det(B')} \right)^{3/2} \equiv M', \tag{15}$$

where the matrix B' is a rank-1 update of the matrix $B = A' + A$, $B' = B + \gamma w w^{\mathsf{T}}$.

If the determinant and the inverse of the matrix B are known, their rank-1 updates can be calculated efficiently using the update formulas,

$$\det(B + ab^{\mathsf{T}}) = (1 + b^{\mathsf{T}} B^{-1} a) \det(B), \tag{16}$$

---

[2] We first calculate two integrals,

$$\langle g' | \mathbf{r} | g \rangle = \left(\frac{\partial}{\partial \mathbf{v}^{\mathsf{T}}} \right) e^{\frac{1}{4} \mathbf{v}^{\mathsf{T}} B^{-1} \mathbf{v}} \left(\frac{\pi^N}{\det(B)} \right)^{3/2} = \mathbf{u} e^{\frac{1}{4} \mathbf{v}^{\mathsf{T}} B^{-1} \mathbf{v}} \left(\frac{\pi^N}{\det(B)} \right)^{3/2}, \tag{11}$$

where $\mathbf{u} = \frac{1}{2} B^{-1} \mathbf{v}$, and

$$\left\langle g' \left| \mathbf{r}^{\mathsf{T}} F \mathbf{r} \right| g \right\rangle = \left(\frac{\partial}{\partial \mathbf{v}} F \frac{\partial}{\partial \mathbf{v}^{\mathsf{T}}} \right) e^{\frac{1}{4} \mathbf{v}^{\mathsf{T}} B^{-1} \mathbf{v}} \left(\frac{\pi^N}{\det(B)} \right)^{3/2} = \left(\frac{3}{2} \operatorname{trace}(FB^{-1}) + \mathbf{u}^{\mathsf{T}} F \mathbf{u} \right) e^{\frac{1}{4} \mathbf{v}^{\mathsf{T}} B^{-1} \mathbf{v}} \left(\frac{\pi^N}{\det(B)} \right)^{3/2}, \tag{12}$$

from which the sought integral,

$$\left\langle g' \left| -\frac{\partial}{\partial \mathbf{r}} \Lambda \frac{\partial}{\partial \mathbf{r}^{\mathsf{T}}} \right| g \right\rangle = \left\langle g' \left| (\mathbf{s}' - 2A'\mathbf{r})^{\mathsf{T}} \Lambda (\mathbf{s} - 2A\mathbf{r}) \right| g \right\rangle, \tag{13}$$

follows directly.

$$(B + ab^\mathsf{T})^{-1} = B^{-1} - \frac{B^{-1}ab^\mathsf{T}B^{-1}}{1 + b^\mathsf{T}B^{-1}a}, \tag{17}$$

where a and b are size-N columns of numbers.

Oscillator form-factor Another potential with a simple analytic matrix element is the oscillator potential, $V(w^\mathsf{T}\mathbf{r}) \propto \mathbf{r}^\mathsf{T}ww^\mathsf{T}\mathbf{r}$, relevant for cold atoms in traps. The matrix element directly follows from (12),

$$\left\langle g' \left| \mathbf{r}^\mathsf{T}ww^\mathsf{T}\mathbf{r} \right| g \right\rangle = \left(\frac{3}{2}w^\mathsf{T}B^{-1}w + \mathbf{u}^\mathsf{T}ww^\mathsf{T}\mathbf{u} \right) M. \tag{18}$$

Other analytic form-factors For a potential with a general form-factor, $V \propto f(w^\mathsf{T}\mathbf{r})$, the matrix element reduces to a three-dimensional integral,[3]

$$\left\langle g' \left| f(w^\mathsf{T}\mathbf{r}) \right| g \right\rangle = M \left(\frac{\beta}{\pi} \right)^{\frac{3}{2}} \int d^3\vec{r}\, f(\vec{r})\, e^{-\beta(\vec{r}-\vec{q})^2}, \tag{21}$$

where $\beta = \left(w^\mathsf{T}B^{-1}w \right)^{-1}$ and $\vec{q} = w^\mathsf{T}\mathbf{u}$.

If the potential does not depend on the direction of its argument the integral reduces further to a one-dimensional integral,

$$\left\langle g' \left| f(|w^\mathsf{T}\mathbf{r}|) \right| g \right\rangle = M \left(\frac{\beta}{\pi} \right)^{\frac{3}{2}} 2\pi \frac{e^{-\beta q^2}}{\beta q} \int_0^\infty r\,dr\, f(r)\, e^{-\beta r^2} \sinh(2\beta q r) \tag{22}$$
$$\equiv M J[f],$$

where

$$J[f] \doteq \left(\frac{\beta}{\pi} \right)^{\frac{3}{2}} 2\pi \frac{e^{-\beta q^2}}{\beta q} \int_0^\infty r\,dr\, f(r)\, e^{-\beta r^2} \sinh(2\beta q r). \tag{23}$$

The integral (23) gives relatively simple analytic results for the Coulomb form-factor, $1/r$,

$$J\left[\frac{1}{r} \right] = \frac{\mathrm{erf}(\sqrt{\beta}q)}{q}, \tag{24}$$

and for several short-range form-factors (in addition to the Gaussian form-factor above):

screened Yukawa, $e^{-\gamma r^2 - \mu r}/r$,

$$J\left[\frac{e^{-\gamma r^2 - \mu r}}{r} \right] \tag{25}$$
$$= \frac{\sqrt{\beta}e^{-\beta q^2}}{2q\sqrt{\beta + \gamma}} \left(e^{\frac{(\mu - 2\beta q)^2}{4(\beta + \gamma)}} \left(\mathrm{erf}\left(\frac{2\beta q - \mu}{2\sqrt{\beta + \gamma}} \right) + 1 \right) + e^{\frac{(\mu + 2\beta q)^2}{4(\beta + \gamma)}} \left(\mathrm{erf}\left(\frac{\mu + 2\beta q}{2\sqrt{\beta + \gamma}} \right) - 1 \right) \right),$$

Yukawa, $e^{-\mu r}/r$,

$$J\left[\frac{e^{-\mu r}}{r} \right] = \lim_{\gamma \to 0} J\left[\frac{e^{-\gamma r^2 - \mu r}}{r} \right] \tag{26}$$

---

[3] Suppose the form-factor $f(\vec{r})$ has a Fourier-transform $\mathcal{F}(\vec{k})$, then

$$\left\langle g' \left| f(w^\mathsf{T}\mathbf{r}) \right| g \right\rangle = \int \frac{d^3\vec{k}}{(2\pi)^3} \mathcal{F}(\vec{k}) \left\langle g' \left| e^{ikw^\mathsf{T}\mathbf{r}} \right| g \right\rangle = e^{\frac{1}{4}\mathbf{v}^\mathsf{T}B^{-1}\mathbf{v}} \left(\frac{\pi^N}{\det(B)} \right)^{3/2} \int \frac{d^3\vec{k}}{(2\pi)^3} \mathcal{F}(\vec{k}) e^{-\alpha k^2 + i\vec{k}\vec{q}} \tag{19}$$

where $\alpha = \frac{1}{4}w^\mathsf{T}B^{-1}w$, $\vec{q} = \frac{1}{2}w^\mathsf{T}B^{-1}\mathbf{v}$. Now the last integral can as well be written as

$$\left(\frac{\beta}{\pi} \right)^{\frac{3}{2}} \int d^3\vec{r}\, f(\vec{r})\, e^{-\beta(\vec{r}-\vec{q})^2}, \tag{20}$$

where $\beta = \frac{1}{4\alpha} = (w^\mathsf{T}B^{-1}w)^{-1}$.

$$= \frac{e^{\frac{\mu^2}{4\beta} - \mu q}}{2q} \left(1 - e^{2\mu q} + \mathrm{erf}\left(\frac{2\beta q - \mu}{2\sqrt{\beta}}\right) + e^{2\mu q} \mathrm{erf}\left(\frac{2\beta q + \mu}{2\sqrt{\beta}}\right) \right), \tag{27}$$

screened Coulomb, $e^{-\gamma r^2}/r$,

$$J\left[\frac{e^{-\gamma r^2}}{r}\right] = \lim_{\mu \to 0} J\left[\frac{e^{-\gamma r^2 - \mu r}}{r}\right] = \frac{\sqrt{\beta} e^{-\frac{\beta \gamma q^2}{\beta + \gamma}} \mathrm{erf}\left(\frac{\beta q}{\sqrt{\beta + \gamma}}\right)}{q\sqrt{\beta + \gamma}}, \tag{28}$$

screened exponential, $e^{-\gamma r^2 - \mu r}$,

$$J\left[e^{-\gamma r^2 - \mu r}\right] = \frac{-\partial}{\partial \mu} J\left[\frac{e^{-\gamma r^2 - \mu r}}{r}\right] = \frac{-\sqrt{\beta}}{4q(\beta + \gamma)^{3/2}} e^{\frac{(\mu - 2\beta q)^2}{4(\beta + \gamma)} - \beta q^2} \tag{29}$$

$$\left((\mu - 2\beta q)\mathrm{erf}\left(\frac{2\beta q - \mu}{2\sqrt{\beta + \gamma}}\right) + (\mu + 2\beta q)e^{\frac{2\beta \mu q}{\beta + \gamma}} \mathrm{erf}\left(\frac{\mu + 2\beta q}{2\sqrt{\beta + \gamma}}\right) + \mu - 2\beta q e^{\frac{2\beta \mu q}{\beta + \gamma}} - \mu e^{\frac{2\beta \mu q}{\beta + \gamma}} - 2\beta q \right),$$

exponential, $e^{-\mu r}$,

$$J\left[e^{-\mu r}\right] = \lim_{\gamma \to 0} J\left[e^{-\gamma r^2 - \mu r}\right] = \frac{-e^{\frac{\mu(\mu - 4\beta q)}{4\beta}}}{4\beta q} \tag{30}$$

$$\left((\mu - 2\beta q)\mathrm{erf}\left(\frac{2\beta q - \mu}{2\sqrt{\beta}}\right) + e^{2\mu q}(\mu + 2\beta q)\mathrm{erf}\left(\frac{\mu + 2\beta q}{2\sqrt{\beta}}\right) + \mu - 2\beta q e^{2\mu q} - 2\beta q - \mu e^{2\mu q} \right).$$

Since the error function has analytic derivative, $\mathrm{erf}'(x) = 2e^{-x^2}/\sqrt{\pi}$, the following form-factors also have analytic matrix elements,

$$J\left[r^n \frac{e^{-\gamma r^2 - \mu r}}{r}\right] = \left(\frac{-\partial}{\partial \mu}\right)^n J\left[\frac{e^{-\gamma r^2 - \mu r}}{r}\right]. \tag{31}$$

Although there are few other form-factors with analytic matrix elements, the resulting relatively complicated expressions are not conducive to subsequent analytic calculations of gradients with respect to the elements of matrices A and the shift-vectors $\mathbf{s}$ through the quantities β and q. In such cases it is probably more efficient to represent the potential as a linear combinations of several Gaussians.

3.3.2 Tensor Potential

The tensor potential V_t between two particles with coordinates $\vec{r}_1$, $\vec{r}_2$ and spins $\vec{S}_1$, $\vec{S}_2$ can be written in the form [11]

$$V_t(r) \propto f(r)(\vec{S}_1 \cdot \vec{r})(\vec{S}_2 \cdot \vec{r}), \tag{32}$$

where $\vec{r} = \vec{r}_1 - \vec{r}_2$ is the relative coordinate between the particles, and $f(r)$ is the radial form-factor of the potential. In this form the potential has a central spin-spin component,

$$\frac{1}{3} f(r) r^2 (\vec{S}_1 \cdot \vec{S}_2), \tag{33}$$

which is often subtracted from the above form to make sure the potential contains only the spherical tensor component.

Introducing the size-N column of numbers w,

$$w = \{w_i | w_1 = 1, w_2 = -1, w_{i \neq 1,2} = 0\}, \tag{34}$$

and vector-columns $\mathbf{y}_1 = \vec{S}_1 w$, $\mathbf{y}_2 = \vec{S}_2 w$, the tensor potential can be written in a convenient general form,

$$\hat{V}_t \propto f(w^\mathsf{T} \mathbf{r})(\mathbf{y}_1^\mathsf{T} \mathbf{r})(\mathbf{y}_2^\mathsf{T} \mathbf{r}), \tag{35}$$

where

$$\mathbf{y}^{\mathsf{T}}\mathbf{r} \equiv \sum_{i=1}^{N} \vec{y}_i \cdot \vec{r}_i. \tag{36}$$

The tensor matrix element can be represented as a derivative of the central matrix element with the same form-factor,

$$\left\langle g' \left| f(w^{\mathsf{T}}\mathbf{r})(\mathbf{y}_1^{\mathsf{T}}\mathbf{r})(\mathbf{y}_2^{\mathsf{T}}\mathbf{r}) \right| g \right\rangle = \left(\mathbf{y}_1^{\mathsf{T}}\frac{\partial}{\partial \mathbf{v}^{\mathsf{T}}} \right) \left(\mathbf{y}_2^{\mathsf{T}}\frac{\partial}{\partial \mathbf{v}^{\mathsf{T}}} \right) \left\langle g' \left| f(w^{\mathsf{T}}\mathbf{r}) \right| g \right\rangle. \tag{37}$$

The central part (35) of the tensor potential has a similar analytic representation,

$$\left\langle g' \left| f(w^{\mathsf{T}}\mathbf{r})(w^{\mathsf{T}}\mathbf{r})(w^{\mathsf{T}}\mathbf{r}) \right| g \right\rangle = \left(w^{\mathsf{T}}\frac{\partial}{\partial \mathbf{v}^{\mathsf{T}}} \right) \left(w^{\mathsf{T}}\frac{\partial}{\partial \mathbf{v}^{\mathsf{T}}} \right) \left\langle g' \left| f(w^{\mathsf{T}}\mathbf{r}) \right| g \right\rangle. \tag{38}$$

Thus if the matrix element of the central potential with a given form-factor $f(w^{\mathsf{T}}\mathbf{r})$ is analytic, the matrix element of the tensor potential $f(w^{\mathsf{T}}\mathbf{r})(\mathbf{y}_1^{\mathsf{T}}\mathbf{r})(\mathbf{y}_2^{\mathsf{T}}\mathbf{r})$ is also analytic. In particular, the tensor matrix elements of the screened Yukawa form-factor and its descendants (31) are also analytic.

Gaussian form-factor For a Gaussian form-factor the matrix elements (37,38) are readily given as

$$\left\langle g' \left| e^{-\gamma \mathbf{r}^{\mathsf{T}}ww^{\mathsf{T}}\mathbf{r}}(\mathbf{y}_1^{\mathsf{T}}\mathbf{r})(\mathbf{y}_2^{\mathsf{T}}\mathbf{r}) \right| g \right\rangle = \left(\frac{1}{2}\mathbf{y}_1^{\mathsf{T}}B'^{-1}\mathbf{y}_2 + (\mathbf{y}_1^{\mathsf{T}}\mathbf{u}')(\mathbf{y}_2^{\mathsf{T}}\mathbf{u}') \right) M',$$

$$\left\langle g' \left| e^{-\gamma \mathbf{r}^{\mathsf{T}}ww^{\mathsf{T}}\mathbf{r}}(\mathbf{r}^{\mathsf{T}}ww^{\mathsf{T}}\mathbf{r}) \right| g \right\rangle = \left(\frac{3}{2}w^{\mathsf{T}}B'^{-1}w + \mathbf{u}'^{\mathsf{T}}ww^{\mathsf{T}}\mathbf{u}' \right) M',$$

where $B' = B + \gamma ww^{\mathsf{T}}$, $B = A' + A$, $\mathbf{u}' = \frac{1}{2}B'^{-1}\mathbf{v}$, $\mathbf{v} = \mathbf{s}' + \mathbf{s}$, and where

$$M' = e^{\frac{1}{4}\mathbf{v}^{\mathsf{T}}B'^{-1}\mathbf{v}} \left(\frac{\pi^N}{\det(B')} \right)^{3/2} \tag{39}$$

is the central Gaussian matrix element.

Again the updates $\det(B + \gamma ww^{\mathsf{T}})$ and $(B + \gamma ww^{\mathsf{T}})^{-1}$ can be efficiently calculated using rank-1 update formulas.

Other form-factors The tensor matrix element (37) can be written in component form as

$$\left\langle g' \left| f(w^{\mathsf{T}}\mathbf{r})(\mathbf{y}_1^{\mathsf{T}}\mathbf{r})(\mathbf{y}_2^{\mathsf{T}}\mathbf{r}) \right| g \right\rangle = \sum_{i,j=1}^{N}\sum_{a,b=1}^{3} (\vec{y}_1)_{ia}(\vec{y}_2)_{jb} \left\langle g' \left| f(w^{\mathsf{T}}\mathbf{r})\vec{r}_{ia}\vec{r}_{jb} \right| g \right\rangle$$

$$= \sum_{i,j=1}^{N}\sum_{a,b=1}^{3} (\vec{y}_1)_{ia}(\vec{y}_2)_{jb}\frac{\partial}{\partial \vec{v}_{ia}}\frac{\partial}{\partial \vec{v}_{jb}} \left\langle g' \left| f(w^{\mathsf{T}}\mathbf{r}) \right| g \right\rangle. \tag{40}$$

where $\vec{r}_{ia}$ is the number-a component of the vector $\vec{r}_i$, and where

$$\left\langle g' \left| f(w^{\mathsf{T}}\mathbf{r}) \right| g \right\rangle = MJ, \tag{41}$$

$$M = e^{\frac{1}{4}\mathbf{v}^{\mathsf{T}}B^{-1}\mathbf{v}} \left(\frac{\pi^N}{\det(B)} \right)^{\frac{3}{2}}, \tag{42}$$

$$J = \left(\frac{\beta}{\pi} \right)^{\frac{3}{2}} 2\pi \frac{e^{-\beta q^2}}{\beta q} \int_0^{\infty} r\,dr\,f(r)\,e^{-\beta r^2}\sinh(2\beta qr). \tag{43}$$

The derivative of the central matrix element at the right-hand side of (40) can now be evaluated as

$$\frac{\partial}{\partial \vec{v}_{ia}}\frac{\partial}{\partial \vec{v}_{jb}}(MJ) = \left(\frac{\partial}{\partial \vec{v}_{ia}}\frac{\partial}{\partial \vec{v}_{jb}}M \right) J + M \left(\frac{\partial}{\partial \vec{v}_{ia}}\frac{\partial}{\partial \vec{v}_{jb}}J \right)$$

$$+ \left(\frac{\partial}{\partial \vec{v}_{ia}}M \right) \left(\frac{\partial}{\partial \vec{v}_{jb}}J \right) + \left(\frac{\partial}{\partial \vec{v}_{jb}}M \right) \left(\frac{\partial}{\partial \vec{v}_{ia}}J \right), \tag{44}$$

where

$$\frac{\partial}{\partial \vec{v}_{ia}} M = \sum_{k=1}^{N} \frac{1}{2} B_{ik}^{-1} \vec{v}_{ka} M, \tag{45}$$

$$\frac{\partial}{\partial \vec{v}_{jb}} \frac{\partial}{\partial \vec{v}_{ia}} M = \frac{1}{2} B_{ji}^{-1} \delta_{ab} M + \sum_{k,l=1}^{N} \frac{1}{2} B_{ik}^{-1} \vec{v}_{ka} \frac{1}{2} B_{jl}^{-1} \vec{v}_{lb} M, \tag{46}$$

$$\frac{\partial}{\partial \vec{v}_{ia}} J = \frac{\partial J}{\partial q} \frac{\partial q}{\partial \vec{v}_{ia}}, \tag{47}$$

$$\frac{\partial}{\partial \vec{v}_{jb}} \frac{\partial}{\partial \vec{v}_{ia}} J = \frac{\partial^2 J}{\partial q^2} \frac{\partial q}{\partial \vec{v}_{jb}} \frac{\partial q}{\partial \vec{v}_{ia}} + \frac{\partial J}{\partial q} \frac{\partial^2 q}{\partial \vec{v}_{jb} \partial \vec{v}_{ia}}, \tag{48}$$

$$\frac{\partial q}{\partial \vec{v}_{ia}} = \frac{1}{q} \sum_{k=1}^{N} h_i h_k \vec{v}_{ka}, \tag{49}$$

$$\frac{\partial^2 q}{\partial \vec{v}_{jb} \partial \vec{v}_{ia}} = \frac{1}{q} h_i h_j \delta_{ab} - \frac{2}{q^2} \sum_{k=1}^{N} h_i h_k \vec{v}_{ka} \sum_{l=1}^{N} h_j h_l \vec{v}_{lb}. \tag{50}$$

where $h = w^\mathsf{T} \frac{1}{2} B^{-1}$.

Now to finish the calculation one only needs to calculate $\partial J/\partial q$ and $\partial^2 J/\partial q^2$ for the given potential. For the screened Yukawa form-factor and its descendants (31) the derivatives $\partial J/\partial q$ and $\partial^2 J/\partial q^2$ are analytic although the actual calculations are relatively tedious and should be best performed by a computer algebra software like Maxima [12]. For example, the expression for

$$\frac{\partial^2}{\partial q^2} J \left[\frac{e^{-\gamma r^2 - \mu r}}{r} \right] \tag{51}$$

can be readily obtained by the following Maxima script,

```
assume(beta>0,gamma>0,mu>0,q>0);
J(f) := (beta/%pi)^(3/2)*2*%pi/beta/q*exp(-beta*q^2)
*integrate(r*f*exp(-beta*r^2)*sinh(2*beta*q*r),r,0,inf);
fortran(diff(J(exp(-gamma*r^2-mu*r)/r), q, 2));
```

which analytically calculates $\frac{\partial^2 J}{\partial q^2}$ for the screened Yukawa potential and outputs the corresponding Fortran code.

Even if J is not analytic one can evaluate its derivatives by calculating numerically a few extra integrals,

$$\frac{\partial J}{\partial q} = -\frac{2\sqrt{\beta} e^{-\beta q^2} \int_0^\infty r e^{-\beta r^2} f(r) \sinh(2\beta q r) \, dr}{\sqrt{\pi} q^2}$$
$$-\frac{4\beta^{\frac{3}{2}} e^{-\beta q^2} \int_0^\infty r e^{-\beta r^2} f(r) \sinh(2\beta q r) \, dr}{\sqrt{\pi}}$$
$$+\frac{4\beta^{\frac{3}{2}} e^{-\beta q^2} \int_0^\infty r^2 e^{-\beta r^2} f(r) \cosh(2\beta q r) \, dr}{\sqrt{\pi} q}, \tag{52}$$

$$\frac{\partial^2 J}{\partial q^2} = \frac{8\beta^{\frac{5}{2}} e^{-\beta q^2} \int_0^\infty r^3 e^{-\beta r^2} f(r) \sinh(2\beta q r) \, dr}{\sqrt{\pi} q}$$
$$+\frac{8\beta^{\frac{5}{2}} q e^{-\beta q^2} \int_0^\infty r e^{-\beta r^2} f(r) \sinh(2\beta q r) \, dr}{\sqrt{\pi}}$$
$$+\frac{4\beta^{\frac{3}{2}} e^{-\beta q^2} \int_0^\infty r e^{-\beta r^2} f(r) \sinh(2\beta q r) \, dr}{\sqrt{\pi} q}$$

$$+\frac{4\sqrt{\beta}\,e^{-\beta q^2}\int_0^\infty r\,e^{-\beta r^2}f(r)\sinh(2\beta q r)\,dr}{\sqrt{\pi}\,q^3}$$

$$-\frac{8\beta^{\frac{3}{2}}e^{-\beta q^2}\int_0^\infty r^2\,e^{-\beta r^2}f(r)\cosh(2\beta q r)\,dr}{\sqrt{\pi}\,q^2}$$

$$-\frac{16\beta^{\frac{5}{2}}e^{-\beta q^2}\int_0^\infty r^2\,e^{-\beta r^2}f(r)\cosh(2\beta q r)\,dr}{\sqrt{\pi}}. \tag{53}$$

3.3.3 Spin-Orbit Potential

The spin-orbit potential between two particles with coordinates $\vec{r}_1$ and $\vec{r}_2$ and spins $\vec{S}_1$ and $\vec{S}_2$ can be written in the form [11]

$$V_{\text{so}} \propto f(r)\left(\vec{S}\cdot\vec{L}\right), \tag{54}$$

where $\vec{r} = \vec{r}_1 - \vec{r}_2$ is the relative coordinate between the particles; $f(r)$ is the radial form-factor of the potential; $\vec{S} = \vec{S}_1 + \vec{S}_2$ is the total spin of the two particles; and $\vec{L}$ is the relative orbital momentum between the two particles,

$$\vec{L} = (\vec{r}_1 - \vec{r}_2) \times \frac{-i}{2}\left(\frac{\partial}{\partial\vec{r}_1} - \frac{\partial}{\partial\vec{r}_2}\right), \tag{55}$$

where "$\times$" denotes vector-product of two vectors.

The orbital momentum operator can be written, using the size-N column of numbers w,

$$w = \{w_i | w_1 = 1, w_2 = -1, w_{i\neq1,2} = 0\}, \tag{56}$$

in the general form,

$$\vec{L} = \frac{-i}{2}\left(w^{\mathsf{T}}\mathbf{r} \times w^{\mathsf{T}}\frac{\partial}{\partial\mathbf{r}^{\mathsf{T}}}\right), \tag{57}$$

where

$$w^{\mathsf{T}}\mathbf{r} \equiv \sum_{i=1}^N w_i\vec{r}_i, \quad w^{\mathsf{T}}\frac{\partial}{\partial\mathbf{r}^{\mathsf{T}}} \equiv \sum_{i=1}^N w_i\frac{\partial}{\partial\vec{r}_i}. \tag{58}$$

For a given form-factor $f(w^{\mathsf{T}}\mathbf{r})$ the spin-orbit matrix element can be represented through the central matrix element with the same form-factor,

$$\left\langle g'\left|f(w^{\mathsf{T}}\mathbf{r})\left(w^{\mathsf{T}}\mathbf{r} \times w^{\mathsf{T}}\frac{\partial}{\partial\mathbf{r}^{\mathsf{T}}}\right)\right|g\right\rangle = \left(w^{\mathsf{T}}\frac{\partial}{\partial\mathbf{v}^{\mathsf{T}}}\right) \times w^{\mathsf{T}}\left(\mathbf{s} - 2A\frac{\partial}{\partial\mathbf{v}^{\mathsf{T}}}\right)\left\langle g'\left|f(w^{\mathsf{T}}\mathbf{r})\right|g\right\rangle, \tag{59}$$

so that if the central matrix element is analytic—as is the case for the screened Yukawa form-factor and its descendants (31)—the spin-orbit matrix element is also analytic.

Gaussian form-factor For a Gaussian form-factor the spin-orbit matrix element is given as

$$\left\langle g'\left|e^{-\gamma\mathbf{r}^{\mathsf{T}}ww^{\mathsf{T}}\mathbf{r}}\left(w^{\mathsf{T}}\mathbf{r} \times w^{\mathsf{T}}\frac{\partial}{\partial\mathbf{r}^{\mathsf{T}}}\right)\right|g\right\rangle = \frac{1}{2}w^{\mathsf{T}}B'^{-1}\mathbf{v} \times \left(w^{\mathsf{T}}\mathbf{s} - w^{\mathsf{T}}AB'^{-1}\mathbf{v}\right)M', \tag{60}$$

where $B' = B + \gamma ww^{\mathsf{T}}$, $B = A' + A$, $\mathbf{u}' = \frac{1}{2}B'^{-1}\mathbf{v}$, $\mathbf{v} = \mathbf{s}' + \mathbf{s}$, and where

$$M' = e^{\frac{1}{4}\mathbf{v}^{\mathsf{T}}B'^{-1}\mathbf{v}}\left(\frac{\pi^N}{\det(B')}\right)^{3/2} \tag{61}$$

is the central Gaussian matrix element.

Again $\det(B')$ and B'^{-1} can be efficiently calculated using rank-1 update formulas.

General form-factor The number-a component of the spin-orbit matrix element (59) can be written in component form as

$$\left\langle g'\left|f(w^{\mathsf{T}}\mathbf{r})\left(w^{\mathsf{T}}\mathbf{r} \times w^{\mathsf{T}}\frac{\partial}{\partial\mathbf{r}^{\mathsf{T}}}\right)_a\right|g\right\rangle \tag{62}$$

Springer

$$= \sum_{b,c=1}^{3} \epsilon_{abc} \sum_{k=1}^{N} w_k \frac{\partial}{\partial \vec{v}_{kb}} \sum_{l=1}^{N} w_l \left(\vec{s}_{lc} - 2 \sum_{j=1}^{N} A_{lj} \frac{\partial}{\partial \vec{v}_{jc}} \right) \langle g' \left| f(w^\mathsf{T} \mathbf{r}) \right| g \rangle . \tag{63}$$

It is clearly a linear combination of the first, $\frac{\partial}{\partial \vec{v}_{kb}} \langle g' \left| f(w^\mathsf{T} \mathbf{r}) \right| g \rangle$, and second, $\frac{\partial}{\partial \vec{v}_{kb}} \frac{\partial}{\partial \vec{v}_{jc}} \langle g' \left| f(w^\mathsf{T} \mathbf{r}) \right| g \rangle$, derivatives of the corresponding central matrix element $\langle g' \left| f(w^\mathsf{T} \mathbf{r}) \right| g \rangle = MJ$. These quantities have been calculated in the previous chapter.

3.4 Many-Body Forces

Many-body potentials in nuclear physics have form-factors which depend on the coordinates of several nucleons. In addition they may have tensor and spin-orbit post-factors.

For a many-body potential with a Gaussian form-factor, $e^{-\mathbf{r}^\mathsf{T} W \mathbf{r}}$, where W is a symmetric positive-definite matrix, the central matrix element is given as

$$\left\langle g' \left| e^{-\mathbf{r}^\mathsf{T} W \mathbf{r}} \right| g \right\rangle = e^{\frac{1}{4} \mathbf{v}(B+W)^{-1} \mathbf{v}} \left(\frac{\pi^N}{\det(B+W)} \right)^{3/2} . \tag{64}$$

The simple tensor and spin-orbit post-factors can be calculated in the same way as has been done in the previous two chapters.

4 Gradients of the Matrix Elements

The direct optimization strategy for the correlated Gaussian method minimizes the variational energy,

$$E = \frac{c^\mathsf{T} \mathcal{H} c}{c^\mathsf{T} \mathcal{N} c}, \tag{65}$$

(where $\mathcal{H}$ and $\mathcal{N}$ are the Hamilton and the overlap matrices in the given Gaussian basis) with respect to the non-linear parameters of the Gaussians, under conditions

$$\mathcal{H}c = E\mathcal{N}c , \ c^\mathsf{T} \mathcal{N} c = 1 . \tag{66}$$

Many popular multidimensional minimization algorithms require evaluations of the gradient,

$$\frac{\partial E}{\partial \alpha} = c^\mathsf{T} \left(\frac{\partial \mathcal{H}}{\partial \alpha} - E \frac{\partial \mathcal{N}}{\partial \alpha} \right) c, \tag{67}$$

where α is a given non-linear parameter of a Gaussian. The gradient involves evaluations of the derivatives of the matrix elements $\mathcal{H}$ and $\mathcal{N}$ with respect to the non-linear parameters of the Gaussians – the matrix A and the shift-vector $\mathbf{s}$.

4.1 Gradient with Respect to Shift Vector

4.1.1 Overlap and Kinetic Energy

The gradient of the overlap,

$$\langle g'|g \rangle \equiv M = e^{\frac{1}{4} \mathbf{v}^\mathsf{T} B^{-1} \mathbf{v}} \left(\frac{\pi^N}{\det(B)} \right)^{3/2} , \tag{68}$$

with respect to the shift vector $\mathbf{s}$ is given in matrix notation as

$$\frac{\partial}{\partial \mathbf{s}^\mathsf{T}} M = \frac{1}{2} B^{-1} \mathbf{v} M . \tag{69}$$

In component notation the gradient is given as

$$\frac{\partial}{\partial \vec{s}_{ia}} M = \frac{1}{2} \sum_{j=1}^{N} B_{ij}^{-1} \vec{v}_{ja} M. \tag{70}$$

The gradient of the kinetic energy,

$$\left\langle g' \left| -\frac{\partial}{\partial \mathbf{r}} \Lambda \frac{\partial}{\partial \mathbf{r}^\mathsf{T}} \right| g \right\rangle = \left(6 \operatorname{trace}(A' \Lambda A B^{-1}) + (\mathbf{s}' - 2A'\mathbf{u})^\mathsf{T} \Lambda (\mathbf{s} - 2A\mathbf{u}) \right) M, \tag{71}$$

(where $\mathbf{u} = \frac{1}{2} B^{-1}(\mathbf{s} + \mathbf{s}')$) with respect to the shift vector is given in the matrix notation as

$$\frac{\partial}{\partial \mathbf{s}^\mathsf{T}} \left\langle g' \left| -\frac{\partial}{\partial \mathbf{r}} \Lambda \frac{\partial}{\partial \mathbf{r}^\mathsf{T}} \right| g \right\rangle = \left(6 \operatorname{trace}(A' \Lambda A B^{-1}) + (\mathbf{s}' - 2A'\mathbf{u})^\mathsf{T} \Lambda (\mathbf{s} - 2A\mathbf{u}) \right) \frac{\partial M}{\partial \mathbf{s}^\mathsf{T}} \tag{72}$$

$$+ \left(\Lambda \mathbf{s}' - 2\Lambda A'\mathbf{u} - B^{-1} A' \Lambda \mathbf{s} - B^{-1} A \Lambda \mathbf{s}' + 2B^{-1} A' \Lambda A \mathbf{u} + 2B^{-1} A \Lambda A' \mathbf{u} \right) M. \tag{73}$$

4.1.2 Gaussian Form-Factor

For the central Gaussian matrix element,

$$\left\langle g' \left| e^{-\mathbf{r}^\mathsf{T} w w^\mathsf{T} \mathbf{r}} \right| g \right\rangle \equiv M' = e^{\frac{1}{4} \mathbf{v}^\mathsf{T} B'^{-1} \mathbf{v}} \left(\frac{\pi^N}{\det(B')} \right)^{3/2}, \tag{74}$$

the gradient with respect to the shift vector is given as

$$\frac{\partial}{\partial \mathbf{s}^\mathsf{T}} M' = \frac{1}{2} B'^{-1} \mathbf{v} M', \tag{75}$$

or, in component notation,

$$\frac{\partial}{\partial \vec{s}_{ia}} M' = \frac{1}{2} \sum_{j=1}^{N} B_{ij}'^{-1} \vec{v}_{ja} M'. \tag{76}$$

The gradient of the Gaussian tensor matrix element (39) is given as

$$\frac{\partial}{\partial \mathbf{s}^\mathsf{T}} \left\langle g' \left| e^{-\mathbf{r}^\mathsf{T} w w^\mathsf{T} \mathbf{r}} (\mathbf{y}_1^\mathsf{T} \mathbf{r})(\mathbf{y}_2^\mathsf{T} \mathbf{r}) \right| g \right\rangle = \left(\frac{1}{2} \mathbf{y}_1^\mathsf{T} B'^{-1} \mathbf{y}_2 + (\mathbf{y}_1^\mathsf{T} \mathbf{u}')(\mathbf{y}_2^\mathsf{T} \mathbf{u}') \right) \frac{\partial}{\partial \mathbf{s}^\mathsf{T}} M'$$

$$+ \left(\frac{1}{2} B'^{-1} \mathbf{y}_1 (\mathbf{y}_2^\mathsf{T} \mathbf{u}') + (\mathbf{y}_1^\mathsf{T} \mathbf{u}') \frac{1}{2} B'^{-1} \mathbf{y}_2 \right) M', \tag{77}$$

The spin-orbit matrix element (59) is in fact a linear combination of factors

$$\vec{s}_{ib} \vec{s}_{jc} M', \ \vec{s}_{ib} M' \tag{78}$$

whose gradients with respect to $\vec{s}_{kd}$ are given as

$$\frac{\partial}{\partial \vec{s}_{kd}} \vec{s}_{ib} \vec{s}_{jc} M' = \delta_{ik} \delta_{bd} \vec{s}_{jc} M' + \vec{s}_{ib} \delta_{jk} \delta_{cd} M' + \vec{s}_{ib} \vec{s}_{jc} \frac{\partial M'}{\partial \vec{s}_{kd}}, \tag{79}$$

$$\frac{\partial}{\partial \vec{s}_{kd}} \vec{s}_{ib} M' = \delta_{ik} \delta_{bd} M' + \vec{s}_{ib} \frac{\partial M'}{\partial \vec{s}_{ib}}. \tag{80}$$

4.1.3 General Form-Factor

The central matrix element of a general form-factors has an additional factor J,

$$\left\langle g' \left| f(w^{\mathsf{T}} \mathbf{r}) \right| g \right\rangle = M J, \tag{81}$$

which depends upon the shift vector $\mathbf{s}$ via the quantity $q = |\frac{1}{2} w^{\mathsf{T}} B^{-1} \mathbf{v}|$ whose gradient is given as

$$\frac{\partial q}{\partial \vec{s}_{ia}} = \frac{1}{q} \sum_{k=1}^{N} h_i h_k \vec{v}_{ka}, \tag{82}$$

where $h = w^{\mathsf{T}} \frac{1}{2} B^{-1}$. The gradient of the general central matrix element with respect to the shift-vector is then given as

$$\frac{\partial}{\partial \vec{s}_{ia}} (MJ) = \frac{\partial M}{\partial \vec{s}_{ia}} J + M \frac{\partial J}{\partial \vec{s}_{ia}}, \tag{83}$$

where $\frac{\partial M}{\partial \vec{s}_{ia}}$ is given above (70) and where

$$\frac{\partial J}{\partial \vec{s}_{ia}} = \frac{\partial J}{\partial q} \frac{\partial q}{\partial \vec{s}_{ia}}. \tag{84}$$

The derivative $\partial J / \partial q$ has been considered in section 3.3.2. It is analytic for the screened Yukawa form-factor and its descendants (31) and can be calculated with a Maxima script like the following,

```
assume(beta>0,gamma>0,mu>0,q>0);
J(f) := (beta/%pi)^(3/2)*2*%pi/beta/q*exp(-beta*q^2)
*integrate(r*f*exp(-beta*r^2)*sinh(2*beta*q*r),r,0,inf);
f: exp(-gamma*r^2-mu*r)/r;
diff(J(f),q);
```

The general tensor and spin-orbit matrix elements contain extra factors of the type ∂J and $\partial^2 J$ whose gradients with respect to s_{ib},

$$\frac{\partial}{\partial \vec{s}_{ib}} \left(\frac{\partial J}{\partial q} \right)^n = \left(\frac{\partial J}{\partial q} \right)^{n+1} \frac{\partial q}{\partial \vec{s}_{ib}}, \tag{85}$$

are analytic for the screened Yukawa potential and its descendants (31) and can be calculated by suitable Maxima scripts.

4.2 Gradient with Respect to Correlation Matrix

The matrix elements depend on the correlation matrix A via the quantities $\det(B)$ and B^{-1}, where $B = A' + A$ (or $B' = A' + A + \gamma w w^{\mathsf{T}}$) and therefore one first needs to find the gradients of these quantities. The derivative of the determinant is given by the Jacobi's formula,

$$\frac{\partial \det(B)}{\partial \alpha} = \det(B) \operatorname{trace} \left(B^{-1} \frac{\partial B}{\partial \alpha} \right), \tag{86}$$

and the derivative of the inverse matrix is given as

$$\frac{\partial B^{-1}}{\partial \alpha} = -B^{-1} \frac{\partial B}{\partial \alpha} B^{-1}. \tag{87}$$

Note that this formulas do not require time-consuming recalculation of $\det(B)$ and B^{-1}.

One cannot treat the individual elements of the correlation matrix as variational parameters because of the constraint that the correlation matrix must be positive-definite. Therefore the gradient $\partial B / \partial \alpha$ depends on the chosen parametrization of the correlation matrix.

One popular parametrization is the representation of a correlated Gaussian in the following form,[4]

$$\langle \mathbf{r} \mid g \rangle = \exp\left(-\sum_{i<j=1}^{N} \alpha_{ij}(\vec{r}_i - \vec{r}_j)^2\right) = \exp\left(-\sum_{i<j=1}^{N} \alpha_{ij}\mathbf{r}^{\mathsf{T}}\omega_{ij}\omega_{ij}^{\mathsf{T}}\mathbf{r}\right), \tag{88}$$

where α_{ij} are positive numbers and ω_{ij} is a size-N column of numbers which are all equal zero except for number-i which equals 1 and number-j which equals -1. In this representation the correlation matrix is a linear combination of positive-definite rank-1 matrices,

$$A = \alpha\omega\omega^{\mathsf{T}} + \ldots, \tag{89}$$

with the derivative

$$\frac{\partial B}{\partial \alpha} = \omega\omega^{\mathsf{T}}. \tag{90}$$

The gradients of the determinant and the inverse of the matrix B (and B') are then given as[5]

$$\frac{\partial \det(B)}{\partial \alpha} = \det(B)\left(\omega^{\mathsf{T}}B^{-1}\omega\right). \tag{91}$$

$$\frac{\partial B^{-1}}{\partial \alpha} = -B^{-1}\omega\omega^{\mathsf{T}}B^{-1}. \tag{92}$$

The gradient of the overlap M is given as

$$\frac{\partial}{\partial \alpha}M = \frac{1}{4}\mathbf{v}^{\mathsf{T}}\frac{\partial B^{-1}}{\partial \alpha}\mathbf{v}M - \frac{3}{2}\frac{1}{\det(B)}\frac{\partial \det(B)}{\partial \alpha}M. \tag{93}$$

The gradient of the kinetic energy is given as

$$\frac{\partial}{\partial \alpha}\left\langle g' \left| -\frac{\partial}{\partial \mathbf{r}}\Lambda\frac{\partial}{\partial \mathbf{r}^{\mathsf{T}}} \right| g \right\rangle = \left(6\operatorname{trace}(A'\Lambda AB^{-1}) + (\mathbf{s}' - 2A'\mathbf{u})^{\mathsf{T}}\Lambda(\mathbf{s} - 2A\mathbf{u})\right)\frac{\partial}{\partial \alpha}M$$

$$+ \left(6\operatorname{trace}\left(A'\Lambda A\frac{\partial B^{-1}}{\partial \alpha}\right) + \left(-2A'\frac{\partial \mathbf{u}}{\partial \alpha}\right)^{\mathsf{T}}\Lambda(\mathbf{s} - 2A\mathbf{u}) + (\mathbf{s}' - 2A'\mathbf{u})^{\mathsf{T}}\Lambda\left(-2A\frac{\partial \mathbf{u}}{\partial \alpha}\right)\right)M, \tag{94}$$

where $\mathbf{u} = \frac{1}{2}B^{-1}(\mathbf{s} + \mathbf{s}')$ and $\frac{\partial \mathbf{u}}{\partial \alpha} = \frac{1}{2}\frac{\partial B^{-1}}{\partial \alpha}(\mathbf{s} + \mathbf{s}')$.

4.2.1 Gaussian Form-Factor

The gradient of the central Gaussian matrix element M' is given as

$$\frac{\partial}{\partial \alpha}M' = \frac{1}{4}\mathbf{v}^{\mathsf{T}}\frac{\partial B'^{-1}}{\partial \alpha}\mathbf{v}M' - \frac{3}{2}\frac{1}{\det(B')}\frac{\partial \det(B')}{\partial \alpha}M'. \tag{95}$$

The gradient of the tensor Gaussian matrix element (39) is given as

$$\frac{\partial}{\partial \alpha}\left\langle g' \left| e^{-\mathbf{r}^{\mathsf{T}}ww^{\mathsf{T}}\mathbf{r}}(\mathbf{y}_1^{\mathsf{T}}\mathbf{r})(\mathbf{y}_2^{\mathsf{T}}\mathbf{r}) \right| g \right\rangle = \left(\frac{1}{2}\mathbf{y}_1^{\mathsf{T}}B'^{-1}\mathbf{y}_2 + (\mathbf{y}_1^{\mathsf{T}}\mathbf{u}')(\mathbf{y}_2^{\mathsf{T}}\mathbf{u}')\right)\frac{\partial}{\partial \alpha}M' \tag{96}$$

$$+ \left(\frac{1}{2}\mathbf{y}_1^{\mathsf{T}}\frac{\partial B'^{-1}}{\partial \alpha}\mathbf{y}_2 + \left(\mathbf{y}_1^{\mathsf{T}}\frac{\partial \mathbf{u}'}{\partial \alpha}\right)(\mathbf{y}_2^{\mathsf{T}}\mathbf{u}') + (\mathbf{y}_1^{\mathsf{T}}\mathbf{u}')\left(\mathbf{y}_2^{\mathsf{T}}\frac{\partial \mathbf{u}'}{\partial \alpha}\right)\right)M', \tag{97}$$

where $\mathbf{u}' = \frac{1}{2}B'^{-1}(\mathbf{s} + \mathbf{s}')$ and $\frac{\partial \mathbf{u}'}{\partial \alpha} = \frac{1}{2}\frac{\partial B'^{-1}}{\partial \alpha}(\mathbf{s} + \mathbf{s}')$.

The gradient of the spin-orbit Gaussian matrix element is given as

$$\frac{\partial}{\partial \alpha}\left\langle g' \left| e^{-\gamma\mathbf{r}^{\mathsf{T}}ww^{\mathsf{T}}\mathbf{r}}\left(w^{\mathsf{T}}\mathbf{r} \times w^{\mathsf{T}}\frac{\partial}{\partial \mathbf{r}^{\mathsf{T}}}\right) \right| g \right\rangle = \frac{1}{2}w^{\mathsf{T}}B'^{-1}\mathbf{v} \times \left(w^{\mathsf{T}}\mathbf{s} - w^{\mathsf{T}}AB'^{-1}\mathbf{v}\right)\frac{\partial}{\partial \alpha}M'$$

$$+ \left(\frac{1}{2}w^{\mathsf{T}}\frac{\partial B'^{-1}}{\partial \alpha}\mathbf{v} \times \left(w^{\mathsf{T}}\mathbf{s} - w^{\mathsf{T}}AB'^{-1}\mathbf{v}\right) - \frac{1}{2}w^{\mathsf{T}}B'^{-1}\mathbf{v} \times \left(w^{\mathsf{T}}A\frac{\partial B'^{-1}}{\partial \alpha}\mathbf{v}\right)\right)M'. \tag{98}$$

---

[4] One has to remember however that in this parametrization the correlation matrix has one zero eigenvalue which corresponds to the center-of-mass motion, therefore this degree of freedom has to be removed by first a suitable coordinate transformation and then disregarding the center-of-mass coordinate.

[5] For other parametrization of the correlation matrix one has to use the actual $\partial B/\partial \alpha$.

4.2.2 Other Form-Factors

Compared to Gaussian form-factors the matrix elements of the other form-factors have additional factors J and their derivatives over q. These factors depends on the matrix B via the quantity $\beta = (w^\mathsf{T} B^{-1} w)^{-1}$ and $\vec{q} = w^\mathsf{T} \frac{1}{2} B^{-1} \mathbf{v}$. For a differentiable function $F(\beta, q)$ its gradient with respect to α is given as

$$\frac{\partial}{\partial \alpha} F = \frac{\partial F}{\partial \beta} \frac{\partial \beta}{\partial \alpha} + \frac{\partial F}{\partial q} \frac{\partial q}{\partial \alpha}, \tag{99}$$

$$\frac{\partial \beta}{\partial \alpha} = -\beta^2 \left(w^\mathsf{T} \frac{\partial B^{-1}}{\partial \alpha} w \right), \tag{100}$$

$$\frac{\partial q}{\partial \alpha} = \frac{1}{q} \vec{q} \cdot w^\mathsf{T} \frac{1}{2} \frac{\partial B^{-1}}{\partial \alpha} \mathbf{v}. \tag{101}$$

The derivatives $\left(\left(\frac{\partial}{\partial q} \right)^n \left(\frac{\partial}{\partial \beta} \right)^m J \right)$ can be conveniently calculated by Maxima scripts similar the one in chapter 3.3.2.

5 Conclusion

In quantum few-body physics the analytic matrix elements and their gradients are of importance for the correlated Gaussians method as they facilitate extensive numerical optimizations of the variational wave-functions. In this paper it has been shown that potentials in the form

$$f(w^\mathsf{T}\mathbf{r}), \quad f(w^\mathsf{T}\mathbf{r})(\mathbf{y}^\mathsf{T}\mathbf{r})(\mathbf{z}^\mathsf{T}\mathbf{r}), \quad f(w^\mathsf{T}\mathbf{r})(a^\mathsf{T}\mathbf{r})(b^\mathsf{T}\mathbf{r}), \quad f(w^\mathsf{T}\mathbf{r}) \left(w^\mathsf{T}\mathbf{r} \times w^\mathsf{T} \frac{\partial}{\partial \mathbf{r}^\mathsf{T}} \right), \tag{102}$$

have analytic matrix elements and gradients between shifted correlated Gaussians for the following class of form-factors,

$$f(r) = r^n \frac{e^{-\gamma r^2 - \mu r}}{r}, \quad n = 0, 1, \ldots. \tag{103}$$

Of these form-factors the Gaussian form-factors produce particularly simple and concise analytic expressions. For other form-factors the gradients—although straightforward in principle—in practice require computer algebra software to actually perform the differentiations. Therefore an efficient strategy could be to represent the potentials at hand as linear combinations of Gaussians and then use the analytic expressions for the Gaussians.

References

1. J. Mitroy et al., Theory and applications of explicitly correlated Gaussians. Rev. Mod. Phys. **85**, 693 (2013)
2. S. Bubin, M. Pavanello, W.-C. Tung, K.L. Sharkey, Ludwik Adamowicz, Born-Oppenheimer and non-born-oppenheimer atomic and molecular calculations with explicitly correlated Gaussians. Chem. Rev. **113**, 36 (2013)
3. Mauricio Cafiero, Ludwik Adamowicz, Analytical gradients for Singers multicenter n-electron explicitly correlated Gaussians. Int. J. Quantum Chem. **82**, 151159 (2001)
4. L. Keeper, S.B. Sharkey, L. Adamowicz, An algorithm for calculating atomic D states with explicitly correlated Gaussian functions. J. Chem. Phys. **134**, 044120 (2011)
5. K.M. Daily, H.C. Greene, Extension of the correlated Gaussian hyperspherical method to more particles and dimensionsb. Phys. Rev. A **89**, 012503 (2014)
6. Wei-Cheng Tung, Ludwik Adamowicz, Accurate potential energy curve of the LiH+ molecule calculated with explicitly correlated Gaussian functions. J. Chem. Phys. **140**, 124315 (2014)
7. Sergiy Bubin, Martin Formanek, Ludwik Adamowicz, Universal all-particle explicitly-correlated Gaussians for non-BornOppenheimer calculations of molecular rotationless states. Chem. Phys. Lett. **647**, 122 (2016)
8. X.Y. Yin, D. Blume, Trapped unitary two-component Fermi gases with up to ten particles. Phys. Rev. A **92**, 013608 (2015)
9. M. Galassi et al, GNU Scientific Library Reference Manual–Third Edition (January 2009), ISBN 0954612078
10. Y. Suzuki, K. Varga, *Stochastic Variational Approach to Quantum-Mechanical Few-Body Problems* (Springer, Berlin, 1998)
11. P. Ring, P. Schuck, The Nuclear Many-Body Problem, ISBN 3-540-09820-8 (Springer, Berlin, 1980)
12. Maxima.sourceforge.net. Maxima, a Computer Algebra System. Version 5.34.1 (2014). http://maxima.sourceforge.net/

Few-Body Syst (2016) 57:1227–1241
DOI 10.1007/s00601-016-1157-2

E. Garrido · A. Kievsky · M. Viviani

Three-Body Coulomb Functions in the Hyperspherical Adiabatic Expansion Method

Received: 2 August 2016 / Accepted: 9 September 2016 / Published online: 4 October 2016
© Springer-Verlag Wien 2016

Abstract In this work we describe a numerical method devised to compute continuum three-body wave functions. The method is implemented using the hyperspherical adiabatic expansion for the three-body wave function imposing a box boundary condition. The continuum energy spectrum results discretized and, for specific quantum number values, all the possible incoming and outgoing channels are simultaneously computed. For a given energy, the hyperradial continuum functions form a matrix whose ij-term refers to specific incoming and outgoing channels. When applied to three-body systems interacting only through the Coulomb potential, this method provides the adiabatic representation of the regular three-body Coulomb wave function. The computation of the irregular Coulomb wave function representation is also discussed. These regular and irregular Coulomb functions can be used to extract the $\mathcal{S}$-matrix for those reactions where, together with some short-range potential, the Coulomb interaction is also present. The method is illustrated in the case of the $3 \rightarrow 3$ process of three alpha particles.

1 Introduction

The description of a system in the continuum is equivalent to describing the collision between the particles involved in the system under investigation. In other words, one has to keep track on how the particles approach to each other, how they interact, and how they move far apart. This information is in fact kept in the $\mathcal{S}$-matrix of the reaction which, in a partial wave decomposition of the process, has a dimension equal to the number of open channels for specific values of the quantum number selected. Given a total energy E, the square of each term, $|S_{ij}(E)|^2$, gives the probability of going out through channel j when the particles approach through channel i. Explicitly

$$\sum_j |S_{ij}(E)|^2 = 1 \tag{1}$$

with the number of open channels being finite or, as many times happens, infinite. In this last case the above relation can be used as a criterium for truncation. From a theoretical point of view, a complete solution of the scattering problem corresponds to the knowledge of the $\mathcal{S}$-matrix. In principle this could be done by computing the continuum wave function up to sufficiently large distances such that the wave function can be matched to

---

This article belongs to the special issue "30th anniversary of Few-Body Systems".

This work was supported by funds provided by DGI of MINECO (Spain) under contract No. FIS2014-51971-P.

E. Garrido (✉)
Instituto de Estructura de la Materia, CSIC, Serrano 123, 28042 Madrid, Spain
E-mail: e.garrido@csic.es

A. Kievsky · M. Viviani
Istituto Nazionale di Fisica Nucleare, Largo Pontecorvo 3, 56100 Pisa, Italy

its known analytical behaviour. This strategy is frequently used in configuration space methods and requires knowledge of the asymptotic solutions of the problem. For example, in the case of two particles interacting through a short-range potential, the analytical form could be given in terms of ingoing and outgoing Bessel functions with their relative weight determined by the S-matrix. When the long-range Coulomb interaction is present the Bessel functions are replaced by Coulomb functions. The extension of this strategy to treat scattering processes of three particles is a delicate task. The Faddeev equations have been largely used in momentum as well as in configuration space to describe $N - d$ collisions below and above the deuteron breakup threshold [1]. The $p - d$ case has been a challenge for many years due to long range Coulomb interaction between the two protons. A solution to this problem has been given in momentum space solving the AGS equations and using a screening renormalization procedure [2]. In configuration space different approaches have been used as the Faddeev equations [3], the Kohn variational principle [4,5], and complex scaling [6].

The mentioned references are examples of the difficulties and huge efforts done to solve the scattering problem for three particles. New methods appeared in recent years trying to avoid the explicit description of the wave function for large distances [7] or using a surface-integral approach [8]. Using similar ideas, a method based on two integral relations, derived from the Kohn variational principle, has been proposed in order to obtain the S-matrix [9–11]. The advantage of this method is that the S-matrix is computed from integrals in which only the internal part of the continuum wave function enters, avoiding the explicit knowledge of the wave function in the asymptotic region. As a consequence of this, the required numerical effort is drastically reduced. However, it is important to keep in mind that, although the large distance part of the wave function is not needed, application of the integral relations still requires the knowledge of the regular and irregular asymptotic solutions of the problem, since these functions enter explicitly in the integral relations. When only short-range interactions are involved, these free solutions can be given in terms of regular and irregular Bessel (or Hankel) functions, whose orders are well defined. The problem comes when the long-range Coulomb interaction enters. In this case the asymptotic solutions are known to be regular and irregular Coulomb functions, but for processes involving more than two particles only approximate analytical solutions exist [12]. An additional problem concerns the fact that, in order to obtain the complete S-matrix using a partial wave decomposition and for specific values of the quantum number selected, the continuum wave functions for all the possible incoming channels are necessary. In practice, this means that, for a given energy, the solution is degenerate and the corresponding full set of orthogonal wave functions are needed. How to compute all of them, being consistent with their normalization, and fulfilling the necessary orthogonality properties is not a simple task either.

The purpose of this work is twofold. First, we shall describe an efficient method that permits to compute continuum wave functions keeping all the information concerning the incoming and outgoing channels. The functions for all the open channels will be obtained simultaneously. In particular, this work will focus on three-body systems, and the wave functions will be obtained by means of the hyperspherical adiabatic expansion method [13]. When applied to the case in which only the Coulomb interaction is considered, the method provides a representation of three-body regular Coulomb functions in terms of the hyperspherical adiabatic basis. The second goal of this work will be to show how to construct the three-body irregular Coulomb functions from the previously computed regular Coulomb functions. In this way, with the regular and irregular partners at hand, the matching procedure can be performed allowing for a first order estimate of the S-matrix. A second order estimate is obtained by applying the integral relations as discussed for example in Ref. [14]. The three-alpha system will be taken as an example in order to illustrate an application of the method.

In the next two sections we briefly describe the main aspects of the hyperspherical adiabatic expansion method. Special emphasis is made on the asymptotic properties of the wave functions for the short-range case, Sect. 2, and the Coulomb case, Sect. 3. In Sect. 4 we describe the method used to compute the full continuum wave functions, and therefore the regular Coulomb functions. In the same section we also give the details of the procedure that permits to obtain the irregular Coulomb functions from the previously computed regular ones. As an illustration, the results obtained for the three-alpha system are shown in Sect. 5. The computed regular and irregular Coulomb functions are tested in Sect. 6. Finally, in Sect. 7 a Summary and the Conclusions are given.

2 The Adiabatic Expansion Method: The Short-Range Asymptotics

Three-body systems are usually described by use of the Jacobi coordinates, from which the hyperspherical coordinates $\{\rho, \Omega\}$ can be constructed [13]. The only radial coordinate is the hyperradius ρ, and Ω collects the five hyperangles. The three-body wave function describing the system is in general expanded in terms of some complete basis set $\{|\varphi_i\rangle\}$, like for instance the hyperspherical harmonics, which contains the whole dependence on the hyperangles. The coefficients in the expansion are the radial wave functions, which are obtained as the solution of a coupled set of differential equations where some effective radial potentials, $V_{ij}^{(eff)}(\rho) = \langle\varphi_i|\hat{V}|\varphi_j\rangle_\Omega$, enter. In this expression $\hat{V}$ represents the interaction between the three particles and $\langle||\rangle_\Omega$ indicates integration over the hyperangles.

When dealing with three particles in the continuum the procedure is exactly the same, and the indeces i and j, which in the definition of $V_{ij}^{(eff)}(\rho)$ are associated to the different terms in the expansion of the three-body wave function, can also be understood as the different channels through which the three particles can approach each other or move far apart from each other after the collision. The quantum numbers associated to the state $|\varphi_i\rangle$ determine the quantum numbers associated to channel i. In this way, if the three-body energy E is such that $E > V_{ii}^{(eff)}(\infty)$, then the channel i is open, meaning that there exist a three-body wave function, Ψ_i, that describes the three-body system when the three particles approach each other through that channel. In other words, the three-body continuum problem is equivalent to a multichannel problem, in such a way that the full three-body wave function Ψ contains as many components, Ψ_i, as open channels are accessible at the energy under consideration.

In the particular case of the adiabatic expansion method, the different channels are associated to the adiabatic potentials used in the expansion of the three-body wave function. More precisely, the full continuum three-body wave function can be written as:

$$
\Psi = \begin{pmatrix} \Psi_1 \\ \Psi_2 \\ \vdots \\ \Psi_{n_0} \end{pmatrix} = \frac{1}{\rho^{5/2}} \begin{pmatrix} f_{11} & f_{21} & \cdots & f_{n_A 1} \\ f_{12} & f_{22} & \cdots & f_{n_A 2} \\ \vdots & \vdots & \vdots & \vdots \\ f_{1n_0} & f_{2n_0} & \cdots & f_{n_A n_0} \end{pmatrix} \begin{pmatrix} \Phi_1 \\ \Phi_2 \\ \vdots \\ \Phi_{n_A} \end{pmatrix}
\tag{2}
$$

where Ψ_i is the three-body wave function corresponding to the incoming channel i, n_0 is the number of open channels, and n_A is the number of terms included in adiabatic expansion. From the expression above it is obvious that the wave function Ψ_i is written as:

$$
\Psi_i = \frac{1}{\rho^{5/2}} \sum_{n=1}^{n_A} f_{ni}(\rho)\Phi_n(\rho, \Omega),
\tag{3}
$$

which is the usual form of the three-body wave function after an adiabatic expansion, and the complete basis set of angular functions $\{\Phi_n(\rho, \Omega)\}$ is obtained as the eigenfunctions of the angular part of the hamiltonian $\hat{\mathcal{H}}_\Omega$ (see [13] for details):

$$
\hat{\mathcal{H}}_\Omega \Phi_n(\rho, \Omega) = \frac{\hbar^2}{2m} \frac{1}{\rho^2} \lambda_n(\rho) \Phi_n(\rho, \Omega),
\tag{4}
$$

where m is normalization mass used to define the Jacobi coordinates.

As shown in Ref.[13], the asymptotic behavior of the λ_n-functions in Eq. (4) permits a clean distinction between those channels associated to a particle-dimer structure (1+2 channels) and those corresponding to having the three particles in the continuum (breakup channels). In this way, for three-body systems containing 1+2 channels and three-body energies below the three-body breakup threshold, we have that the breakup channels are closed, and therefore $n_0 < n_A$ in Eq. (2). Furthermore, n_0, which in this case is the number of 1+2 channels open, is typically a very modest number [10]. Also, when n refers to one of the closed channels, the radial functions f_{ni} in Eq. (4) go to zero asymptotically, and only the n_0 open channels survive at large distances. For energies above the breakup threshold all the channels are open, and $n_0 = n_A$.

The radial wave functions f_{ni}, where i and n indicate the incoming and outgoing channels, respectively, are obtained by solving the set of coupled radial equations [13]:

$$
\left(-\frac{\partial^2}{\partial\rho^2} - Q_{nn}(\rho) + \frac{\lambda_n(\rho) + \frac{15}{4}}{\rho^2} - \frac{2mE}{\hbar^2}\right) f_{ni}(\rho) - \sum_{n' \neq n} \left(2P_{nn'}(\rho)\frac{\partial}{\partial\rho} + Q_{nn'}(\rho)\right) f_{n'i}(\rho) = 0,
\tag{5}
$$

where E is the total three-body energy, the $\lambda_n(\rho)$-functions are the eigenvalues of the angular part of the hamiltonian (Eq. 4), and the coupling functions $P_{nn'}(\rho)$ and $Q_{nn'}(\rho)$ can be found for instance in [13].

For a given incoming channel i, the asymptotic behaviour of the Ψ_i wave function can be written as [10]:

$$\Psi_i \to \sum_{n=1}^{n_0} (A_{in} F_n + B_{in} G_n),$$ (6)

where, for short-range interactions, the regular and irregular functions F_n and G_n are known to be given in terms of the regular, j, and irregular, η, spherical Bessel functions as [11]:

$$F_n = \sqrt{k_y^{(n)}} \, j_{\ell_y} \left(k_y^{(n)} y_n \right) \frac{1}{\rho^{3/2}} \Phi_n(\rho, \Omega),$$ (7)

$$G_n = \sqrt{k_y^{(n)}} \, \eta_{\ell_y} \left(k_y^{(n)} y_n \right) \frac{1}{\rho^{3/2}} \Phi_n(\rho, \Omega),$$ (8)

when n corresponds to a $1 + 2$ channel, and by

$$F_n = \sqrt{\kappa} \, j_{K+\frac{3}{2}}(\kappa\rho) \frac{1}{\rho^{3/2}} \Phi_n(\rho, \Omega),$$ (9)

$$G_n = \sqrt{\kappa} \, \eta_{K+\frac{3}{2}}(\kappa\rho) \frac{1}{\rho^{3/2}} \Phi_n(\rho, \Omega),$$ (10)

when n corresponds to a breakup channel. In Eqs. (7) and (8) ℓ_y is the relative orbital angular momentum between the dimer and the third particle, y_n is the y-Jacobi coordinate when the x-coordinate is constructed between the two particles in the dimer, and $k_y^{(n)}$ is the momentum associated to the Jacobi coordinate y_n, which except for some mass factors corresponds to the relative momentum between the dimer and the third particle [16]. In Eqs. (9) and (10) the index K is the hypermomentum associated to the adiabatic eigenvalue λ_n (which goes asymptotically to $K(K+4)$), and $\kappa = \sqrt{2mE}/\hbar$.

Using Eqs. (6)–(10) we then have that the asymptotic form of the full three-body wave function Ψ can be written in a compact matrix form as:

$$\Psi \to AF + BG$$ (11)

where A and B are $n_0 \times n_0$ matrices made by the A_{in} and B_{in} elements in Eq. (6) and where F and G are column vectors with n_0 terms. For short-range interactions the nth term in the F and G vectors is given by:

$$F_n = \frac{1}{\rho^{5/2}} f_n^{reg} \Phi_n(\rho, \Omega) \quad \text{with} \quad f_n^{reg} = \frac{k_y^{(n)} y_n}{\sqrt{k_y^{(n)}}} j_{\ell_y}(k_y^{(n)} y_n),$$ (12)

and

$$G_n = \frac{1}{\rho^{5/2}} f_n^{irr} \Phi_n(\rho, \Omega) \quad \text{with} \quad f_n^{irr} = \frac{k_y^{(n)} y_n}{\sqrt{k_y^{(n)}}} \eta_{\ell_y}(k_y^{(n)} y_n),$$ (13)

respectively, when n is a 1+2 channel. If n is a breakup channel ℓ_y, y_n, and $k_y^{(n)}$ should be replaced by $K + 3/2$, ρ, and κ, respectively, in the expressions above. In other words, the column vectors F and G, which describe the regular and irregular free solutions, can be written as:

$$F = \begin{pmatrix} F_1 \\ F_2 \\ \vdots \\ F_{n_0} \end{pmatrix} = \frac{1}{\rho^{5/2}} \begin{pmatrix} f_1^{reg} & 0 & \cdots & 0 \\ 0 & f_2^{reg} & \cdots & 0 \\ \vdots & \vdots & \ddots & \vdots \\ 0 & 0 & \cdots & f_{n_0}^{reg} \end{pmatrix} \begin{pmatrix} \Phi_1 \\ \Phi_2 \\ \vdots \\ \Phi_{n_0} \end{pmatrix},$$ (14)

and the same for G, but replacing the diagonal elements f_i^{reg} by their irregular partners f_i^{irr} given in Eq. (13).

Summarizing, for short-range interactions, the adiabatic expansion of the regular and irregular free solutions takes the form given in Eq. (14), and the corresponding radial matrix is diagonal. The radial diagonal elements are given in terms of regular and irregular Bessel functions, and they are known analytically. The complete solution to the scattering problem is obtained from the knowledge of the matrices A and B. In fact the $\mathcal{K}$-matrix of the reaction is

$$\mathcal{K} = A^{-1} B$$ (15)

and the $\mathcal{S}$-matrix is obtained from the relation $\mathcal{S} = (1 + i\mathcal{K})(1 - i\mathcal{K})^{-1}$. As described in Refs. [10, 11, 15], a second order estimate of the matrices A and B is provided by the following relations

$$A_{ij} = -\frac{2m}{\hbar^2} \langle \Psi_i | \hat{\mathcal{H}} - E | G_j \rangle \qquad (16)$$

$$B_{ij} = \frac{2m}{\hbar^2} \langle \Psi_i | \hat{\mathcal{H}} - E | F_j \rangle. \qquad (17)$$

Applications to atom-dimer and $n - d$ collisions have been given in the mentioned references.

3 The Adiabatic Expansion Method: The Coulomb Asymptotics

When the Coulomb potential is present, the $\lambda_n(\rho)$-functions in the radial equations (5) behave at large distances as $\lambda_n(\rho) \to a_n\rho + b_n\sqrt{\rho} + c_n + \mathcal{O}(1/\sqrt{\rho})$. As a consequence, the effective adiabatic potentials are long-range potentials, going asymptotically as $\lambda_n/\rho^2 \to 1/\rho$. On top of this, some of the coupling $P_{nn'}$ functions show the same $1/\rho$ behavior. All this implies that the different radial functions $f_{ni}(\rho)$ do not decouple at large distances. Therefore, when solving the free Coulomb three-body hamiltonian the regular and irregular solutions can still be written as in Eq. (14), but with the important difference that now the radial matrix is not diagonal anymore. That is, the regular solutions can be written as:

$$F = \begin{pmatrix} F_1 \\ F_2 \\ \vdots \\ F_{n_0} \end{pmatrix} = \frac{1}{\rho^{5/2}} \begin{pmatrix} f_{11}^{reg} & f_{21}^{reg} & \cdots & f_{n_01}^{reg} \\ f_{12}^{reg} & f_{22}^{reg} & \cdots & f_{n_02}^{reg} \\ \vdots & \vdots & \vdots & \vdots \\ f_{1n_0}^{reg} & f_{2n_0}^{reg} & \cdots & f_{n_0n_0}^{reg} \end{pmatrix} \begin{pmatrix} \Phi_1 \\ \Phi_2 \\ \vdots \\ \Phi_{n_0} \end{pmatrix} = \frac{1}{\rho^{5/2}} R_{reg} \Phi, \qquad (18)$$

and the same for G, but in terms of the corresponding irregular radial matrix R_{irr} which contains the irregular functions f_{ni}^{irr}. The functions F and G, when written as in Eq. (18), are the adiabatic expansion of the three-body regular and irregular Coulomb functions.

It is known that any $M \times N$ matrix, with $M \geq N$, can be written as the product of an $M \times N$ column-orthogonal matrix U, an $N \times N$ diagonal matrix W, and the transpose of an $N \times N$ orthogonal matrix V. In other words, the radial matrix R_{reg} in Eq. (18) can be written for each ρ as:

$$R_{reg} = U W_{reg} V^T, \qquad (19)$$

where $U^T U = \mathbb{I}$, $V^T V = \mathbb{I}$, and W_{reg} is diagonal. This is the so-called singular value decomposition.

Therefore, the regular Coulomb function (18) can be written in a compact form as:

$$F = \frac{1}{\rho^{5/2}} R_{reg} \Phi = \frac{1}{\rho^{5/2}} U W_{reg} V^T \Phi, \qquad (20)$$

from which we can define

$$F_U = U^T F = \frac{1}{\rho^{5/2}} W_{reg} V^T \Phi = \frac{1}{\rho^{5/2}} W_{reg} \Phi_V, \qquad (21)$$

which, since the W_{reg} matrix is diagonal, is formally identical to Eq. (14).

Equation (21) is therefore a new representation of the three-body regular Coulomb functions. In this representation the wave function is expanded in a new angular basis set $\{\Phi_V = V^T \Phi\}$, in such a way that, when using this angular basis, the corresponding radial equations decouple. Each term in the vector function F_U corresponds to a free, uncoupled, three-body Coulomb solution, whose radial wave functions (which are the different terms in the diagonal matrix W_{reg}) must correspond to regular Coulomb functions. However, the order and the Sommerfeld parameters entering in these regular Coulomb functions are not known, and they should in principle be determined numerically. A more rigorous demonstration of the fact that the diagonal functions in W_{reg} are Coulomb functions is given in Appendix.

When this is done, we of course immediately have the irregular partner of F, which has the form:

$$G = \frac{1}{\rho^{5/2}} U W_{irr} V^T \Phi = \frac{1}{\rho^{5/2}} R_{irr} \Phi, \qquad (22)$$

and where W_{irr} is analogous to W_{reg} but containing in the diagonal the irregular Coulomb functions whose order and Sommerfeld parameters are the same as the ones obtained for the regular part.

The irregular Coulomb functions contained in the diagonal matrix W_{irr}, each of them with some order ξ and Sommerfeld parameter η, can be regularized as done for instance in [11] for the irregular Bessel functions, i.e., by solving the Schrödinger equation with the Coulomb potential together with some arbitrary short-range interaction. The solutions of this equation are known to behave asymptotically as $\Psi_\xi(\eta, z) \rightarrow f_\xi(\eta, z) + \tan\delta g_\xi(\eta, z)$, where $f_\xi(\eta, z)$ and $g_\xi(\eta, z)$ are the regular and irregular Coulomb functions. It is then possible to construct the regularized irregular Coulomb function $\tilde{g}_\xi(\eta, z)$ as:

$$\tilde{g}_\xi(\eta, z) = \frac{\Psi_\xi(\eta, z) - f_\xi(\eta, z)}{\tan\delta}, \tag{23}$$

which by construction is regular at the origin and goes asymptotically to $g_\xi(\eta, z)$.

These irregular Coulomb functions can be used, by means of Eq. (22), to construct the regularized version of the irregular matrix G, which will denoted by $\tilde{G}$.

4 Calculation of Full Three-Body Coulomb Functions

In order to get the radial matrix R_{reg} in Eq. (18) corresponding to the regular three-body Coulomb function for a given three-body energy $E > 0$, one has to solve the three-body problem for that energy including only the Coulomb interaction between the three particles. In the adiabatic expansion method this means that, after solving the angular eigenvalue problem (4), one has to solve the coupled set of radial equations (5). A simple way of doing so is to impose a box boundary condition. In other words, the radial wave functions $f_{ni}(\rho)$ are imposed to be zero for some large value L of the hyperradius. When doing so, the continuum spectrum is automatically discretized.

Each of the discrete states obtained in this way can in principle be normalized to 1 inside the box. In other words, the radial matrix R_{reg} can be normalized such that $\int R_{reg}^T R_{reg} d\rho = \mathbb{I}$, where $\mathbb{I}$ is the identity matrix. However, as shown in [17], when using this normalization for the discrete continuum solutions of the free Schrödinger equation, the radial wave functions obtained are the Coulomb functions (if the Coulomb interaction is present) or the Bessel functions (if no Coulomb interaction is present) but multiplied by the factor $\sqrt{2/L}$. For this reason it is more convenient to normalize the radial matrix R_{reg} such that:

$$\int R_{reg}^T R_{reg} d\rho = \frac{L}{2} \mathbb{I}. \tag{24}$$

Although this is not crucial, when using the normalization above, the R_{reg} matrix is a well-normalized Coulomb function, in the sense that, when diagonalized, the functions in the diagonal are precise Coulomb functions, with no additional normalization factors.

It is important to note that when using this simple discretization method no condition is given in order to determine the incoming and outgoing channels associated to each of the discrete states. However, as shown in Ref. [18], for each three-body energy, all the possible combinations between incoming and outgoing channels are automatically obtained. This is because when discretizing the energy spectrum, the discrete energies appear in groups of states, in such a way that each group contains n_0 terms, which, in this case ($n_0 = n_A$), is the number of adiabatic terms included in the expansion (3). As also shown in Ref. [18], the larger the value of the box size L, the closer to degeneracy the states in each group. Eventually, for an infinitely big box, all the n_0 states in the group merge to the same energy, and they correspond to each of the F_i functions given in Eq. (18), which, in turn, contain all the possible incoming and outgoing channels for that energy.

In practice, the discretization has to be done, of course, using a finite value of L. This means that the computed states in a group of energies, which in principle describe the same three-body state, do actually have slightly different energies. We solve this problem by modifying the value of L around the initially chosen value for each of the n_0 states in the energy group, in such a way that we get all of them at exactly the same energy. The price to pay when doing this, is that since each state in the energy group is normalized to $L/2$ (as shown in Eq. 24), we then get that the normalization of each of them is not exactly the same. We have therefore changed slightly different energies and the same normalization by the same energy and slightly different normalization. However, this second choice is much more convenient. For a typical box size of $L = 300$ fm variations in L of no more than 1 or 2 fm (very usually clearly less that this) are enough to fine tune the energy of the state to the

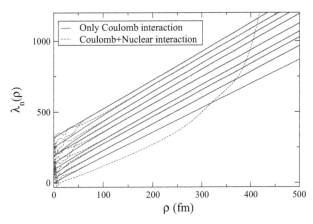

Fig. 1 (color online) The *solid lines* show the λ_n-functions obtained for three α-particles interacting only via the Coulomb potential when coupled to total spin and parity 0^+. The *dashed curves* are the corresponding λ_n-functions when some nuclear α-α interaction (Ali-Bodmer potential) is also included

desired value. The change in the normalization from $L/2$ to $(L+2)/2$ amounts to multiplying the three-body wave function by the factor $\sqrt{(L+2)/L}$, which for $L = 300$ fm is a factor of about 1.003, whose effect on the final results is inappreciable.

Summarizing, the procedure in order to compute the regular three-body Coulomb functions in the adiabatic expansion method is as follows: (i) Solve the angular part of the three-body hamiltonian, Eq. (4), for the three-body system including only the Coulomb potential, (ii) solve the coupled set of radial equations (5) with a box boundary condition at a value L of the hyperradius, (iii) choose the group of n_0 discrete states whose energy is closer to the desired three-body energy E, and (iv) fine tune the value of L such that all the n_0 states are obtained at the same energy E. These states form the radial matrix R_{reg} in Eq. (18), and they are the regular three-body Coulomb wave function for the three-body system under investigation at the energy E. Of course, if the angular part of the three-body hamiltonian in step (i) is solved for arbitrary interactions between the particles, this same procedure obviously permits to obtain the full three-body wave function of that particular system.

Once the regular function has been computed, its irregular partner can be obtained by (i) decomposing the radial matrix R_{reg} for each ρ as given in Eq. (19), (ii) matching the functions in the diagonal matrix W_{reg} with regular Coulomb functions using the order and the Sommerfeld parameter as fitting parameters, iii) constructing the W_{irr} diagonal matrix in Eq. (22) formed by the irregular partners of the regular Coulomb functions obtained in the previous step, and iv) getting the irregular radial matrix R_{irr} as $U W_{irr} V^T$, in analogy with Eq. (19). If the irregular Coulomb functions in the diagonal matrix W_{irr} are regularized, for instance as given in Eq. (23), the new radial matrix $\tilde{W}_{irr}$ provides the regularized irregular partner of R_{reg}, which is given by $\tilde{R}_{irr} = U \tilde{W}_{irr} V^T$.

5 The Three-Alpha Case

Let us illustrate the method described in the previous sections by taking the three-alpha system as an example. In this case there are no bound two-body subsystems, although the two-alpha system, ^{8}Be, has an extremely narrow (only a few ev wide) 0^+ resonance very close to the two-alpha threshold. The case where bound two-body subsystems exist (giving rise to the appearance of the so-called 1+2 channels) will be considered in a forthcoming work.

The fact that the system does not contain any 1+2 channel implies that, asymptotically, only states with the three particles in the continuum are possible. This is illustrated in Fig. 1, where the solid (red) lines show the $\lambda_n(\rho)$ functions obtained after solving the eigenvalue problem in Eq. (4) for a system of three bosons with charge $2e$, mass equal to the mass of the alpha-particle, spin and parity 0^+, and interacting only via the Coulomb potential. Obviously, due to the Coulomb repulsion, these lambdas must be associated to situations where all the three particles are in the continuum. When some nuclear interaction is included (for instance the Ali-Bodmer 'd' potential given in Ref. [19]), the corresponding λ_n-functions are given by the dashed (blue) curves. Very soon these λ's overlap with the ones obtained when including only the Coulomb interaction. The

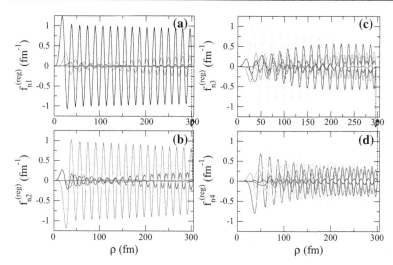

Fig. 2 (color online) First four rows of the radial matrix R_{reg} defined in Eq. (18) for a three-alpha system at an energy of 3.5 MeV. In **a–d** the incident channels correspond to the first, second, third, and fourth adiabatic channels, respectively. The *arrows* show the difference in the box size required in order to fix the total three-body energy at 3.5 MeV

only exception is the lowest λ-function, which is associated to the geometry where two of the alpha-particles populate the low-lying 0^+ resonance in ^{8}Be. As seen in the figure, this λ-function crosses all the other λ's, and eventually, only the ones obtained with only the Coulomb potential (associated to pure three-body continuum asymptotics) survive.

5.1 Calculation of the Regular Three-Body Coulomb Functions

After having solved the angular equation (4) with only the Coulomb interaction (the corresponding λ_n-functions are the solid (red) curves in Fig. 1), we obtain the regular Coulomb functions by solving the coupled set of radial equations (5) with a box boundary condition. We have done it with a box size of $L = 300$ fm and we have included eight adiabatic terms in the expansion (3). As already mentioned, this procedure discretizes the continuum spectrum, in such a way the discrete energies appear in well-separated groups, where each group contains as many discrete states as adiabatic terms are included in the expansion, i.e., eight terms in our choice. Among these groups of states we should then choose the one whose energies appear in the vicinity of the energy of interest. For instance, if we choose a three-body energy of $E = 3.5$ MeV, we should pick the group of eight states that, for $L = 300$ fm, appear with energies ranging between 3.40 and 3.53 MeV.

As already discussed, in order to fine tune the energy to the desired value (3.5 MeV in our example), an efficient way is to slightly modify the value of L for each of the (eight) states in the group. This can be done with relative small variations of L (no more than 2 fm in our calculation using $L = 300$ fm). Note that the larger the value of L, the closer to degeneracy the states belonging to the same group, and therefore the smaller the required variation of L in order to fit the energy of all the states to the same value. As an illustration we show in Fig. 2 the eight radial wave functions corresponding to the incoming channels 1 (Fig. 2a), 2 (Fig. 2b), 3 (Fig. 2c), and 4 (Fig. 2d). They have been obtained for a three-body energy of 3.5 MeV. As we can see, the diagonal terms $f_{11}^{(reg)}$ [black curve in panel (a)] and $f_{22}^{(reg)}$ [red curve in panel (b)] dominate, although the coupling with the other adiabatic channels is evident. In the last two panels the dominance of the diagonal terms, $f_{33}^{(reg)}$ [green curve in panel (c)] and $f_{44}^{(reg)}$ [blue curve in panel (d)], it is not so clear. These couplings would be zero in the case of short-range interactions. In the figure, the dashed arrows show how much the size of the box changes from one case to another in order to fit the energy for each of them to the chosen value of $E = 3.5$ MeV. As seen in the figure, this change is very small compared to the total size of the box, which makes the effect of the different normalization completely negligible.

The curves shown in Fig. 2 correspond to the first four rows of the radial matrix R_{reg} given in Eq. (19). The remaining four rows look similar to the ones shown in the figure. When computed, the radial matrix R_{reg} automatically satisfies that $\int R_{reg}^T R_{reg} d\rho$ is proportional to the identity matrix $\mathbb{I}$, and therefore it can be easily normalized as given in Eq. (24). The computed non-diagonal terms in (24) take values of typically $\sim 10^{-4}$.

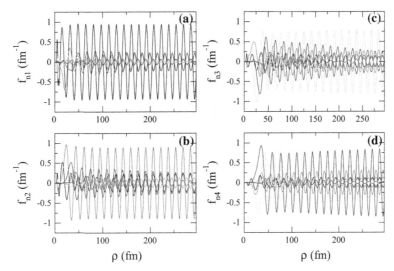

Fig. 3 (color online) First four rows of the full radial matrix wave function defined in Eq. (2) for a three-alpha system at an energy of 3.5 MeV. In **a**–**d** the incident channels correspond to the first, second, third, and fourth adiabatic channels, respectively. The Ali-Bodmer potential (version 'd') in Ref. [19] has been used for the nuclear alpha–alpha interaction

Table 1 Values of the order (ξ) and Sommerfeld (η) parameters obtained when fitting the eight computed eigenvalue functions of the regular matrix R_{reg} by means of the regular Coulomb functions $f_{\xi}(\eta, \rho)$

$f_{\xi}(\eta, \rho)$	1	2	3	4	5	6	7	8
ξ	1.75	5.53	7.82	9.50	11.60	13.46	13.76	15.33
η	2.26	2.45	2.65	2.55	2.67	2.53	2.91	2.88

As already mentioned, the same procedure permits to obtain the full three-body wave function for the system provided that the full (nuclear and Coulomb) two-body interactions are included. For the three-alpha system this corresponds to using in Eq. (5) the λ_n-functions given by the dashed (blue) curves in Fig. 1. In Fig. 3 we show the corresponding first four rows of the full radial wave function given in Eq. (2). Again the three-body wave function energy has been chosen to be $E = 3.5$ MeV.

5.2 Calculation of the Irregular Three-Body Coulomb Functions

In order to compute the irregular partner, R_{irr}, of the regular radial matrix R_{reg}, the first step is to diagonalize R_{reg} for each value of ρ. This is done by the singular value decomposition given in Eq. (19). The diagonal matrix W_{reg} is then adjusted as a function of ρ with regular Coulomb functions whose order (ξ) and Sommerfeld parameters (η) are employed for the fitting. The parameters obtained after the fitting of the eight computed eigenfunctions of the R_{reg} matrix are given in Table 1.

For $\rho = 0$ the λ_n functions entering in the effective adiabatic potentials in Eq. (5) take values equal to $K(K + 4)$, where K is the hypermomentum (see Ref. [13]). In this way, each λ_n is associated to a particular value of K. For the three-alpha system considered in this work, the $\rho = 0$ values of the λ_n-functions plotted in Fig. 1 follow the hypermomentum sequence $K = 0, 4, 6, 8, 10, 12, 12, 14$. In case of dealing with short-range interactions only, the computed regular Coulomb functions would actually be regular Bessel functions, whose order would be given exactly by $K + 3/2$. It is interesting to note that the values given in Table 1 for the order parameter ξ are not dramatically far from $K + 3/2$.

In Fig. 4 we show the absolute value of the eight computed eigenfunctions of the regular matrix R_{reg}, which are contained in the diagonal matrix W_{reg} in Eq. (19). These eigenfunctions are given by the red circles. The corresponding regular Coulomb functions obtained with the parameters given in Table 1 are shown by the solid (blue) curves. As seen in the figure the agreement is rather good.

Using the order and Sommerfeld parameters given in Table 1, one can immediately construct the corresponding irregular functions in the diagonal matrix W_{irr} in Eq. (22). They are given by the thick solid (blue) curves in Fig. 5. These irregular functions can be regularized as given in Eq. (23), giving rise to the regularized

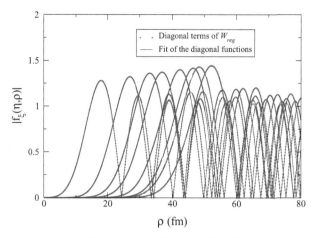

Fig. 4 (color online) The *red circles* show the absolute value of the eigenfunctions contained in the diagonal matrix W_{reg} in Eq. (19). The *blue curves* are the regular Coulomb functions obtained using the order and Sommerfeld parameters given in Table 1

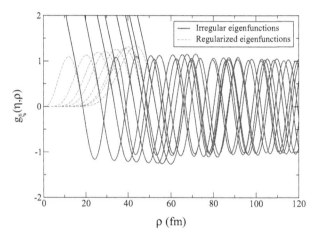

Fig. 5 (color online) The *solid (blue)* curves show the irregular Coulomb functions contained in the diagonal matrix W_{irr} (Eq. (22)) when using the parameters in Table 1. The corresponding regularized functions, $\tilde{W}_{irr}$, are given by the *dashed (red) curves*

irregular Coulomb functions $\tilde{W}_{irr}$, which are shown by the dashed (red) curves in Fig. 5. Replacing now in Eq. (19) the diagonal matrix W_{reg} by W_{irr} (or $\tilde{W}_{irr}$) we finally obtain the irregular partner, R_{irr} (or $\tilde{R}_{irr}$) of the regular matrix R_{reg} shown in Fig. 2. The first four rows of $\tilde{R}_{irr}$ are shown in Fig. 6.

Summarizing, we have computed the regular, R_{reg}, and irregular, R_{irr}, radial Coulomb functions for the three-alpha system at the energy $E = 3.5$ MeV. The full three-alpha function using the nuclear Ali-Bodmer potential and the regularized version of R_{irr} have also been obtained. These matrices account for all the possible incoming and outgoing channels for the chosen three-body energy. The radial matrices R_{reg} and R_{irr} (or $\tilde{R}_{irr}$) permit to obtain, via Eqs. (20) and (22), the adiabatic expansion the three-body Coulomb wave functions F and G (or $\tilde{G}$), which, according to Eq. (11), are needed in order to extract the $\mathcal{S}$-matrix of the process. In particular, F and $\tilde{G}$ can be used in combination with the integral relations described in [9–11], which permit calculation of the $\mathcal{S}$-matrix from the internal part of the three-body wave function. This fact implies that knowledge of F and $\tilde{G}$ is required only up to affordable distances.

6 Test of the Solutions

In the procedure used to obtain the regular and irregular matrices, R_{reg} is directly calculated as the regular solution of the coupled set of radial equations (5). The matrix $\int R_{reg}^T R_{reg} d\rho$ is automatically obtained to be proportional to $\mathbb{I}$, and it can then be normalized according to Eq. (24).

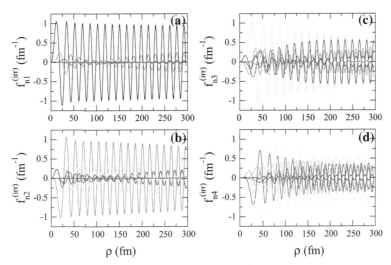

Fig. 6 (color online) First four rows of the (regularized) irregular partner, $\tilde{R}_{irr}$, of the regular radial matrix R_{reg} shown in Fig. 2

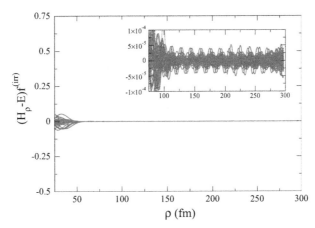

Fig. 7 (color online) *Outer part*: result obtained when applying the set of differential radial equations (5) to the computed (regularized) irregular radial functions in $\tilde{R}_{irr}$. The *inset* shows the detail from $\rho = 50$ fm to $\rho = 300$ fm

Let us check now that the irregular matrix R_{irr}, although not obtained directly as a solution of Eq. (5), is in fact a solution of the radial differential equations. We have done it using the regularized version $\tilde{R}_{irr}$, which is the radial matrix to be used in practice when computing the S-matrix. In this way we also get rid of the numerical inaccuracies arising at small values of ρ due to the divergence of R_{irr}. The first four rows of $\tilde{R}_{irr}$ are shown in Fig. 6. When inserted into the set of Eq. (5) we obtain the result shown in Fig. 7. We show simultaneously the result obtained for the 8 possible incoming channels included in our three-alpha calculation. Since we are using the regularized irregular functions, they are not expected to be a solution of the differential equations at short distances, but, as we can see in the figure, they are good solutions for ρ larger than about 40 fm. In the inset we show a zoom of the result in the large distance region, just to illustrate that for $\rho \gtrsim 75$ fm the result is always close to or smaller than 10^{-4}.

The fact that the irregular matrix is found to be a solution of the three-body equation implies that the orthogonality condition (24) is also satisfied. In fact, this condition is automatically fulfilled by the irregular radial matrix without need of any additional renormalization. Again, the non-diagonal terms of $\int \tilde{R}_{irr}^T \tilde{R}_{irr} d\rho$ are found to be of about 10^{-4}.

An additional test for the computed regular and irregular Coulomb matrices can be made having in mind that, asymptotically, the regular and irregular Coulomb functions behave as the sinus and cosinus, respectively, of the same argument. Therefore, at large distances, the sum of the square of the regular and irregular Coulomb functions has to approach 1. In our matricial language this property implies that the matrix $R_{reg}^T R_{reg} + \tilde{R}_{irr}^T \tilde{R}_{irr}$ should asymptotically approach the identity matrix $\mathbb{1}$. This is what we show in Fig. 8. The outer part shows,

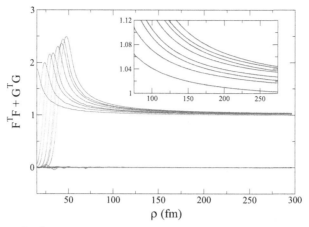

Fig. 8 (color online) $R^T_{reg}R_{reg} + \tilde{R}^T_{irr}\tilde{R}_{irr}$ as a function of ρ. In the inner part a closer plot of the diagonal terms is shown

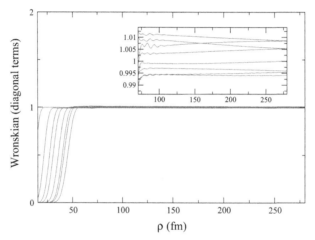

Fig. 9 (color online) Diagonal terms of the Wronskian when using the regular and (regularized) irregular radial matrices R_{reg} and $\tilde{R}_{irr}$. The *inset* shows the detail from $\rho = 70$ fm to $\rho = 300$ fm

as a function of ρ, the 64 functions obtained when computing $R^T_{reg}R_{reg} + \tilde{R}^T_{irr}\tilde{R}_{irr}$. As we can see, eight of the functions (which are the diagonal ones) go to 1, and the remaining ones (non-diagonal terms) go to zero. In the inner part the detail of the long-distance behaviour of the diagonal terms is shown.

Also, the computed regular and irregular Coulomb functions must satisfy the Wronskian relation, which translated again into our matricial language means that

$$R'^T_{reg}\tilde{R}_{irr} - R^T_{reg}\tilde{R}'_{irr} = \mathbb{I}, \tag{25}$$

where R' represents the derivative of the functions contained in the matrix R with respect to $z = \kappa\rho$. This relation has to be satisfied for any value of ρ, although when using the regularized irregular matrix $\tilde{R}_{irr}$ this must be true only at large distances. This is in fact what shown in Fig. 9, where the diagonal terms of the matrix in Eq. (25) are shown. As we can see, for large ρ's the Wronskian approaches pretty much the value of 1. This is better seen in the inset, where the detail from $\rho = 50$ fm to $\rho = 300$ fm is shown. As we can see, all the diagonal terms deviate from 1 by less than 1 %.

Finally, the regular and irregular matrices R_{reg} and $\tilde{R}_{irr}$ permit us to obtain the complete regular and irregular Coulomb functions F and $\tilde{G}$ given in Eqs. (20) and (22). In the same way, the full three-alpha radial matrix (containing the nuclear and Coulomb interaction and whose first four rows are shown in Fig. 3) leads to the total three-body wave function Ψ given in Eq. (2). The asymptotic behaviour of Ψ takes the form given in Eq. (11), where the matrices A and B can be obtained by means of the two integral relations [9–11]:

$$A = -\frac{2m}{\hbar^2}\langle\Psi|\hat{\mathcal{H}} - E|\tilde{G}\rangle \qquad (26)$$

$$B = \frac{2m}{\hbar^2}\langle\Psi|\hat{\mathcal{H}} - E|F\rangle, \qquad (27)$$

where $\hat{\mathcal{H}}$ is the total three-body hamiltonian. The $\mathcal{K}$-matrix of the process is then given by $\mathcal{K} = -A^{-1}B$, from which the $\mathcal{S}$-matrix can obtained as $\mathcal{S} = (1 + i\mathcal{K})(1 - i\mathcal{K})^{-1}$.

After the calculation shown in this paper, making use of the radial matrices partially shown in Figs. 2, 3, and 6, we can obtain the $\mathcal{S}$-matrix for $E = 3.5\,\mathrm{MeV}$, which automatically satisfies the unitarity condition $\mathcal{S}^\dagger\mathcal{S} = \mathbb{1}$. From the $\mathcal{S}$-matrix one can compute the elastic $3 \to 3$ cross section of the process, which is given by [20,21]:

$$\sigma_{el}(E) = 3!\frac{32\pi^2}{\kappa^5}\sum_{J^\pi}\sum_{nn'}(2J + 1)|S^{J^\pi}_{nn'}(E) - \delta_{nn'}|^2, \qquad (28)$$

where the summation of J^π indicates summation over all the possible three-body states with angular momentum J and parity π. However, the expression for the cross section given above is ill-defined, since it depends on the arbitrary choice of the normalization mass m through the three-body momentum $\kappa = \sqrt{2mE}/\hbar$. This is related to the fact that the incoming flux of particles is basically given by κ/m [22], and it therefore depends on the choice made for the normalization mass. Nevertheless, this problem disappears if we consider the reaction rate instead, which for the three-alpha system takes the form [23]:

$$R_{3\alpha}(E) = \frac{\hbar^5}{m_\alpha^3}\frac{144\sqrt{3}}{E^2}\sum_{J^\pi}\sum_{nn'}(2J + 1)|S^{J^\pi}_{nn'}(E) - \delta_{nn'}|^2, \qquad (29)$$

where m_α is the mass of the alpha particle. For the energy considered in this work, $E = 3.5\,\mathrm{MeV}$, and including 8 adiabatic terms, we have obtained that the contribution of the 0^+ three-alpha state to the reaction rate of the process takes the value $R_{3\alpha} = 18900\,\mathrm{fm}^6/s$.

7 Summary and Conclusions

In this paper we have solved a three-body problem using the hyperspherical adiabatic expansion method and considering the Coulomb interaction. Imposing a box boundary condition at sufficiently large distances, we observed that the solutions appear in groups corresponding to discretized energy values and with the number of components in each group equal to the number of adiabatic channels. Selecting a particular group, or equivalently selecting a particular energy, the matrix R_{reg} has been constructed in terms of the various hyperradial functions f_{ij}^{reg}. Differently to the case in which the Coulomb interaction is absent, this matrix is not diagonal indicating the coupling of the adiabatic channels in the asymptotic region. Using a singular-value decomposition, it can be casted in a product of a diagonal matrix, W_{reg}, and two unitary matrices U and V used to rotate the original adiabatic basis. In the new basis the hyperradial functions decouple and, moreover, they result in Coulomb functions with order and Sommerfeld parameter well defined. In a second step the diagonal matrix W_{irr} has been introduced by replacing the regular Coulomb functions in W_{reg} by the corresponding irregular Coulomb functions. In a final step these functions have been regularized at the origin. After this numerical treatment the regular and (regularized) irregular vectors F and $\tilde{G}$ can be used to compute the integral relations given in Eq. (27) for a problem in which these vectors represent the asymptotic solution. As an example here the case of three interacting α particles has been investigated using the short-range potential of Ali-Bodmer. A restricted adiabatic space of 8 basis vectors has been used.

This work can be seen as a first step in the solution of charged three-body reactions. In the present application we have used a fix number of adiabatic channels. It should be noticed that changing the number of adiabatic channels the numerical procedure to determine F and G has to be repeated. However this treatment is independent of the particular short-range interaction under consideration and can be done beforehand and efficiently stored. In the present example the case of three charged particles has been considered, however the method can be applied in a similar way to the case of a neutral plus two charged particles as $p - d$ scattering. In the $n - d$ case is known that a large number of hyperspherical functions and adiabatic states are needed [15]. Preliminary results for the $p - d$ case show a similar pattern of convergence. Studies along this line are under way.

Coulomb Character of the Diagonal Radial Functions

The regular Coulomb functions can be written as given in Eq. (20), where the R_{reg}-matrix contains all the radial regular functions f_{ni}^{reg} introduced in Eq. (18). These radial functions are obtained as solutions of the coupled radial equations (5), which can be written in a compact matrix form as:

$$\left(\frac{\partial^2}{\partial z^2} - \frac{\lambda + \frac{15}{4}}{z^2} + \mathbb{I}\right) R_{reg} + \frac{2}{\kappa} P \frac{\partial}{\partial z} R_{reg} + \frac{1}{\kappa^2} Q R_{reg} = 0, \tag{30}$$

where $z = \kappa\rho$, P and Q are the matrices formed by the $P_{nn'}$ and $Q_{nn'}$ coupling functions, λ is a diagonal matrix containing the $\lambda_n(\rho)$ functions, and $\mathbb{I}$ is the identity matrix.

We define now $\tilde{\lambda}_n = \lambda_n/z$, $\tilde{P} = P\rho$, and $\tilde{Q} = Q\rho^2$, in such a way that the matrices $\tilde{\lambda}$, $\tilde{P}$, and $\tilde{Q}$ go to a constant value asymptotically. In this way the expression above becomes:

$$\left(\frac{\partial^2}{\partial z^2} - \frac{\tilde{\lambda}}{z} - \frac{15/4}{z^2} + \mathbb{I}\right) R_{reg} + \frac{2}{z} \tilde{P} \frac{\partial}{\partial z} R_{reg} + \frac{1}{z^2} \tilde{Q} R_{reg} = 0. \tag{31}$$

Let us write now the radial matrix R_{reg} as:

$$R_{reg} = z^\alpha e^{iz} = e^{iz + \alpha \log z} = \mathbb{I} + \alpha \log z + \frac{1}{2}(\alpha \log z)^2 + \cdots, \tag{32}$$

where α is a general complex matrix.

From the expression above it is clear that diagonalization of the matrix α is equivalent to diagonalization of the radial matrix R_{reg}. Furthermore, if we call α_i to each of the eigenvalues of α, we have that the radial eigenvalue functions of R_{reg} take the form $e^{iz + \alpha_i \log z} = e^{i\kappa\rho + \alpha_i \log \kappa\rho}$. In other words, if we are able to prove that the eigenvalues α_i are pure imaginary ($\alpha_i = -i\eta_i$ with η_i real), then the eigenvalues of R_{reg} will behave as $e^{i(\kappa\rho - \eta_i \log \kappa\rho)}$, which is the behaviour of the Coulomb functions.

In order to prove that the eigenvalues of the matrix α are in fact pure imaginary, let us replace R_{reg} in Eq. (31) by its form given in Eq. (32), i.e. $z^\alpha e^{iz}$. After some algebra we obtain that Eq. (31) can be written as:

$$\frac{1}{z}\left(-\tilde{\lambda} + 2i\tilde{P} + 2i\alpha\right) + \frac{1}{z^2}\left(\alpha(\alpha - \mathbb{I}) - \frac{15}{4} + 2\tilde{P}\alpha + \tilde{Q}\right) = 0. \tag{33}$$

At very large distances the term in $1/z$ dominates, which means that for a sufficiently large value of z, where $\tilde{\lambda}$ and $\tilde{P}$ are constant matrices, we have that:

$$\left(-\tilde{\lambda} + 2i\tilde{P} + 2i\alpha\right) = 0 \implies \alpha = -i\frac{\tilde{\lambda}}{2} - \tilde{P}. \tag{34}$$

Let us keep in mind that the matrices $\tilde{\lambda}$ and $\tilde{P}$ are both real, and whereas $\tilde{\lambda}$ is diagonal, $\tilde{P}$ has the property that $\tilde{P}_{nn'} = -\tilde{P}_{n'n}$ (and therefore $\tilde{P}_{nn} = 0$). As a consequence, from the expression above it clear that the matrix α is an antihermitian matrix ($\alpha^\dagger = -\alpha$). The antihermitian character is of course preserved when α is diagonalized, which implies that the α-eigenvalues have to be pure imaginary. As a consequence, the eigenvalue functions of the radial matrix R_{reg} behave as Coulomb functions.

References

1. Glöckle, W., Wiłala, H., Hüber, D., Kamada, H., Golak, J.: The three-nucleon continuum: achievements, challenges and applications. Phys. Rep. **274**, 107 (1996)
2. Deltuva, A., Fonseca, A.C., Sauer, P.U.: Momentum-space treatment of the Coulomb interaction in three-nucleon reactions with two protons. Phys. Rev. C **71**, 054005 (2005)
3. Kievsky, A., Friar, J.L., Payne, G.L., Rosati, S., Viviani, M.: Phase shifts and mixing parameters for low-energy proton-deuteron scattering. Phys. Rev. C **63**, 064004 (2001)

4. Kievsky, A., Rosati, S., Viviani, M.: Proton-deuteron elastic scattering above the deuteron breakup. Phys. Rev. Lett. **82**, 3759 (1999)
5. Kievsky, A., Viviani, M., Rosati, S.: Polarization observables in $p - d$ scattering below 30 MeV. Phys. Rev. C **64**, 024002 (2001)
6. Lazauskas, R., Carbonell, J.: Application of the complex-scaling method to few-body scattering. Phys. Rev. C **84**, 034002 (2011)
7. Carbonell, J., Deltuva, A., Fonseca, A.C., Lazauskas, R.: Bound state techniques to solve the multiparticle scattering problem. Prog. Part. Nucl. Phys. **74**, 55 (2014)
8. Kadyrov, A.S., Bray, I., Mukhamedzhanov, A.M., Stelbovics, A.T.: Coulomb breakup problem. Phys. Rev. Lett. **101**, 230405 (2008)
9. Barletta, P., Romero-Redondo, C., Kievsky, A., Viviani, M., Garrido, E.: Integral relations for three-body continuum states with the adiabatic expansion. Phys. Rev. Lett. **103**(090402–1), 090402–090404 (2009)
10. Romero-Redondo, C., Garrido, E., Barletta, P., Kievsky, A., Viviani, M.: General integral relations for the description of scattering states using the hyperspherical adiabatic basis. Phys. Rev. A **83**(022705–1), 022705–022712 (2011)
11. Garrido, E., Romero-Redondo, C., Kievsky, A., Viviani, M.: Integral relations and the adiabatic expansion method for $1 + 2$ reactions above the breakup threshold: Helium trimers with soft-core potentials. Phys. Rev. A **86**(052709–1), 052709 (2012)
12. Mukhamedzhanov, A.M., Kadyrov, A.S., Pirlepesov, F.: Leading asymptotic terms of the three-body Coulomb scattering wave function. Phys. Rev. A **73**, 012713 (2006)
13. Nielsen, E., Fedorov, D.V., Jensen, A.S., Garrido, E.: The three-body problem with short-range interactions. Phys. Rep. **347**, 373–459 (2001)
14. Kievsky, A., Viviani, M., Barletta, P., Romero-Redondo, C., Garrido, E.: Variational description of continuum states in terms of integral relations. Phys. Rev. C **81**, 034002 (2010)
15. Garrido, E., Kievsky, A., Viviani, M.: Breakup of three particles within the adiabatic expansion method. Phys. Rev. C **90**, 014607 (2014)
16. Garrido, E., Fedorov, D.V., Jensen, A.S.: Three-body halos. IV. Momentum distributions after fragmentation. Phys. Rev. C **55**, 1327 (1997)
17. Garrido, E., Jensen, A.S., Fedorov, D.V.: Techniques to treat the continuum applied to electromagnetic transitions in ^{8}Be. Few-body Syst. **55**, 101–119 (2014)
18. Garrido, E.: Three-body continuum wave functions with a box boundary condition. Few-body Syst. **56**, 829–836 (2015)
19. Ali, S., Bodmer, A.R.: Phenomenological α-α potentials. Nucl. Phys. A **80**, 99–112 (1966)
20. Danilin, B.V., Thompson, I.J., Vaagen, J.S., Zhukov, M.V.: Three-body continuum structure and response functions of halo nuclei (I): ^{6}He. Nucl. Phys. A **632**, 383–416 (1998)
21. Suno, H., Esry, B.D.: Adiabatic hyperspherical study of triatomic helium systems. Phys. Rev. A **78**, 062701 (2008)
22. Garrido, E., Fedorov, D.V., Jensen, A.S.: Three-body bremsstrahlung and the rotational character of the ^{12}C spectrum. Phys. Rev. C **91**, 054003 (2015)
23. Platter, L.: Low-energy universality in atomic and nuclear physics. Few-body Syst. **46**, 139–171 (2009)

Few-Body Syst (2017) 58:80
DOI 10.1007/s00601-017-1238-x

CrossMark

Takashi Watanabe · Yasuhisa Hiratsuka · Shinsho Oryu ·
Yoshio Togawa

A New Feature of the Screened Coulomb Potential in Momentum Space

Received: 1 October 2016 / Accepted: 18 January 2017 / Published online: 8 February 2017
© Springer-Verlag Wien 2017

Abstract A Coulomb equivalent screened Coulomb potential is proposed for solving the Schrödinger equation and/or the Calogero first order differential equation, where some critical range bands are obtained. Phase shifts for "any" two-charged particle system (from electron–electron to heavy ion–heavy ion) are reproduced by using the universal critical range bands and the appropriate Sommerfeld parameter over a very wide energy region. A Coulomb-like off-shell amplitude is introduced using two-potential theory without employing the usual Coulomb renormalization method.

1 Introduction

Obtaining an accurate method for treating charged particle scattering is one of the more important problems in nuclear reactions. Almost all approaches in the past half century use a screened Coulomb potentials. However, precise three-body calculations require a more accurate solution for the Coulomb problem. In the three-body configuration (r-) space calculation, the three-body wave function [$\Psi(\mathbf{x}, \mathbf{y})$ with the Jacobi coordinates $\mathbf{x}$ and $\mathbf{y}$] is approximately expanded in a separable manner $\sum_i \psi_x^i(\mathbf{x}) \psi_y^i(\mathbf{y})$. In the p-space calculation, the three-body equations, with the nuclear potential and the screened Coulomb potential, are solved assuming convergence is achieved by increasing the range. However, we still have a concern as to whether those converged values exist or not, or whether the method can be used to predict a physical properties.

Usually, in order to obtain the LS equation from the Schrödinger equation, the boundary condition:

$$\lim_{r \to \infty} r V(r) \to 0 \tag{1}$$

This article belongs to the Topical Collection "30th anniversary of Few-Body Systems".

T. Watanabe · Y. Hiratsuka · S. Oryu (✉)
Department of Physics, Tokyo University of Science, 2641 Yamazaki, Noda, Chiba 278-8510, Japan
E-mail: oryu@rs.noda.tus.ac.jp

T. Watanabe
E-mail: watanabe@ph.noda.tus.ac.jp

Y. Hiratsuka
E-mail: mwhyh216@hb.tp1.jp

Y. Togawa
Department of Information Science, Tokyo University of Science, 2641 Yamazaki, Noda, Chiba 278-8510, Japan
E-mail: togawa@is.noda.tus.ac.jp

must be satisfied. The nuclear potential satisfies this condition, but the Coulomb potential does not because the Coulomb potential is

$$V^C(r) = 2k\eta(k)/r \tag{2}$$

so that

$$\lim_{r \to \infty} r V^C(r) \to 2k\eta(k) \neq 0, \tag{3}$$

where $\eta(k) = \nu Z Z' e^2 / k$ is the Sommerfeld parameter, with the reduced mass $\nu = m_1 m_2 / (m_1 + m_2)$ and charges Ze and $Z'e$ in units of $\hbar = c = 1$, and $k = \sqrt{2\nu E}$ denotes the on-(energy-)shell momentum.

On the other hand, for the p-space approach, the "usual Fourier transform" of the Coulomb potential is performed by using a screened Coulomb potential such as a Yukawa-type potential with a dimension-less "universal range": $\mathcal{R} = kR$,

$$V^{\mathcal{R}}(r; k) = \frac{2k\eta(k)}{r} e^{-kr/\mathcal{R}}. \tag{4}$$

The Fourier transform of Eq. (4) is given by

$$V^{\mathcal{R}}(\mathbf{p}, \mathbf{p}'; k) = \frac{8\pi k\eta(k)}{(\mathbf{p} - \mathbf{p}')^2 + (k/\mathcal{R})^2}. \tag{5}$$

Because the Coulomb potential in r-space is obtained in the limit $\mathcal{R} \to \infty$ in Eq. (4), we expect that the Coulomb potential in momentum space is obtained by taking $\mathcal{R} \to \infty$ in Eq. (5),

$$V^C(\mathbf{p}, \mathbf{p}'; k) = \lim_{\mathcal{R} \to \infty} V^{\mathcal{R}}(\mathbf{p}, \mathbf{p}'; k) = \frac{8\pi k\eta(k)}{(\mathbf{p} - \mathbf{p}')^2}, \tag{6}$$

where $V^{\mathcal{R}}(\mathbf{p}, \mathbf{p}'; k)$ and $V^C(\mathbf{p}, \mathbf{p}'; k)$ are the screened Coulomb and the Coulomb potentials in momentum space, respectively.

The screened Coulomb potential $V^{\mathcal{R}}(\mathbf{p}, \mathbf{p}'; k)$ satisfies the Lippmann–Schwinger (LS) equation condition as required by Eq. (1),

$$t^{\mathcal{R}}(\mathbf{p}, \mathbf{p}'; E) = V^{\mathcal{R}}(\mathbf{p}, \mathbf{p}'; k)$$
$$+ \int V^{\mathcal{R}}(\mathbf{p}, \mathbf{p}''; k) G_0(\mathbf{p}''; E) t^{\mathcal{R}}(\mathbf{p}'', \mathbf{p}'; E) \frac{d\mathbf{p}''}{(2\pi)^3}. \tag{7}$$

However, by increasing the range $\mathcal{R}$ (having $1/k \ll R$), the kernel suffers an overlapping singularity between the Green's function and the potential of Eq. (6). In other words, Eq. (6) has a serious divergence at $p = p'' = k \equiv \sqrt{2\nu E}$ in the kernel. We conclude that such a long range limit for the screened Coulomb potential does not converge Eq. (7) to the LS equation for the Coulomb problem or to the Coulomb solution. A condition similar to Eq. (1) in p-space is given by

$$\lim_{p' \to p} V(\mathbf{p}, \mathbf{p}') < \infty. \tag{8}$$

It should be noted that both of a long range and the short range potentials can not be calculated in the *same bases* in the LS equation.

Recently, we introduced a new method for solving the proton-deuteron three-body problem with a rigorous treatment of the Coulomb interaction [1]. To accomplish this the most important breakthrough was demonstrated in the two-charged particle problem for the proton–proton system (using the LS equation) [2–4].

In order to verify the condition Eq. (8), we introduced a modified Coulomb potential V^{MCP} in [3,4] which is defined by using the Eq. (6)-type Coulomb potential and a screened Coulomb potential $V^{\mathcal{R}}(\mathbf{p}, \mathbf{p}'; k)$ with a finite universal range $\mathcal{R}$,

$$V^{MCP}(\mathbf{p}, \mathbf{p}'; k) = V^C(\mathbf{p}, \mathbf{p}'; k)\bar{\delta}_{p,p'} + V^{\mathcal{R}}(\mathbf{p}, \mathbf{p}'; k)\delta_{p,p'}, \tag{9}$$

where we define the delta function $\delta_{p,p'} = 1$ for $p = p'$, and 0 for $p \neq p'$, and $\bar{\delta}_{p,p'} = (1 - \delta_{p,p'}) = 1$ for $p \neq p'$ and 0 for $p = p'$, respectively. The new potential has no diagonal divergence in p-space; therefore V^{MCP}

satisfies the condition in Eq. (8). Furthermore, it was proved that the Fourier transform of V^{MCP} becomes the pure Coulomb potential in r-space [2], which means that the V^{MCP} potential is equivalent to the pure Coulomb potential when it serves as an integrand in momentum space. Therefore, V^{MCP} is an alternative Coulomb potential in the integral sense. The kernel of the LS equation for V^{MCP} is free from the overlapping singularity in principle, where the screening *range* of V^{MCP} in Eq. (9) defined by a critical range $R = e^{a\gamma}/2k$ for energy larger than 1 keV in the proton–proton case [1–4] and a is defined in "Appendix A".

In order to solve the problem, we required an auxiliary potential: $V^\phi \equiv V^C - V^{\mathcal{R}}$ and a *Lemma* in ref. [5,6]. Using the *Lemma* is the only way to obtain a "critical range (or decisive range)" [2–4], and the LS equation for the auxiliary potential was numerically solved by two methods, the first one adopted "special Gaussian points" with a critical range [7], and the second one solved by a contour deformation method with the critical range [2,8].

The above mentioned critical range should be generalized to the wider energy region: $0 < E < \infty$, especially for the very low energy region of less than 1 keV for the proton–proton case. In this paper for convenience, we obtain a critical range which represents the auxiliary phase shift becoming zero (or the *Lemma* in terms of the phase shift): $\phi(k) = \sigma_l(k) - \delta_l^{\mathcal{R}}(k) \to 0$ to high accuracy using the r-space calculation.

On the other hand, we can choose a screening range for the Coulomb potential which can directly determine the Coulomb phase shift by solving a differential equation for the phase shift [9–11] or also by solving the Schrödinger equation. We obtain a very high precision with more than six digits. In momentum space, the LS equation gives the same accurate phase shift.

In our past several papers, we demonstrated that two potentials: $V^C = V^{\mathcal{R}} + V^\phi$ can be summarized by selecting V^ϕ to obtain $T^C = T^{\mathcal{R}\phi} + T^\phi$ where T^ϕ is the LS amplitude, and $T^{\mathcal{R}\phi}$ is a modified amplitude with respect to the $V^{\mathcal{R}}$ by T^ϕ. Let us call it the *Lemma* method where we require a *Lemma*: $T^\phi(k, k) = T^\phi(k, p') = T^\phi(p, k) = 0$ for the on- and half on-shell amplitudes which is proved in our papers. In this paper, we present another method which is given in "Appendix B" by using the GSE method [12–17]. For this purpose, we have to introduce an innocuous range parameter $\mathcal{R}'$, which is very close to the critical range $\mathcal{R}' \approx \mathcal{R}$. The details are shown in "Appendix B".

As a summary, we assert that the Coulomb problem in momentum space is solved by using the Coulomb phase shift "renormalization" method which was introduced firstly by W. Tobocman et al. in the 1950s [18], by A. M. Veselova in early 1970s [19–24] and by E. O. Alt et al in late 1970s [25–28]. Some improvements have been proposed by A. Deltuva, et al with a benchmark test [29–33], and also by ourselves[34]. However, our method is not the same as the renormalization method, but requires "zero renormalization" to obtain a critical range (to solve the LS equation) which is presented in the references [1–5,7,8].

Unfortunately, almost all approaches can not be solved analytically. Our former method [5,6] analytically reaches this goal only by assistance of the *Lemma*. In order to obtain good accuracy, the renormalization method requires a numerical calculation with many digit precision to avoid a loss of trailing digits (or a cancellation of significant digits). Such a loss of information in the trailing term could propagate to the leading term through the renormalization, and finally affect the higher wave interactions in the three-body problem. We have emphasized, in three-body problems, that the reproduction of the two-body Coulomb phase shifts is important, because the Coulomb higher waves do not converge and interfere with the non-Coulomb higher waves. Therefore, we have to take care regarding the propagation of errors in the "renormalization" method in the three-body problem.

In this context, we found a counterpart of the two-potential formalism by selecting $V^{\mathcal{R}}$ instead of V^ϕ to make $T^C = T^{\phi\mathcal{R}} + T^{\mathcal{R}}$ where $T^{\mathcal{R}}$ is the LS amplitude for $V^{\mathcal{R}}$, and the remainder term is an amplitude for V^ϕ modified by the distortion of $T^{\mathcal{R}}$. Let us call this case *direct* method.

In this paper, we will investigate the "direct" method in the Sect. 2. The numerical calculation for the LS equation is done by using the GSE method [12–17] in momentum space: the Schrödinger equation and the phase shift differential equation are solved in r-space. The calculated results for the universal critical range bands are shown together with the Coulomb phase shifts for scattering from "electron–electron" to "heavy ion–heavy ions" in Sect. 3. Our conclusion and a discussion will be given in Sect. 4.

2 A New Direct Method to Obtain a Coulomb Equivalent Phase Shift

In this paper, we would like to present a new method which is practical and applicable to few-body problems and in general.

Let us start from Eq. (9), that is,

$$
\begin{aligned}
V_l^{MCP}(p, p') &= V_l^{\mathcal{R}}(p, p') + \overline{V}_l^{\phi}(p, p') \\
&= \begin{cases} V_l^{\mathcal{R}}(p, p') & (p = p') \\ V_l^{C}(p, p') & (p \neq p'). \end{cases}
\end{aligned} \tag{10}
$$

Contrary to two-potential theory for the *Lemma* method which is "selecting T_l^{ϕ}", and we adopt the two-potential theory for the *direct* method: "selecting $T_l^{\mathcal{R}}$", we obtain the off-shell Coulomb amplitude which is given in "Appendix C",

$$
\begin{aligned}
T_l^{C}(p, p'; E) &= \int_0^{\infty} \int_0^{\infty} \frac{p''^2 p'''^2 dp'' dp'''}{(2\pi^2)^2} \overline{\omega}_l^{\mathcal{R}}(p, p''; E) \\
&\quad \times t_l^{\phi\mathcal{R}}(p'', p'''; E) \omega_l^{\mathcal{R}}(p''', p'; E) + T_l^{\mathcal{R}}(p, p'; E)
\end{aligned} \tag{11}
$$

with the core amplitude,

$$
\begin{aligned}
t_l^{\phi\mathcal{R}}(p, p'; E) &= V_l^{\phi}(p, p') + \int_0^{\infty} \int_0^{\infty} \frac{p''^2 p'''^2 dp'' dp'''}{(2\pi^2)^2} \\
&\quad \times V_l^{\phi}(p, p'') G_l^{\mathcal{R}}(p'', p'''; E) t_l^{\phi\mathcal{R}}(p''', p'; E).
\end{aligned} \tag{12}
$$

The Møller functions for the screened Coulomb potential,

$$
\overline{\omega}_l^{\mathcal{R}}(p, p''; E) = \frac{2\pi^2}{p^2}\delta(p - p'') + T_l^{\mathcal{R}}(p, p''; E)G_0(p''; E), \tag{13}
$$

$$
\omega_l^{\mathcal{R}}(p''', p'; E) = \frac{2\pi^2}{p'^2}\delta(p''' - p') + G_0(p'''; E)T_l^{\mathcal{R}}(p''', p'; E), \tag{14}
$$

the Green's function for the screened Coulomb potential,

$$
\begin{aligned}
G_l^{\mathcal{R}}(p'', p'''; E) &= \frac{2\pi^2}{p''^2}\delta(p'' - p''')G_0(p''; E) \\
&\quad + G_0(p''; E)T_l^{\mathcal{R}}(p'', p'''; E)G_0(p'''; E),
\end{aligned} \tag{15}
$$

and the screened Coulomb amplitude,

$$
\begin{aligned}
T_l^{\mathcal{R}}(p, p'; E) &= V_l^{\mathcal{R}}(p, p') + \frac{1}{2\pi^2}\int_0^{\infty} p''^2 dp'' \\
&\quad \times V_l^{\mathcal{R}}(p, p'')G_0(p''; E)T_l^{\mathcal{R}}(p'', p'; E).
\end{aligned} \tag{16}
$$

$T_l^{\mathcal{R}}$ is the dominant term in Eq. (11), and the first term of the r.h.s. in Eq. (11) is a secondary term with a small auxiliary potential, although Eq. (12) may produce a numerical difficulty in the kernel $V^{\phi}(p, p'')G_0(p''; E) \equiv \overline{V}^{\phi}(p, p'')G_0(p''; E)$. However, the kernel is regular at $p = p' = k$ by Eq. (10). Since we can obtain the on-shell $T_l^{\mathcal{R}}$ precisely by using the universal range $\mathcal{R}$ in Eq. (50), then the amplitude $t_l^{\phi\mathcal{R}}$ could be minimized. The universal range is obtained through the r space calculation with the Schrödinger equation in "Appendix A". Therefore $T_l^{\mathcal{R}}$ can represent the on- and half-off Coulomb amplitude; however, the off-shell part of T_l^{C} could be compensated by the first term of the r.h.s. in Eq. (11).

Therefore, we confirm that if the secondary term $T_l^{\phi\mathcal{R}}$ is obtained, then the off-shell amplitude $T_l^{C}(p', p''; E)$ can be calculated together with $T_l^{\mathcal{R}}$. This result guarantees the existence of the off-shell Coulomb-like t-matrix by Eq. (11) [35], which was not defined in Ref. [36].

Table 1 The parameters a_m and b_m (with $m = 0, 1, 2, \ldots$) in Eq. (17) for five bands: $\mathcal{R} = \mathcal{R}_n(\eta)$ (with $n=0, 1, 2,\ldots$) are shown as calculated by the direct method

$\mathcal{R}_n$	m	0	1	2	3	4
$\mathcal{R}_0$	a_m	−0.122527	67.0542	33.4284	−33.6583	
$0.00353 \le \eta \le 1.58$	b_m	−0.14194	90.8834	15.8711	12.7285	
$\mathcal{R}_1$	a_m	5.57504	9.02951	12.9823	−3.20748	
$1.58 \le \eta \le 4.22$	b_m	−0.731152	−0.925015	1.87304	0.68854	
$\mathcal{R}_2$	a_m	1.04150	2.62713	5.58478	−0.64538	
$3.95 \le \eta \le 5.60$	b_m	2.11619	3.39619	3.91962	−2.57886	0.397671
$\mathcal{R}_3$	a_m	3.26102	12.2893	43.3765	−5.55534	
$5.60 \le \eta \le 7.90$	b_m	0.419669	−0.514822	−1.45874	0.924641	0.00556185
$\mathcal{R}_4$	a_m	1.1429	1.8394	3.89062	−0.245209	−0.0191509
$7.07 \le \eta \le 9.13$	b_m	2.12168	3.59564	−1.4752	0.186987	

3 Numerical Result for Screening Ranges

3.1 Universal Critical Range Bands

To obtain the range $\mathcal{R}$ by the direct-method, the Schrödinger equation and/or the Calogero differential equation [9–11] can be used. We obtain a very accurate on-shell phase shift by Calogero's differential equation with a finite range potential; therefore, using the same range, the phase shift is reproduced by the LS equation with high accuracy. It seems clear that a proper screening range is obtained by increasing the range in the potential: $\lim_{R\to\infty} e^2 e^{-r/R}/r \to e^2/r$; in the two- and three-body problems, however, that idea is wrong. Because the long range singularity can not be covered with a closed neighborhood in the potential theory, the limiting procedure fails. It was confirmed that a critical screening range exists [2–4].

We found that five discrete screening range bands are necessary to reproduce the Coulomb phase shift for $\sigma_0 \le 4\pi$.

1. The first (lowest) band (black circle in Fig. 1) can represent the higher energy region better than the energy which satisfies $\sigma_0(k) = 0$, and the range should be $\mathcal{R} = \mathcal{R}_0(k) = 0$ fm at $\sigma_0(k) = 0$.
2. The second lower band (black nabla symbol) can be applied in the phase shift region: $0 \le \sigma_0(k) \le \pi$ and the range must be $\mathcal{R} = \mathcal{R}_1(k) = 0$ fm at $\sigma_0(k) = \pi$.
3. The next band (black square symbol) covers the phase shift region: $\pi \le \sigma_0(k) \le 2\pi$, and the range should be $\mathcal{R} = \mathcal{R}_2(k) = 0$ fm at $\sigma_0(k) = 2\pi$.
4. The next band (white circle symbol) applies for: $2\pi \le \sigma_0(k) \le 3\pi$, and the range $\mathcal{R} = \mathcal{R}_3(k) = 0$ fm at $\sigma_0(k) = 3\pi$.

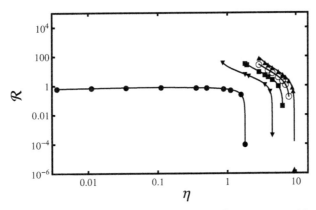

Fig. 1 The range bands by the *direct* fitting method. The universal ranges $\mathcal{R}(k)$ are denoted by five different symbols: *black circle, black nabla, black square, white circle,* and *black delta.* The universal ranges $\mathcal{R} = \mathcal{R}(k) = \mathcal{R}_n(k)$ $(n = 0, 1, 2, 3, 4)$ are fitted as functions of $\eta = \eta(k)$. 1) $\mathcal{R}_0 = 0$ at $\sigma_0(k) = 0$, 2) $\mathcal{R}_1 = 0$ at $\sigma_1(k) = \pi$, 3) $\mathcal{R}_2 = 0$ at $\sigma_2(k) = 2\pi$, 4) $\mathcal{R}_3 = 0$ at $\sigma_3(k) = 3\pi$, 5) $\mathcal{R}_4 = 0$ at $\sigma_0(k) = 4\pi$. All bands become zero at $\sigma_0(k) = n\pi$ $(n = 0, 1, 2, \ldots)$; however, one could replace them by the finite range of the upper band. Each band makes it possible to derive the Coulomb phase shifts for the corresponding energy region

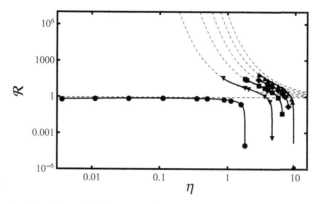

Fig. 2 The direct fitting range bands and extension by our model range. The legends are the same as in Fig. 1. *Dashed lines* are given by Eq. (50) in the case: $m = 1$ of the Yukawa type from the lowest line $n = 0$ to the highest line $n = 5$

5.   The highest band (black delta symbol) applies for: $3\pi \leq \sigma_0(k) \leq 4\pi$ and the range $\mathcal{R} = \mathcal{R}_4(k) = 0$ fm at $\sigma_0(k) = 4\pi$.

We find that $\sigma_0(k) = n\pi$ $(n = 0, 1, 2, \ldots)$ leads to $\mathcal{R}_n(k) = 0$ which denotes $V^{\mathcal{R}}(r) = 0$. However, the zero range screened Coulomb property seems to be unusual; then we adopt the finite range of the upper band instead. These bands are fitted by the N/D form

$$\mathcal{R} = \frac{a_0 + a_1\eta + a_2\eta^2 + a_3\eta^3 + \cdots}{b_0 + b_1\eta + b_2\eta^2 + b_3\eta^3 + \cdots} \tag{17}$$

with the universal ranges $\mathcal{R} \equiv kR$, and the Sommerfeld parameter $\eta = \eta(k)$. Parameters $a_0, a_1, \ldots$ and $b_0, b_1, \ldots$ are shown in the Table 1. Therefore, the Coulomb phase shifts of the each system are obtained by taking the individual range $R = \mathcal{R}/k$ and incorporating the individual values of the Sommerfeld parameters.

Especially for the cases: $\sigma_l(k) = n\pi$ $(n = 0, 1, 2, \ldots)$, the range converges to a zero value. The given ranges provide the nine-digit results with the Calogero equation.

Obviously, the phase shifts for the higher partial waves require the individual range-bands for the higher waves, because the screened Coulomb potential plus the centrifugal term $l(l + 1)/r^2$ are used to obtain the range. Therefore, Fig. 1 is available only for the S-wave. However, the longer range $\mathcal{R}$ which is given by Eq. (50) is the partial wave independent, because it is satisfied in the asymptotic region where the centrifugal potential is negligible.

Let us call the critical range bands which are defined by Eq. (50) "asymptotic" bands, while the former critical range bands which are obtained numerically, "numerical" bands. Therefore, the universal critical range bands are given by the smooth continuation between both range bands in Fig. 2. It is seen that universal critical ranges of the asymptotic bands are generally very large except for $n = 0$, therefore it is rather difficult to obtain the numerical solution by the LS equation because of cancellation of significant digits. On the other hand, universal critical ranges within numerical bands are smaller, therefore the LS equation is successfully solved.

3.2 Coulomb Phase Shifts

By using our universal critical range bands, we can easily obtain the Coulomb phase shifts by using the LS equation in momentum space. We found that phase shifts for "any" two-charged particle systems (from electron-electron to heavy ion-heavy ion) are automatically reproduced by using the universal critical range bands and the appropriate Sommerfeld parameter over a very wide energy region. Figures 3 and 4 are the Coulomb phase shifts for many different two-charged particle systems for S-wave. The higher partial waves are shown for reference in Fig. 5. The Coulomb-like off-shell amplitude could be introduced using the two-potential theory. The numerical results will be shown on another occasion.

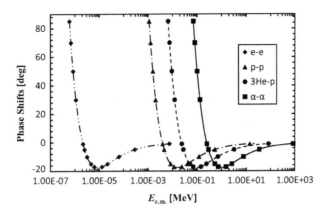

Fig. 3 The Coulomb phase shifts in the S-wave are illustrated for several systems from e$^-$-e$^-$ to $\alpha - \alpha$ by the *diamond, triangle, circle*, and *square symbols*, respectively. The results are compared with the analytic solutions by the *double-dashed, dotted-dashed, dashed*, and *solid lines*, respectively. These are calculated using the universal ranges which are given in Fig. 1. The present range is obtained for the Yukawa-type screened Coulomb potential. The results are illustrated only for $0 \leq \sigma_0 \leq \pi$; however, our results are fitted with the analytic solution in an energy region greater than that

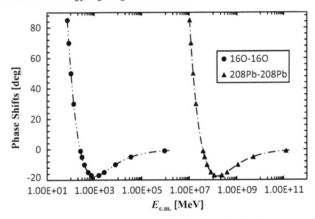

Fig. 4 The Coulomb phase shifts in the S-wave are illustrated for ^{16}O-^{16}O and ^{208}Pb-^{208}Pb. Our calculated results are shown by the *solid-circle* and the *triangle*. These are compared with the analytic results by the *double-dashed*, and the *dashed-dotted lines*, respectively. The other aspects are the same as in Fig. 3

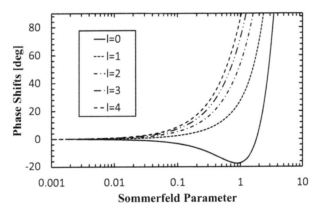

Fig. 5 Our screened Coulomb potential in Eq. (4) which satisfies the *Lemma* is applicable for the higher partial wave cases. The calculated results are fitted to the analytic partial wave phase shifts: $\sigma_l(k) = \arg \Gamma(l + 1 + i\eta(k))$

4 Conclusion and Discussion

In this paper, we derived a critical screening range which is used to calculate the Coulomb phase shift in the energy region: $0 < E < \infty$ with high accuracy without using the Coulomb phase shift renormalization method but instead the direct method. The direct method gives not only the on- and half off-shell amplitudes very accurately, but also the off-shell part is analytically generated where the same asymptotic behavior as the Coulomb wave function is in effect.

In this paper, the quantized critical range in the V^{MCP}-potential is introduced with the aid of the r-space calculation for convenience in the relatively short range region, but in the longer range region with an analytic form shown in "Appendix A". The quantized critical range includes the long range Coulomb information as shown in "Appendix A". By using the critical range, we proposed a new direct method instead of the historical renormalization method.

Our Coulomb treatment by the dimensionless universal range and appropriate Sommerfeld parameter is general for any two-charged particle system from "electron"–"electron" to the "heavy ion"–"heavy ion". Because the Sommerfeld parameter, as a universal variable is determined by the charges and the reduced mass, then our method may be applicable to systems with non-integer charges such as quarks.

Finally, we conclude that the Coulomb potential can be treated on the same basis of the usual scattering theory for the short range potential by the aid of the critical range bands, where the pinching singularity, by the dispersion theoretical terminology, does not occur in our theory. We believe that the two-body Coulomb problems in momentum space can be solved not only theoretically but also numerically with high precision in this paper. A partial wave-independent asymptotic critical range calculation is ready for performing with very high precision. A practical three-body calculation is now in progress.

A An Auxiliary Phase Shift $\phi(k)$ and a Critical Range $\mathcal{R}$

The auxiliary phase shift $\phi(k)$ was introduced in [5,6]. In order to obtain the phase shift, let us start from the Schrödinder equation for the screened Coulomb potential and the pure Coulomb potential.

$$\left[\frac{d^2}{dr^2} - \frac{L(L+1)}{r^2} - V^C(r) + k^2\right] w_L^{(\pm)}(r) = 0 \tag{18}$$

$$\left[\frac{d^2}{dr^2} - \frac{L(L+1)}{r^2} - V^{\mathcal{R}}(r) + k^2\right] h_L^{(\pm)}(r) = 0 \tag{19}$$

Therefore in Eq. (18), using the auxiliary potential: $V^\phi(r) = \{V^C(r) - V^{\mathcal{R}}(r)\}$, we have

$$\left[\frac{d^2}{dr^2} - \frac{L(L+1)}{r^2} - V^{\mathcal{R}}(r) - V^\phi(r) + k^2\right] w_L^{(\pm)}(r) = 0 \tag{20}$$

The potentials are defined by using the two-body reduced mass and the charges with Ze and $Z'e$,

$$V^C(r) = 2v\frac{ZZ'e^2}{r} = 2k\frac{\eta(k)}{r} \tag{21}$$

$$V^{\mathcal{R}}(r) = 2v\frac{ZZ'e^2}{r}e^{-(r/R)^m} = 2k\frac{\eta(k)}{r}e^{-(r/R)^m} \tag{22}$$

with the Sommerfeld parameter $\eta(k) = ZZ'e^2v/k$. The ratio between the Coulomb wave function $w_L^{(\pm)}(r)$ and the screened Coulomb wave function $h_L^{(\pm)}(r)$ is given by a new function $y_L^{(\pm)}(r)$,

$$h_L^{(\pm)}(r) \equiv y_L^{(\pm)}(r) \cdot w_L^{(\pm)}(r). \tag{23}$$

Substituting Eq. (23) into Eq. (19), we have the following relation by using Eq. (18),

$$\frac{y_L''(r)}{y_L(r)} + 2\frac{w_L'(r)}{w_L(r)} \cdot \frac{y_L'(r)}{y_L(r)} = V^{\mathcal{R}}(r) - V^C(r) \equiv -V^\phi(r); \tag{24}$$

equivalently we have a differential equation for the auxiliary potential,

$$y_L''(r) + 2\frac{w_L'(r)}{w_L(r)} \cdot y_L'(r) + V^\phi(r)y_L(r) = 0. \tag{25}$$

The asymptotic wave functions based on these potentials are given by

$$w_L^{(\pm)}(r) \sim \exp\left[\pm i\left(kr - \frac{\pi L}{2} - \eta(k)\ln 2kr + \sigma_L(k)\right)\right] \tag{26}$$

$$h_L^{(\pm)}(r) \sim \exp\left[\pm i\left(kr - \frac{\pi L}{2} + \delta_L^R(k)\right)\right] \tag{27}$$

where $\sigma_L(k)$ and $\delta_L^R(k)$ are the Coulomb and the screened Coulomb phase shifts, respectively. It is obvious that the main difference in the asymptotic wave functions between the long range and the short range interactions appears in the r dependent phase: $\ln 2kr$ in the Coulomb wave function Eq. (26).

By using the asymptotic forms of Eqs. (26), (27), the first and the second terms of Eq. (24) are given by where the asymptotic form of the second term of Eq. (24) is given by

$$\frac{w_L'(r)}{w_L(r)} = \frac{d}{dr}[\ln w_L(r)] = \left(k - \frac{\eta(k)}{r}\right) i \approx ik \tag{28}$$

$$\frac{y_L'(r)}{y_L(r)} = \frac{d}{dr}[\ln y_L(r)] \equiv \frac{d\mathcal{Y}_L(r)}{dr} \tag{29}$$

with a new function $\mathcal{Y}_L(r)$,

$$\mathcal{Y}_L(r) \equiv \ln y_L(r), \tag{30}$$

and with the asymptotic phase

$$y_L^{(\pm)}(r) \sim \exp\left[\pm i\left(\eta(k)\ln 2kr - \sigma_L(k) + \delta_L^R(k)\right)\right]. \tag{31}$$

The first term of Eq. (24) is rewritten using Eq. (29) as

$$\frac{y_L''(r)}{y_L(r)} = \frac{d}{dr}\left[\frac{y_L'(r)}{y_L(r)}\right] + \left[\frac{y_L'(r)}{y_L(r)}\right]^2 = \frac{d^2\mathcal{Y}_L(r)}{dr^2} + \left[\frac{d\mathcal{Y}_L(r)}{dr}\right]^2 \tag{32}$$

where the terms are assumed to be very small or vanishing in the asymptotic region. Because, $y_L(r) \to 1$, or $\mathcal{Y}_L(r) \to 0$ is verified and are expected to be smooth function due to Eqs. (23) and (30). Consequently, Eq. (24) is rewritten, by using (28) and (29), as a simple differential equation in the asymptotic region,

$$2ik\mathcal{Y}_L'(r) = V^R(r) - V^C(r) = -V^\phi(r). \tag{33}$$

Because the integral of Eq. (33) should be performed in the region between r and ∞, which is the Coulomb boundary, then the solution is given by using an arbitrary constant b,

$$\mathcal{Y}_L(r) = \int_b^r \frac{\{V^R(r) - V^C(r)\}}{2ik} dr = i\eta(k)\int_b^r \frac{1 - e^{-(r/R)^m}}{r} dr. \tag{34}$$

Here, $\mathcal{Y}_L(r)$ could be separated by using the screening range R

$$\mathcal{Y}_L(r) = i\eta(k)\left[\int_b^0 \frac{1 - e^{-(r/R)^m}}{r} dr + \int_0^R \frac{1 - e^{-(r/R)^m}}{r} dr + \int_R^r \frac{1 - e^{-(r/R)^m}}{r} dr\right]$$

$$= i\eta(k)\left[C + \left\{\int_0^R \frac{1 - e^{-(r/R))^m}}{r} dr + \int_R^r \frac{1}{r} dr - \int_R^r \frac{e^{-(r/R)^m}}{r} dr\right\}\right]$$

$$= i\eta(k)\left[C + \left\{\int_0^R \frac{1 - e^{-(r/R)r^m}}{r} dr - \int_R^\infty \frac{e^{-(r/R)^m}}{r} dr\right\}\right.$$

$$\left. + \int_R^r \frac{1}{r} dr + \int_r^\infty \frac{e^{-(r/R)^m}}{r} dr\right] \tag{35}$$

with

$$C = \int_b^0 \frac{1 - e^{-(r/R)^m}}{r} dr.$$ (36)

The first part $\{\}$ in Eq. (35) is the Euler constant $\gamma = 0.57721\ldots$ which is defined by,

$$\gamma \equiv \int_0^1 \frac{1 - e^{-t}}{t} dr - \int_1^\infty \frac{e^{-t}}{t} dt.$$ (37)

That is, by putting $(r/R)^m = t$ and $dt = m(r/R)^{m-1} dr/R = mtdr/R$, and also $t = 1$ at $r = R$, then we obtain

$$\left\{ \int_0^R \frac{1 - e^{-(r/R)^m}}{r} dr - \int_R^\infty \frac{e^{-(r/R)^m}}{r} dr \right\}$$

$$= \int_0^1 \frac{1 - e^{-t}}{t} \frac{dt}{m} - \int_R^1 \frac{e^{-t}}{t} \frac{dt}{m} = \left\{ \frac{\gamma}{m} \right\}.$$ (38)

Therefore, Eq. (35) becomes

$$\mathcal{Y}_L(r) = i\eta(k) \left[C + \left\{ \frac{\gamma}{m} \right\} + (\ln 2kr - \ln 2kR) + \int_r^\infty \frac{e^{-(r/R)^m}}{r} dr \right]$$

$$= i\eta(k) \left[a\gamma + (\ln 2kr - \ln 2kR) + \int_r^\infty \frac{e^{-(r/R)^m}}{r} dr \right]$$ (39)

with $a = (C/\gamma + 1/m)$, and $\mathcal{Y}_L(r)$ is an r-dependent function.

The asymptotic phase in Eq. (31) can be compared with Eq. (39) by neglecting the last term of Eq. (39), and also by adopting $b = 0$ or $C = 0$. We obtain

$$\eta(k) \left[\left\{ \frac{\gamma}{m} \right\} + (\ln 2kr - \ln 2kR) \right] = \left(\eta(k) \ln 2kr - \sigma_L(k) + \delta_L^R(k) \right)$$ (40)

Therefore, we have

$$\eta(k) \left[\left\{ \frac{\gamma}{m} \right\} - \ln 2kR \right] = -\sigma_L(k) + \delta_L^R(k),$$ (41)

and

$$\sigma_L(k) = \delta_L^R(k) + \eta(k) \left(\ln 2kR - \left\{ \frac{\gamma}{m} \right\} \right)$$ (42)

$$\equiv \delta_L^R(k) + \phi(k, R).$$ (43)

If we can choose

$$\phi(k, R) \equiv \eta(k) \left(\ln 2kR - \left\{ \frac{\gamma}{m} \right\} \right) = 2n\pi$$ (44)

with $n = 0, 1, 2, 3, \ldots$, then we have

$$\ln 2kR = \frac{2n\pi}{\eta(k)} + \left\{ \frac{\gamma}{m} \right\}.$$ (45)

By substituting Eq. (45) into Eq. (39), $\mathcal{Y}_L(r)$ becomes

$$
\begin{aligned}
\mathcal{Y}_L(r) &= i\eta(k)\left[\left\{\frac{\gamma}{m}\right\} + (\ln 2kr - \ln 2kR)\right] \\
&= i\eta(k)\left[\left\{\frac{\gamma}{m}\right\} + \left(\ln 2kr - \frac{2n\pi}{\eta(k)} - \left\{\frac{\gamma}{m}\right\}\right)\right] \\
&= i\left[\eta(k)\ln 2kr - 2n\pi\right].
\end{aligned}
\tag{46}
$$

Therefore, Eq. (31) is given by

$$
\begin{aligned}
y_L^{(\pm)}(r) &\sim \exp\left[\pm i\left(\eta(k)\ln 2kr - \sigma_L(k) + \delta_L^R(k)\right)\right] \\
&= \exp\left[\pm i\left(\eta(k)\ln 2kr - 2n\pi\right)\right].
\end{aligned}
\tag{47}
$$

We can confirm that the long range property of the Coulomb wave function can be compensated by this phase of Eq. (47) in the screened Coulomb wave function. Hence, by using Eqs. (47) and (26), Eq. (23) becomes

$$
\begin{aligned}
h_L^{(\pm)}(r) &= y_L^{(\pm)}(r) \cdot w_L^{(\pm)}(r) \\
&= \exp\left[\pm i\left(\eta(k)\ln 2kr - 2n\pi\right)\right] \\
&\quad \times \exp\left[\pm i\left(kr - \frac{\pi L}{2} - \eta(k)\ln 2kr + \sigma_L(k)\right)\right] \\
&= \exp\left[\pm i\left(kr - \frac{\pi L}{2} + \sigma_L(k) - 2n\pi\right)\right] \\
&= \exp\left[\pm i\left(kr - \frac{\pi L}{2} + \delta_L^R(k)\right)\right].
\end{aligned}
\tag{48}
$$

This result is equivalent to Eq. (27).

Finally, we can conclude that the energy dependent "critical range" parameter in the asymptotic region is given by using Eq. (45),

$$
R = \frac{1}{2k}\exp\left[\frac{2n\pi}{\eta(k)} + \left\{\frac{\gamma}{m}\right\}\right],
\tag{49}
$$

and the universal range is

$$
\mathcal{R} = \frac{1}{2}\exp\left[\frac{2n\pi}{\eta(k)} + \left\{\frac{\gamma}{m}\right\}\right].
\tag{50}
$$

B Proof of the *Lemma* for the Modified Coulomb Potential

The *Lemma* for the modified Coulomb potential will be proved. In order to satisfy the *Lemma*: $\overline{T}_l^\phi(k, k; E) = \overline{T}_l^\phi(p, k; E) = \overline{T}_l^\phi(k, p'; E) = 0$ for $\overline{V}_l^\phi(p, p')$, we adopt the GSE method [12–17] where $\overline{V}_l^\phi(k, k)$ should not be zero, for this purpose, we use Eq. (9) by taking $\mathcal{R}' \sim \mathcal{R}$. Therefore Eq. (10) is rewritten by omitting $(; k)$ in the potentials for simplicity,

$$
\begin{aligned}
\overline{V}_l^\phi(p, p') &\equiv \begin{cases} V_l^{\mathcal{R}'}(p, p') - V_l^{\mathcal{R}}(p, p') \sim \varepsilon & (p = p') \\ V_l^C(p, p') - V_l^{\mathcal{R}}(p, p') & (p \neq p') \end{cases} \\
&= V_l^{MCP}(p, p') - V_l^{\mathcal{R}}(p, p'),
\end{aligned}
\tag{51}
$$

where we can choose a very small number ε which verifies Eq. (51) for any partial waves. Therefore, the LS equation for the potential $\overline{V}_l^\phi$ becomes,

$$\overline{T}_l^\phi(p, p'; E) = \overline{V}_l^\phi(p, p') + \frac{1}{2\pi^2} \int_0^\infty p''^2 dp''$$

$$\times \overline{V}_l^\phi(p, p'') G_0(p''; E) \overline{T}_l^\phi(p'', p'; E). \tag{52}$$

The GSE method gives the solutions of $\overline{T}_l^\phi(k, k; E)$, $\overline{T}_l^\phi(p, k; E)$ and $\overline{T}_l^\phi(k, p'; E)$, [12–16]

$$\overline{T}_l^\phi(k, k; E) = \frac{[\overline{V}_l^\phi(k, k)]^2}{A_l^\phi(k, k; E)}, \tag{53}$$

$$\overline{T}_l^\phi(k, p'; E) = \frac{\overline{V}_l^\phi(k, k) \overline{\varphi}_l^\phi(k, p'; E)}{A_l^\phi(k, k; E)}, \tag{54}$$

$$\overline{T}_l^\phi(p, k; E) = \frac{\overline{\varphi}_l^\phi(p, k; E) \overline{V}_l^\phi(k, k)}{A_l^\phi(k, k; E)}, \tag{55}$$

with

$$\overline{\varphi}_l^\phi(p, k; E) = \overline{V}_l^\phi(p, k) + \int_0^\infty \overline{V}_l^{\phi(2)}(p, p') G_0(p'; E) \overline{\varphi}_l^\phi(p', k; E) \frac{p'^2 dp'}{2\pi^2} \tag{56}$$

$$\overline{A}_l^\phi(k, k; E) = \overline{V}_l^\phi(k, k) + \int_0^\infty \overline{V}_l^\phi(k, p') G_0(p'; E) \overline{\varphi}_l^\phi(p', k; E) \frac{p'^2 dp'}{2\pi^2} \tag{57}$$

and,

$$\overline{V}_l^{\phi(2)}(p, p') = \frac{\begin{vmatrix} \overline{V}_l^\phi(k, k) & \overline{V}_l^\phi(k, p') \\ \overline{V}_l^\phi(p, k) & \overline{V}_l^\phi(p, p') \end{vmatrix}}{\overline{V}_l^\phi(k, k)}. \tag{58}$$

In Eq. (51), because of the relation,

$$|\overline{V}_l^\phi(p, k)| \sim |\overline{V}_l^\phi(k, p')| \sim |\overline{V}_l^\phi(p, p')| > |\overline{V}_l^\phi(k, k)|, \tag{59}$$

and $|\overline{V}_l^\phi(k, k)| \sim |\varepsilon|$, we have

$$\overline{V}_l^{\phi(2)}(p, p') \sim |\varepsilon|^{-1}, \tag{60}$$

$$|\overline{\varphi}_l^\phi(p, k; E)| \sim |\varepsilon|^{-1}, \tag{61}$$

$$|\overline{A}_l^\phi(k, k; E)| \sim |\varepsilon|^{-1}. \tag{62}$$

Therefore, we obtain from Eqs. (53), (54) and (55),

$$\overline{T}_l^\phi(k, k; E) \sim |\varepsilon|^3, \tag{63}$$

$$\overline{T}_l^\phi(k, p'; E) \sim |\varepsilon|, \tag{64}$$

$$\overline{T}_l^\phi(p, k; E) \sim |\varepsilon|. \tag{65}$$

Consequently, the *Lemma* is proven by taking the limit $\mathcal{R}' \to \mathcal{R}$ which satisfies $\varepsilon \to 0$ independent of the partial wave.

C Two-Potential Theory Regarding $T^{\mathcal{R}}$

The two-potential theory in operator form is introduced for the LS equation which is given by the potential V^{MCP} in Eq. (9); hereafter, we adopt simply the notation V^{C} instead of V^{MCP},

$$T^{C} = (V^{\mathcal{R}} + V^{\phi}) + (V^{\mathcal{R}} + V^{\phi})G_0 T^{C} \tag{66}$$

$$T^{\mathcal{R}} = V^{\mathcal{R}} + V^{\mathcal{R}}G_0 T^{\mathcal{R}} \equiv V^{\mathcal{R}}\omega^{\mathcal{R}}$$
$$= V^{\mathcal{R}} + T^{\mathcal{R}}G_0 V^{\mathcal{R}} \equiv \overline{\omega}^{\mathcal{R}} V^{\mathcal{R}} \tag{67}$$

$$\omega^{\mathcal{R}} = 1 + G_0 T^{\mathcal{R}}$$
$$\overline{\omega}^{\mathcal{R}} = 1 + T^{\mathcal{R}}G_0. \tag{68}$$

Let us define

$$T^{C} \equiv T^{\phi\mathcal{R}} + T^{\mathcal{R}}. \tag{69}$$

By using Eqs (66),(69)

$$T^{\phi\mathcal{R}} + T^{\mathcal{R}} = (V^{\mathcal{R}} + V^{\phi})$$
$$+ (V^{\mathcal{R}} + V^{\phi})G_0(T^{\phi\mathcal{R}} + T^{\mathcal{R}})$$
$$= (V^{\mathcal{R}} + V^{\phi})$$
$$+ V^{\phi}G_0 T^{C} + V^{\mathcal{R}}G_0(T^{\phi\mathcal{R}} + T^{\mathcal{R}}). \tag{70}$$

Multiply $\overline{\omega}^{\mathcal{R}}$ from the left of Eq. (70),

$$(1 + T^{\mathcal{R}}G_0)T^{\phi\mathcal{R}} = \overline{\omega}^{\mathcal{R}}V^{\phi}\omega^{\mathcal{R}}$$
$$+ (1 + T^{\mathcal{R}}G_0)V^{\phi}G_0 T^{\phi\mathcal{R}}$$
$$+ (1 + T^{\mathcal{R}}G_0)V^{\mathcal{R}}G_0 T^{\phi\mathcal{R}}, \tag{71}$$

$$T^{\phi\mathcal{R}} = \overline{\omega}^{\mathcal{R}}V^{\phi}\omega^{\mathcal{R}} + \overline{\omega}^{\mathcal{R}}V^{\phi}G_0 T^{\phi\mathcal{R}}. \tag{72}$$

Let us define

$$T^{\phi\mathcal{R}} \equiv \overline{\omega}^{\mathcal{R}} t^{\phi\mathcal{R}} \omega^{\mathcal{R}}, \tag{73}$$

substituting Eq. (73) into Eq. (72), we obtain

$$t^{\phi\mathcal{R}} = V^{\phi} + V^{\phi}(G_0 + G_0 T^{\mathcal{R}}G_0)t^{\phi\mathcal{R}}$$
$$= V^{\phi} + V^{\phi}G^{\mathcal{R}}t^{\phi\mathcal{R}} \tag{74}$$

$$G^{\mathcal{R}} = G_0\overline{\omega}^{\mathcal{R}} = G_0 + G_0 T^{\mathcal{R}}G_0. \tag{75}$$

Therefore we have

$$T^{C} = \overline{\omega}^{\mathcal{R}} t^{\phi\mathcal{R}} \omega^{\mathcal{R}} + T^{\mathcal{R}}. \tag{76}$$

The momentum representations of these operator relations are given by the partial wave formula of Eq. (76) (see also [35]),

$$T_l^{C}(p, p'; E) = \int_0^{\infty}\int_0^{\infty} \frac{p''^2 dp''}{2\pi^2} \frac{p'''^2 dp'''}{2\pi^2}$$
$$\times \overline{\omega}_l^{\mathcal{R}}(p, p''; E)t_l^{\phi\mathcal{R}}(p'', p'''; E)\omega_l^{\mathcal{R}}(p''', p'; E) + T_l^{\mathcal{R}}(p, p'; E). \tag{77}$$

In order to verify the relation between the double integral and single integral, the following relations should be defined,

$$G_l^{\mathcal{R}}(p, p'; E) = \frac{2\pi^2}{p^2} \delta(p - p') G_0(p; E) + G_0(p; E) T_l^{\mathcal{R}}(p, p'; E) G_0(p'; E) \tag{78}$$

$$\overline{\omega}_l^{\mathcal{R}}(p, p'; E) = \frac{2\pi^2}{p^2} \delta(p - p') + T_l^{\mathcal{R}}(p, p'; E) G_0(p'; E) \tag{79}$$

$$\omega_l^{\mathcal{R}}(p, p'; E) = \frac{2\pi^2}{p^2} \delta(p - p') + G_0(p; E) T_l^{\mathcal{R}}(p, p'; E). \tag{80}$$

It should be noted that the volume factors $2\pi^2/p^2$ and $2\pi^2/p'^2$ are missing in our former articles [5,6] in $G_l^\phi, \overline{\omega}_l^\phi, \omega_l^\phi$ regarding T^ϕ, because of the different definition of the free Green's function. However, the final results are not changed.

References

1. Y. Hiratsuka, S. Oryu, S. Gojuki, D(p, p)D elastic scattering with rigorous Coulomb treatment. J. Phys. G Nucl. Part. Phys. **40**, 025106 (2013)
2. S. Oryu, Y. Hiratsuka, S. Nishinohara, S. Chiba, Proton-proton phase shifts calculation in momentum space by a rigorous Coulomb treatment. J. Phys. G Nucl. Part. Phys. **39**, 045101 (2012)
3. S. Oryu, S. Nishinohara, N. Shiiki, S. Chiba, Coulomb phase shift calculation in momentum space. Phys. Rev. C **75**, 021001 (2007)
4. S. Nishinohara, S. Chiba, S. Oryu, The Coulomb scattering in momentum space for few-body systems. Nucl. Phys. A **790**, 277c (2007)
5. S. Oryu, Two- and three-charged particle nuclear scattering in momentum space: A two-potential theory and a boundary condition model. Phys. Rev. C **73**, 054001 (2006)
6. S. Oryu, Erratum: Two- and three-charged particle nuclear scattering in momentum space: A two-potential theory and a boundary condition mode. Phys. Rev. **76**, 069901 (2007)
7. S. Nishinohara, A study of Coulomb scattering by the momentum representation (in Japanese), Doctor thesis in Tokyo University of Science (2006)
8. Y. Hiratsuka, A study of the rigorous solution for two- and three-charged nucleons scattering in momentum space, Doctor thesis in Tokyo University of Science (2014)
9. F. Calogero, *Variable Phase Approach to Potential Scattering* (Academic Press, New York, 1967)
10. B. Liu, H. Crater, Two-body Dirac equations for nucleon-nucleon scattering. Phys. Rev. C **67**, 024001-1-36 (2003)
11. V.D. Viterbo, N.H.T. Lemes, J.P. Rraga, Variable phase equation in quantum scattering. Rev. Bras. Ensino Fis. **36**(1), 1310 (2014)
12. S. Oryu, Separable expansion fit to the Reid soft-core fully off-shell t matrices: $^1s_0$ and $^3s_1 - ^3D_1$ states. Phys. Rev. C **27**, 2500–2514 (1983)
13. S. Oryu, An efficient low-rank separable T-matrix formalism and its application to the reid soft core potential. Prog. Theor. Phys. **62**, 847–848 (1979)
14. S. Oryu, Generalized separable potential theory and Bateman's method on the scattering problem. Prog. Theor. Phys. **52**, 550–566 (1974)
15. S. Oryu, On the convergence of generalized separable approximation in the scattering problems. Prog. Theor. Phys. **51**, 1626–1628 (1974)
16. S. Oryu, Convergence property of the generalized separable expansion of the two-body transition matrix and the UPA model. Prog. Theor. Phys. Suppl. **61**, 199–228 (1977)
17. S. Oryu, An Analytic solution of the three-body Amado-Lovelace equation at positive energies. Prog. Theor. Phys. Suppl. **61**, 180–198 (1977)
18. W. Tobocman, M.H. Kalos, Numerical calculation of (d, p) angular distributions. Phys. Rev. **97**, 55 (1955)
19. A.M. Veselova, Separation of two-particle Coulomb singularities in a system of three charged particles. Teor. Mat. Fiz. **3**, 326 (1970)
20. A.M. Veselova, Separation of two-particle Coulomb singularities in a system of three charged particles (in Russian). Theor. Math. Phys. **3**, 542 (1970)
21. A.M. Veselova, Determination of scattering amplitudes in problems with two and three charged particles. Teor. Mat. Fiz. **13**, 368 (1972)
22. A.M. Veselova, Determination of scattering amplitudes in problems with two and three charged particles (in Russian). Theor. Math. Phys. **3**, 1200 (1972)
23. A.M. Veselova, Integral equations for three particles with Coulomb long-range interaction. Teor. Mat. Fiz. **35**, 180 (1978)
24. A.M. Veselova, Theor. Math. Phys. **35**, 395 (1978)
25. E.O. Alt, W. Sandhas, H. Zankel, H. Ziegelmann, Coulomb corrections in proton-deuteron scattering. Phys. Rev. Lett. **37**, 1537 (1976)
26. E.O. Alt, W. Sandhas, H. Ziegelmann, Coulomb effects in three-body reactions with two charged particles. Phys. Rev. C **17**, 1981 (1978)

27. E.O. Alt, W. Sandhas, Scattering amplitudes and integral equations for the collision of two charged composite particles. Phys. Rev. C **21**, 1733 (1980)
28. E.O. Alt, W. Sandhas, H. Ziegelmann, Calculation of proton-deuteron phase parameters including the coulomb force. Nucl. Phys. A **445**, 429 (1985)
29. A. Deltuva, A.C. Fonseca, P.U. Sauer, Momentum-space description of three-nucleon breakup reactions including the Coulomb interaction. Phys. Rev. C **72**, 054004 (2005)
30. A. Deltuva, A.C. Fonseca, A. Kievsky, S. Rosati, P.U. Sauer, M. Viviani, Benchmark calculation for proton-deuteron elastic scattering observables including the Coulomb interaction. Phys. Rev. C **71**, 064003 (2005)
31. A. Deltuva, A.C. Fonseca, P.U. Sauer, Momentum-space treatment of the Coulomb interaction in three-nucleon reactions with two protons. Phys. Rev. C **71**, 054005 (2005)
32. A. Kievsky, S. Rosati, W. Tornow, M. Viviani, Critical comparison of experimental data and theoretical predictions for N-d scattering below the breakup threshold. Nucl. Phys. A **607**, 402–424 (1996)
33. A. Kievsky, S. Rosati, M. Viviani, Proton-Deuteron elastic scattering above the Deuteron breakup. Phys. Rev. Lett. **82**, 3759 (1999)
34. T. Watanabe, Y. Hiratsuka, S. Oryu, Effective screened coulomb potential from $e^- - e^-$ to 208Pb-208Pb systems. EPJ Web Conf. **113**, 03018 (2016)
35. S. Oryu, T. Watanabe, Y. Hiratsuka, Y. Togawa, A Coulomb-like off-shell T-matrix with the correct coulomb phase shift in *Proceedings of EFB23 Conference Few-Body Systems* (to be published). Aahus, Denmark (2016)
36. H. van Haeringen, *Charged Particle Interactions Theory and Formulas* (Coulomb Press Leyden, Leiden, 1985)

Printed in the United States
By Bookmasters